COMPUTERS IN NEUROBIOLOGY
AND BEHAVIOR

COMPUTERS IN NEUROBIOLOGY AND BEHAVIOR

BRANKO SOUČEK
Professor

State University of New York, Stony Brook, New York
and Institute Ruder Bošković, Zagreb, Yugoslavia

ALBERT D. CARLSON
Professor

State University of New York, Stony Brook, New York

A WILEY-INTERSCIENCE PUBLICATION

JOHN WILEY and SONS, New York • London • Sydney • Toronto

Copyright © 1976 by John Wiley & Sons, Inc.

All rights reserved. Published simultaneously in Canada.

No part of this book may be reproduced by any means, nor transmitted, nor translated into a machine language without the written permission of the publisher.

Library of Congress Cataloging in Publication Data:

Souček, Branko.
 Computers in neurobiology and behavior.

 Includes bibliographical references and index.
 1. Neurobiology—Data processing. 2. Animals, Habits and behavior of—Data processing. 3. Human behavior—Data processing. I. Carlson, Albert D., 1930- joint author. II. Title.

QP357.5.S68 591.1'88'02854 75-25677
ISBN 0-471-81389-3

Printed in the United States of America

10 9 8 7 6 5 4 3 2 1

To Erika and Barbara

PREFACE

The industrial and scientific laboratories are more and more interested in specialists with application-oriented training and with interdisciplinary profile. In this book we are trying to bridge the gap between two disciplines: life science and computer science.

The book describes those computer techniques that are crucial to life scientists and engineers: basic programming, signal processing, data collection, analyzing, and simulation. It then concentrates on biological problems that are of interest to both life scientists and computer scientists: biological signals, codes and messages, and their use for the communication between and within living organisms. It is hoped that this volume will be useful to scientists and engineers involved in the study of behavior, neurophysiology, animal communication, physiology, and biomedicine. Also, computer and electrical engineers could benefit by learning communication and information processing systems as designed by nature.

This book is written as a text book for students, as well as a reference for practicing scientists and engineers. The treatment is kept as straightforward as possible, with emphasis on functions and systems. A minimal background in related areas at the undergraduate level is assumed, although many fundamentals are reviewed.

The book is divided into two parts: Part 1 deals with the data acquisition and processing of biological signals. Application of computers to problems in life science is explained. Programming in BASIC language is described in detail. As this language can be learned in only a few days, many biomedical scientists are using it. A simple explanation of how the experiment is interfaced to the computer is given. Computer simulation and modeling techniques are presented in a simple, nonmathematical way. The basic signal processing techniques are explained: measurement of amplitude and latency histograms, correlation functions, and power spectra. Examples with biological and biomedical signals are presented and discussed. This material will provide the basic working knowledge necessary in the modern laboratory with computerized instrumentation. Special attention is given to minicomputers and microprocessors. Although nonexpensive, these machines are quite powerful and are found everywhere.

Part 2 deals with computer models of biological communications. The general features of animal communications are introduced. In biology, communication occurs on two distinct levels: (1) internal communication within living organisms through neural process, and (2) external communication between two living organisms. Computer models of internal communication (neuromuscular transmission) and external communication (insect calls, insect light flashes, bird singing, etc.) are designed and described. Computer modeling has helped to find some basic rules present in biocommunications. A large volume of experimental data has been used to design the theoretical models. Simulated communication patterns and signal sequences are in excellent agreement with field measured data. New theories have been developed that explain solo, alternating and aggressive communication, centrally controlled communication, pulse pattern recognition, and frequency pattern recognition. Models of quantized information transmission on neural terminals are described and discussed.

Concrete flow charts and programs written in BASIC are presented for many data acquisition techniques and for communication models. The student could directly use those programs for data acquisition or for system simulation. The programs could be used to simulate experiments in which the parameters could be easily changed. This is the best way to understand the importance and role of the different elements in behavioral and neurophysiological systems. As the BASIC is a widely used language, one could run the programs in the computer centers, on minicomputers, on programmable desk calculators, or on time-share terminals.

BRANKO SOUČEK
ALBERT D. CARLSON

Stony Brook, New York
February 1975

ACKNOWLEDGMENT

We wish to thank Professor Charles Walcott, Chairman, Department of Cellular and Comparative Biology, S.U.N.Y. Stony Brook for his helpful suggestions. We appreciate the interest and support of Professors B. Katz, G. Korn, and A. R. Martin in our work which is described in this book. We thank our students F. Vencl, J. Copeland, and W. O'Neill for participating in some of the experiments described in the book. We appreciate the cooperation of J. Knapp and H. Kushner of Shering Corporation who provided the opportunity to apply the techniques described to problems in pharmaceutical research. In writing this book, it was necessary to borrow some information from the manuals of computer manufacturers. We thank the following companies for permission to use material from their publications:

>Digital Equipment Company
>Hewlett-Packard Company
>Intel Corporation
>General Automation, Inc.

We are grateful to J. Škvaril for allowing us to quote from his paper.

We greatly appreciate the efforts of Joyce Roe who produced some of the drawings in this book.

B.S.
A.D.C.

CONTENTS

PART 1. DATA ACQUISITION AND PROCESSING OF BIOLOGICAL SIGNALS, B. Souček

Chapter 1. Biomedical and Biocommunication Signals and Their Computer Processing   3

Introduction and Survey, 3

1.1. Communication Between and Within Living Organisms, 4
1.2. Neural Point Processes, 6
1.3. Behavioral or Communication Point Processes, 11
1.4. Neural Continuous Processes, 14
1.5. Behavioral or Communication Continuous Processes, 17
References, 21

Chapter 2. Information Coding   25

Introduction and Survey, 25

2.1. Decimal, Binary, and Octal Number Systems, 25
2.2. Basic Logical Operations, 29
2.3. Flip-Flop, 35
2.4. Logical Neurons, 37
2.5. Information Content of Memory and of Communication Channel, 37
2.6. Redundancy and Message Distances, 41
References, 44

Chapter 3. Interfacing the Experiment to the Computer 45

Introduction and Survey, 45

3.1. Laboratory Mini- and Microcomputers, 45
3.2. Programmed Input–Output Transfer, 51
3.3. Party line for Programmed Input–Output, 53
3.4. Example of Programmed Input–Output Transfer, 58
3.5. Direct Memory Access, 58
References, 60

Chapter 4. Laboratory Computer System 61

Introduction and Survey, 61

4.1. Computerized Experiments, 61
4.2. Digital-to-Analog Decoder, 64
4.3. Analog-to-Digital Convertor, 66
4.4. Time-to-Digital Conversion, 68
4.5. Real-Time Clock, 69
References, 72

Chapter 5. Basic Programming 73

Introduction and Survey, 73

5.1. BASIC Language, 73
5.2. Control Operations, 76
5.3. Loops, 78
5.4. Subroutines, 81
5.5. Input–Output Programming, 83
5.6. Other Features, 86
References, 92

Chapter 6. Real-Time Programming 93

Introduction and Survey, 93

6.1. Real-Time BASIC for Minicomputers, 93
6.2. Real-Time Executive BASIC System, 103
6.3. Real-Time PL/M Language for Microprocessors, 113
References, 115

Contents xiii

Chapter 7. Simulation   **116**

Introduction and Survey, 116

- 7.1. Deterministic Data Simulation, 117
- 7.2. Random Data and Probability Distributions, 122
- 7.3. Monte Carlo Techniques, 127
- 7.4. Simulation of Experimental and Theoretical Data, 130
  References, 139

Chapter 8. Amplitude and Latency Histograms   **140**

Introduction and Survey, 140

- 8.1. Probability Density Function, 140
- 8.2. Direct Recording Analyzers, 145
- 8.3. Basic Minicomputer-Analyzer System, 147
- 8.4. Sample Example, 152
  References, 152

Chapter 9. Correlation Measurement   **155**

Introduction and Survey, 155

- 9.1. Correlation Function, 155
- 9.2. Amplitude Correlations, 158
- 9.3. Interval Correlation, 159
  References, 172

Chapter 10. Fourier Analysis and Power Spectra   **173**

Introduction and Survey, 173

- 10.1. Periodic Data and Fourier Series, 173
- 10.2. Frequency Spectrum Program, 175
- 10.3. Some Typical Spectra, 177
- 10.4. The Sonograph Machine, 184
  References, 184

PART 2. COMPUTER MODELS OF NEURAL ACTIVITIES AND OF ANIMAL COMMUNICATIONS, B. Souček and A. D. Carlson

Chapter 11. General Features of Animal Communication and Neural Processes and their Computer Modeling 187

Introduction and Survey, 187

- 11.1. Description of Animal Communication Process, 188
- 11.2. Functions of Animal Communication, 190
- 11.3. Modalities and Patterns of Communication, 191
- 11.4. Neurological Signals, 194
- 11.5. Receptors and Neural Integration, 195
- 11.6. Evolution of Communication Patterns, 198
- 11.7. Computer Modeling of Neural and Behavioral Activities, 198
  References, 202

Chapter 12. Solo, Alternating, and Aggressive Communication 203

Introduction and Survey, 203

- 12.1. Example of Katydid Chirping, 204
- 12.2. Response Function, 208
- 12.3. Typical Sequence and Noise, 211
- 12.4. Stable Sequence, Sliding Sequence, and Transfer Function, 214
- 12.5. Aggression, 217
- 12.6. Model Based on Response Function and Transfer Function, 221
  References, 223

Chapter 13. Communication Based on Timing and Pulse Pattern Recognition 224

Introduction and Survey, 224

- 13.1. Example of Firefly Flashing, 225
- 13.2. Response Function, 227

Contents

- 13.3. Sensitivity Adjustment and Memory, 232
- 13.4. Locked-In Sequences, 234
- 13.5. Preparation for Model, 236
  References, 237

Chapter 14. Computer Simulation of Firefly Flash Sequences   238

Introduction and Survey, 238

- 14.1. Firefly Flash Sequences, 239
- 14.2. Switching the Conditions, 244
- 14.3. Memory and Locked-In Sequences, 247
- 14.4. Computer Model, 247
- 14.5. Significance of the Computer Modeling, 255
  References, 257

Chapter 15. Neural, Communication, and Behavioral Sequential Patterns   258

Introduction and Survey, 258

- 15.1. Basic Elements or Syllables, 259
- 15.2. Pairs, 262
- 15.3. Triplets, 269
- 15.4. Trees, 269
- 15.5. Significance of the Tree Method, 279
  References, 285

Chapter 16. Communication Based on Frequency Pattern Recognition   286

Introduction and Survey, 286

- 16.1. Example of Bird Duetting, 287
- 16.2. Definitions and Classification of Syllables, 287
- 16.3. Analysis of Syllable Sequences, 290
- 16.4. Dependencies and Independencies Controlling Song Sequences, 296
- 16.5. Discussion of the Model, 297
  References, 298

Chapter 17. Models of Quantized Information Transmission on Neural Terminals — **300**

Introduction and Survey, 300
17.1. Example of the End-Plate Potential, 301
17.2. Computer Model, 303
17.3. Fixed and Variable Latency, 312
References, 316

Appendix — **317**

Index — **321**

COMPUTERS IN NEUROBIOLOGY
AND BEHAVIOR

PART I

DATA ACQUISITION AND PROCESSING OF BIOLOGICAL SIGNALS

B. Souček

Biomedical and Biocommunication Signals and their Computer Processing, 3   Information Coding, 25   Interfacing the Experiment to the Computer, 45   Laboratory Computer System, 61   BASIC Programming, 73   Real-time Programming, 93   Simulation, 116   Amplitude and Latency Histograms 140   Correlation Measurement, 155   Fourier Analysis and Power Spectra, 173

Chapter 1

BIOMEDICAL AND BIOCOMMUNICATION SIGNALS AND THEIR COMPUTER PROCESSING

Introduction and Survey

Behavioral biologists, neurophysiologists, and other life scientists and engineers are no longer restricted to simplistic descriptions of the trains of spikes or of continuous waveforms that constitute the basic signals in biocommunications. The development of a wide variety of computational techniques, the rise of computer models for framing and testing hypothesis and the development of new theories, all have engendered new interpretations, new questions, and new directions in the quantitative investigation of biocommunications. One could feel that trend in the study of both communication between living organisms and communication within living organisms.

Communication is a process of exchanging the messages between two points using some kind of signal.

First of all, it must be mentioned that the communication signal is normally distorted by random fluctuations, that are caused by the on-going biological activity, by the common "black box" approach and great complexity of the biological object, by the environment, and finally by the instrumentation used. Because of these facts, the statistical approach to the biocommunication signals processing must be considered as the main method used in this domain.

This chapter gives an overview of the biomedical and biocommunication signals and their computer processing. It is shown that virtually the same computer techniques could be used to study two fields: communication between living organisms and communication within living organisms.

This chapter reviews a number of typical problems and shows some of the solutions. The purpose of this chapter is to give a bird's eye view of the field, rather than to explain the techniques used. Numerous computer techniques used to solve the problems are then explained in detail in the chapters that follow.

Three functions are frequently used in biological data analysis: probability distribution, correlation function, and power spectra. These functions describe the random data in a similar way, as amplitude, signal shape, and signal frequency describe the deterministic data.

Probability distribution describes the amplitude property of the random data. As the data fluctuate in a random fashion, the probability distribution shows which amplitude range is more probable, and which amplitude range is less probable.

Correlation function describes the dependence of the signal at one instant of time on the signal at another instant of time. The difference between two time instants is called time lag or correlation delay. Correlation function is plotted as a function of the time lag. Maximal value of the correlation function shows the time lag for which the two signals influence each other the most.

Power spectra shows the frequency composition of the signal. Each signal could be approximated with the mixture of sine waves of different frequencies and amplitudes. The power spectra shows the amplitudes of the sine waves as a function of the frequencies.

All these functions and many others could be measured and calculated using laboratory computers. Each of the techniques used is described in a separate chapter.

In this chapter we use the above definitions of the basic functions, and without going into details, we review a number of typical problems and show some of the solutions.

1.1 Communication Between and Within Living Organisms

Communication involves the production of a signal at a source, which stimulates a response in a receiver. In biology communication is carried on two distinct levels.

- Communication within living organisms, through neural processes.
- Communication between living organisms, for example, message exchange between two animals or between two humans.

The purpose and modalities of these two levels of communication are quite different. Many times the researcher has to study both levels of communications to understand the biological system. One of the main tools in communication study is a digital computer. The communication is carried through signals and messages, and the computer can be used for many purposes: data acquisition, processing, system simulation, and comparison between theoretical and experimental results. Fortunately, the same computer techniques that are used in the field of communication between living organisms can be also used in the field of communication within living organisms and vice versa.

A communicatory system involves a number of components which are represented somewhat after the fashion of communication engineers and shown in Figure 1.1.

The emitter sends the signal over the communication channel to the receiver. The signal incorporates the message, which is based on the code held

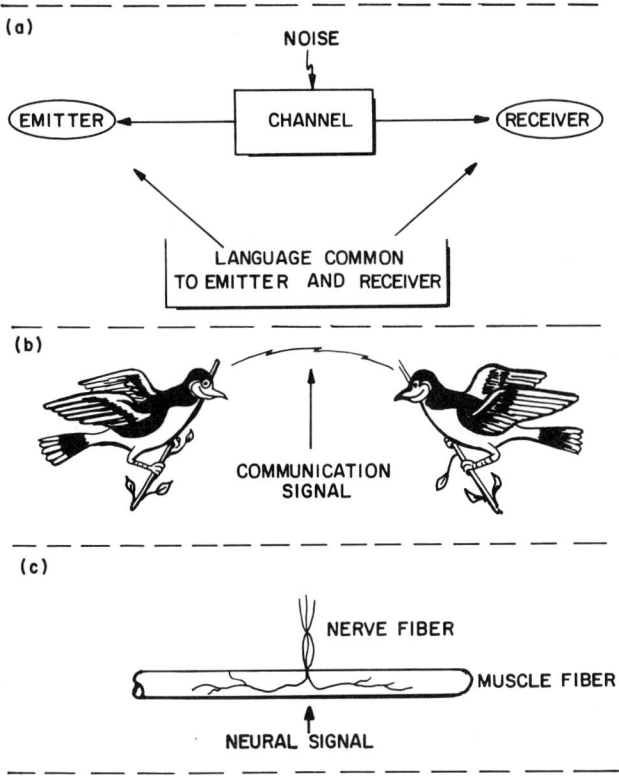

Fig. 1-1 Communication system. (*a*) Information exchange between emitter and a receiver, through a noisy channel. (*b*) External communication between two living organisms. (*c*) Internal communication through neural processes.

in common with the receiver. The meaning to the recipient is dependent on the context of the signal. The meaning can probably be operationally defined as the response selected by the recipient from all responses open to it.

During the transmission of information over the communication channel, the noise is added to the signal. One of the tasks of the biological communication study is to distinguish the basic signal from the noise. The always-present noise in biological systems is a reason that biological messages are coded in a special way. Rarely will the messages be coded using available signal elements in an optimal way. Rather, a large degree of redundancy is used.

From the signal point of view, the biological communications could be divided into two groups: point (discrete) processes and continuous processes.

From the system point of view, the biological communications could also be divided into two groups: internal communications through neural processes and external communications between two living organisms.

Four combinations of processes are described in the following sections. Description and examples of neural processes are quoted from the review by Škvaril.[1]

1.2 Neural Point Processes

The nerve cell is considered as the basic functional element of the nervous system. Its basic property common to all living cells is between other the ability to maintain a concentration gradient for different substances on both sides of membrane surrounding the cell. Often these substances are electrically charged, and therefore a potential difference appears across the membrane.[2] The changes in an environmental condition act on the neuron in such a way that its membrane potential is sometimes decreased to a value known as the threshold. When this is the case, this potential markedly decreases and then returns again to the preceding resting potential (duration of this transient change is about 1 msec, and the voltage of this drop is about 70 mV). This event, known as the action potential or spike, spreads along the whole neuronal body, including its processes. The spike could be transferred to the other neurons being in synaptical contact with the mentioned one. This impulse activity represents the basic mechanism of information transmission in the nerve system.[3,4] Therefore, the analysis of it is of considerable importance.

Action potentials or spikes taken from a single neuron may be considered as a sample function of a point process the basic characteristic of which is the histogram of the intervals between consequent events. It is an estimate of the probability density function of the mentioned intervals considered as a sample of a random variable (See Figure 1.2). If the studied series of events

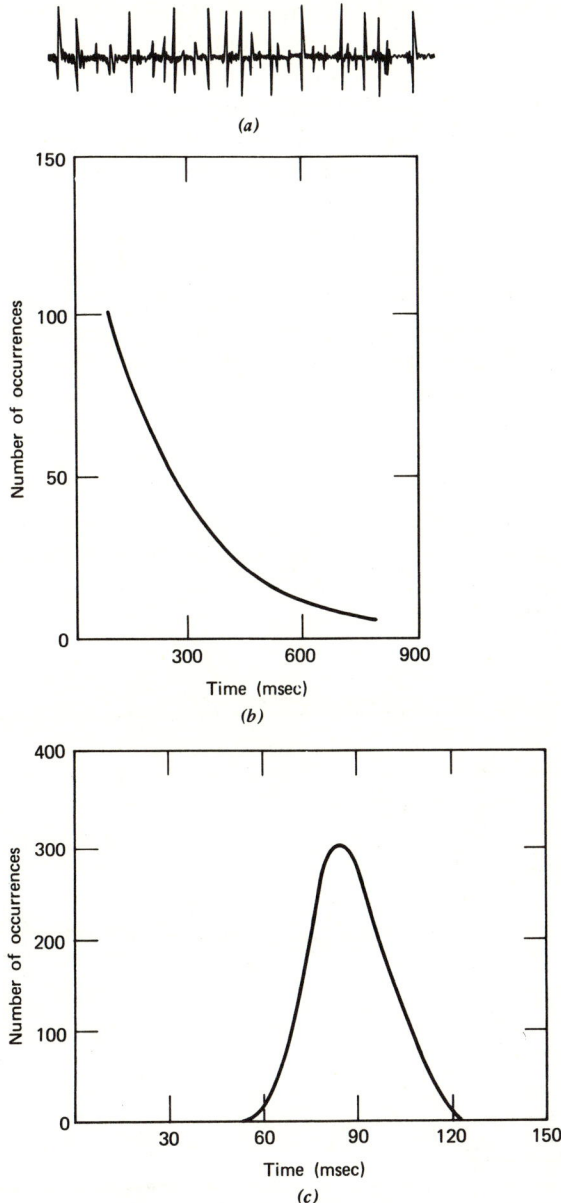

Fig. 1-2 (a) An example of the action potentials of a single neuron recorded by a microelectrode and reproduced from CRT screen. (b) Example of histograms of intervals between consecutive spikes of the spontaneous activity of a single neuron. (c) Another example of an interval histogram.

corresponds to the renewal process, then the knowledge of this density alone is sufficient for the description of the process.[5]

The second stage in the description is represented by the problem of the approximation of this experimental histogram by some general type of distribution like the Gaussian or exponential characterized by only a few parameters. A similar problem is encountered in deciding if two experimental histograms are produced by identical generators (homogeneity of population). Sometimes, simpler statistics are sufficient for the detection of spontaneous changes of the activity observed. The mean value estimated by the arithmetic mean, the variance estimated by the squared deviation, or coefficient of asymmetry or excess could be presented here as typical representatives of these simple statistics (see Figure 1.3).

To detect the influence of the repetitive stimulation of the observed activity the "dwell histogram" is most frequently evaluated. It resembles an

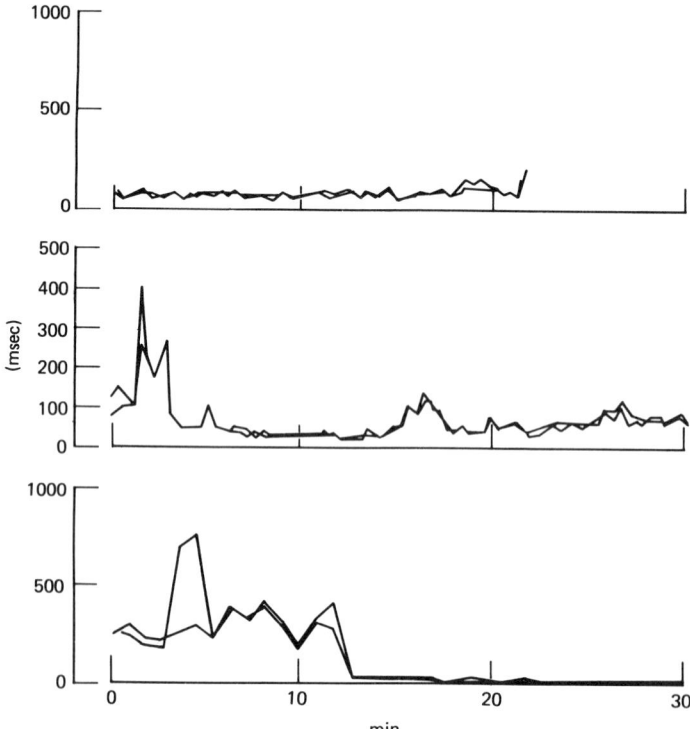

Fig. 1-3 Arithmetic mean (heavy line) and standard deviation (thin line) of intervals between consecutive spikes of the spontaneous unit activity of a single neuron as a function of time after the application of a particular dose of drug. Three examples are given.

Neural Point Processes

estimate of the conditional probability density function that an event occurs in a certain time, if a stimulus is applied in time zero. When the stimulus is not influencing the activity observed and other certain assumptions are fulfilled, the dwell histogram is uniformly distributed over the interstimulation period.[6] Therefore, in some cases the evaluation of the experimental dwell histogram, that is, the acceptance or rejection of the hypothesis about the influence of the given stimulus, may be stated on the basis of the "naked eye" approach only (see Figure 1.4a and b).

To detect the dependences and hidden periodicities in one series of events the estimate of the autocorrelation function (known as the postfiring interval distribution, renewal density function, or expectation density function) is sometimes evaluated.[7] This estimate of a conditional probability density function is like the dwell histogram, but the role of stimulus is fulfilled consequently by each analyzed event in the latter case (see Figure 1.5).

To detect the mutual dependence of two series of events, other functions like the crosscorrelation function or the distribution of forward and backward recurrence times could be used.[7] The first approach is similar to the dwell histogram method described above (events of one process are considered as the stimuli), the second one is described later.

To present other examples of the electrophysiological phenomena reduced to the point process and evaluated by the mentioned methods, the series of the interval histograms between the QRS complex of ECG (Figure 1.6) and the histograms of the intervals between consecutive breathing (Figure 1.7) are shown.

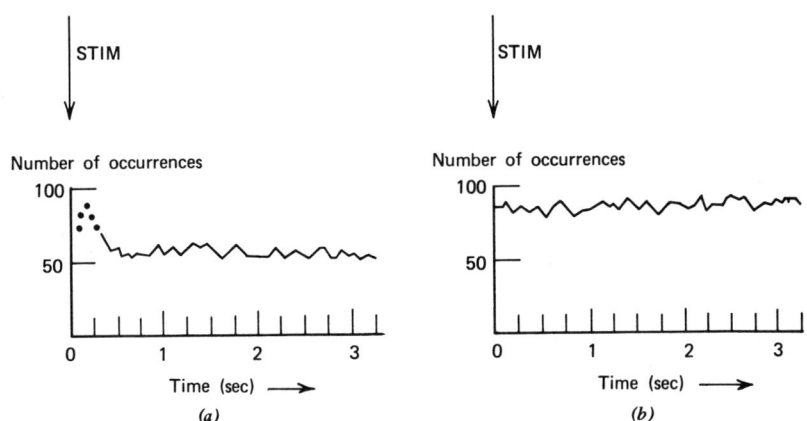

Fig. 1-4 Dwell histograms of two neurons. (a) Shows the strictly defined dependence on the stimulus. (b) Is not influenced.

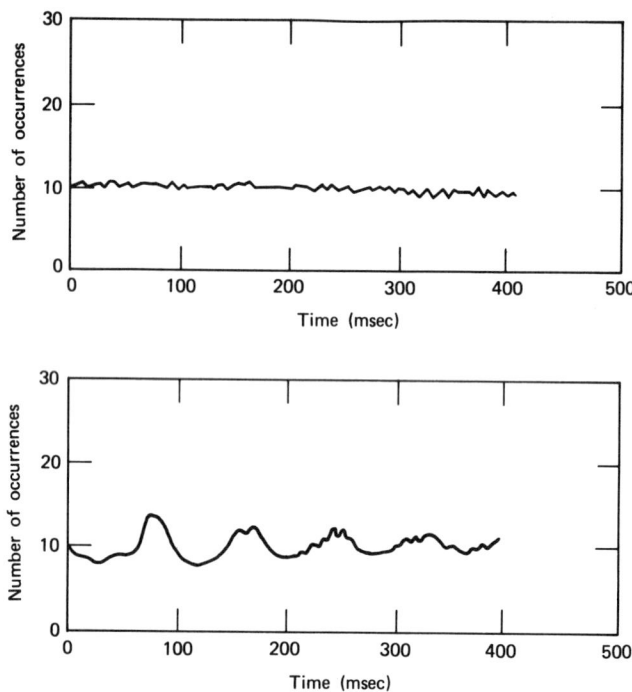

Fig. 1-5 Autocorrelation functions of the spontaneous impulse activity of the same neuron, as used in Fig. 1-2.

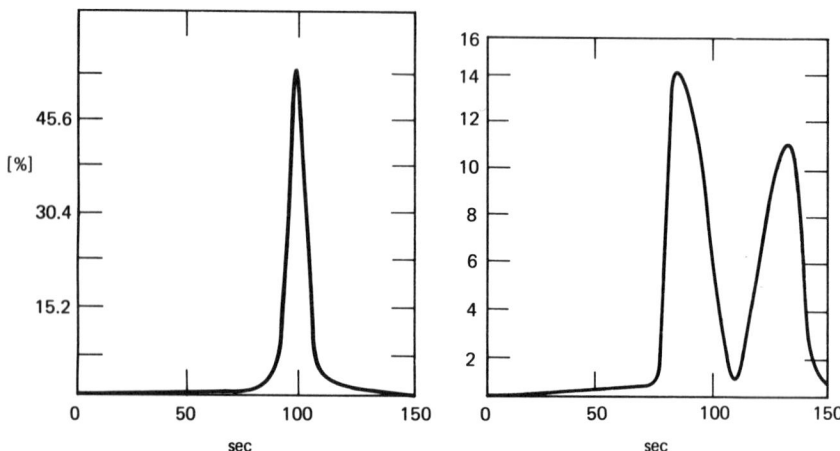

Fig. 1-6 Histograms of the intervals between the consecutive QRS complexes in ECG.

Behavioral or Communication Point Processes

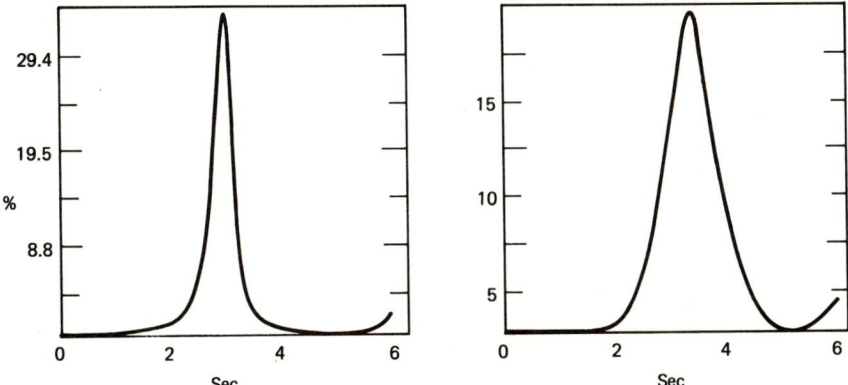

Fig. 1-7 Histograms of the intervals between the consecutive instants of inspiration, moments of which are determined by a chest movement controlled switch.

1.3 Behavioral or Communication Point Processes

The communication is considered as the basic functional element of the animal and human behavior. Its basic property common to all living organisms is the ability between organisms to support a living organism and the species to adapt and survive. By any reasonable definition of the term communication, there can be no doubt that not only human, but also animals communicate with each other.

Most critical from the point of view of natural selection is the presence of the species specific information. Another one is sexual information, used to distinguish between males and females. Next is individual information used to distinguish a particular individual in the social group. Communication messages could be also motivational, for example, the singer is in reproductive condition. Another message could be environmental, for example the singer is within his territory and has no mate.

Communication involves the production of a signal by an individual or a group, which stimulates a response in a receiver. In general, our discussion assumes sender and receiver to be of the same species. Here are a few examples of sensory stimulus comprising the signal:

- Sequence or pattern of sounds.
- Sequence or pattern of light flashes.
- A display of gestures, body position, facial expression.
- An emission of odors, and the like.

The processes described could be either discrete (or point processes) or continuous processes.

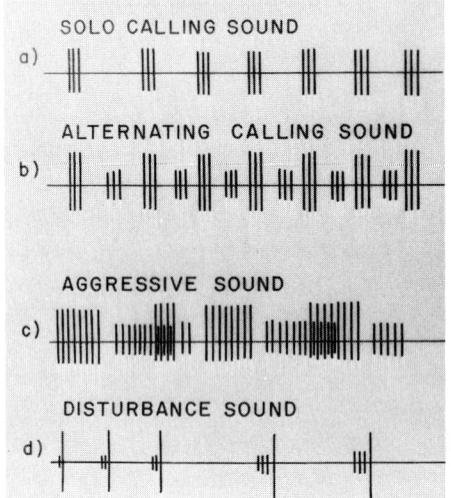

Fig. 1-8 Acoustical communication between two insects (katydid chirping).

Figure 1.8 presents an example of acoustical point processes, katydid chirping. Adult males of the true katydid produce different sequences of chirps. Each chirp consists of two or more pulses.

Figure 1.8a shows the calling sound, whose function is to attract mature females. Figure 1.8b shows the alternating sound: if two male katydid are close enough they will respond to the other's song. One male will be the leader, and it will increase its chirp rate. The other male, the follower will try to sing in such a way that the chirps of the two will be in alternation. Figure 1.8c shows the aggressive sound, produced by the two males who fight the acoustical war over the same territory. Figure 1.8d shows the disturbance sounds produced whenever the katydids are handled.

Computers are becoming the basic tools in measuring and analyzing the communication point processes. Computers are used for direct data acquisition, latency and amplitude histogram generation, correlation analysis, and power spectra analysis. Figure 1.9 shows an example of the correlation analysis of katydid calls. Such techniques can be used to detect the periodic component buried in random noise.

Fig. 1-9 Computer analysis of incest calls (katydid chirping). (a) Interchirp intervals displayed as a function of the chirp serial number. (b) Simple correlation function of interchirp intervals. This example proves that this function is of no use for interval analysis. (c) Modified correlation function of interchirp intervals. One could clearly notice that the analysed process is periodic and also to determine the period of the process.

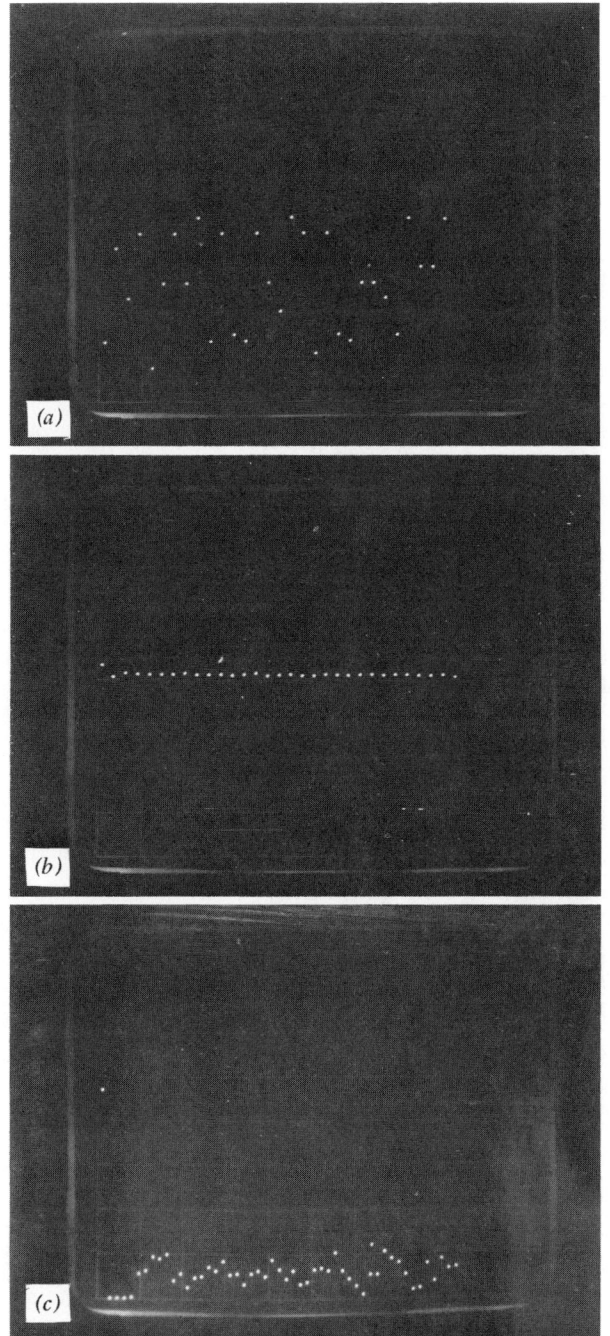

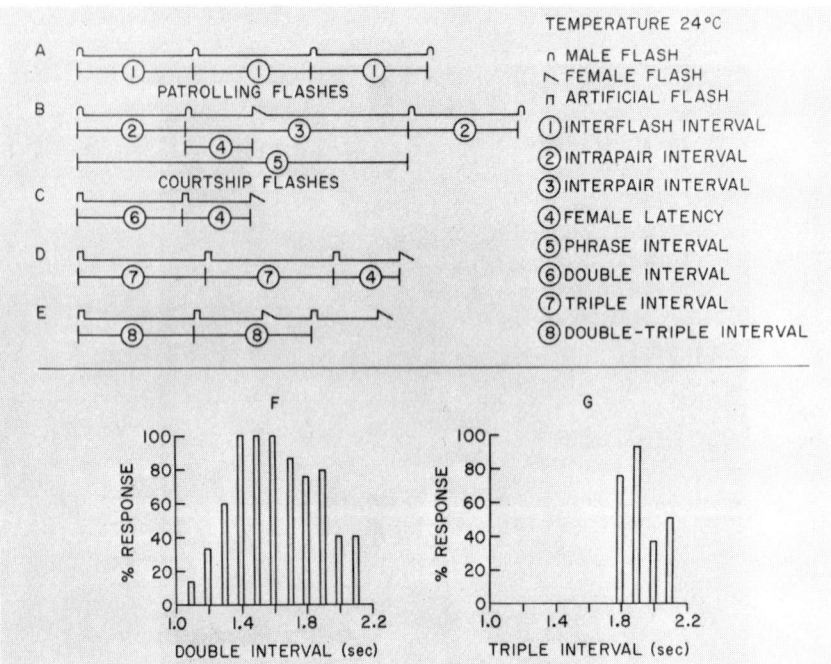

Fig. 1-10 Flashing sequences used for firefly communication. (*a–e*) Different flashing sequences and definitions of intervals. (*f*) Measured probability distribution of "double intervals." (*g*) Measured probability distribution of "triple intervals."

Figure 1.10 presents an example of point process with light flashes used for communication purposes. This exchange of light flashes is used for a courtship communication between male and female fireflies. Much of the basic information transmitted from one member of the species to another member is contained in the interflash interval of male flashes and in the female's flash response latency. Precise time discrimination is needed to recognize the species specific signal and to identify the correct partner.

Computers are used not only to measure and analyze the communication signals, but also to simulate the communication processes. The computer modeling presents the significant step in closing the gap between animal communication and behavior on one side and neural structure study on the other side.

1.4 Neural Continuous Processes

As a typical example of the continuous neurophysiological signal the EEG should be mentioned. Contrary to the neuronal unit activity recorded by the

Neural Continuous Processes

microelectrodes taking the electric potentials from very small neighborhood of the cell, the EEG activity is taken from the surface of the scalp or from the depth of the brain by relatively large electrodes. In our review the term EEG has broader meaning; it is used for electrical activity of the brain recorded by the macroelectrode. The presence of different tissues between the local generators of the electrical activity in the brain influences the potential observed at the surface.[8] Thus the activity of single neurons is not more discernable on EEG record representing the filtered sum of the activity of large set of individual generators. These generators must be synchronized to produce the periodical activity that is sometimes observed.

The evaluation of this activity is the domain of very skilled specialists, and the procedures are very subjective and always time consuming. The application of computers in this task is therefore important but difficult.

Processing of the continuous signals is illustrated on the processing of EEG.

The amplitude histogram representing under some conditions the estimate of the first-order probability density function of the process analyzed is one of the simplest first-order statistics used in connection with both spontaneous and evoked EEG activity (see Figure 1.11).[9] Some parameters characterizing

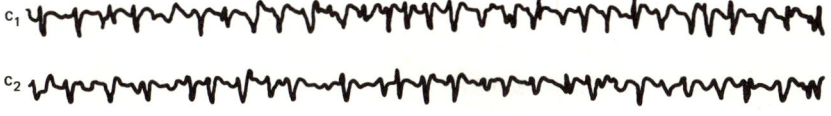

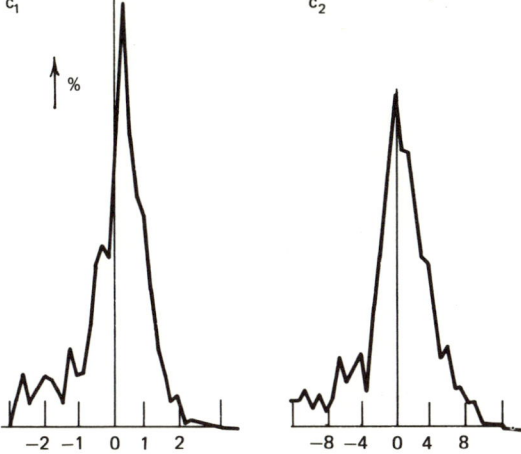

Fig. 1-11 Amplitude histograms of EEG together with the parts of the original EEG records. Frequency of sampling is 60 samples/second.

the form of this histogram are often evaluated. For example, on the basis of the coefficient of asymmetry the typical waveform in EEG could be automatically detected.[10] In further processing we may try to approximate such histograms with some theoretical distributions (for instance, with normal one) to simplify its description.[11]

Estimation of the ensemble mean in chosen time relative to the stimulus represents the most famous statistic evaluated in the domain of the evoked activity in EEG. It is a powerful tool for the detection of the so-called evoked response, that is, the changes of the slow electrical activity of the chosen part of the brain to the given stimuli. This method enables one to extract the response of very small amplitude (signal) hidden in the additive random spontaneous activity (noise) (see Figure 1.12). This technique is useful for the detection of some irregularities in the transmission of the signal in the brain as it is caused by a tumor or to differentiate the sensory function[12] or is useful as a clinical method.[13,14] Theoretical approach to the problem of the evoked response detection is presented in comprehensive form in Bentad[15] and Syers[16] and partially in Ruchkin.[17] In connection with the evoked response, processing the evaluation of the standard deviation in all points of the evoked response could be presented as an example of simple second-order

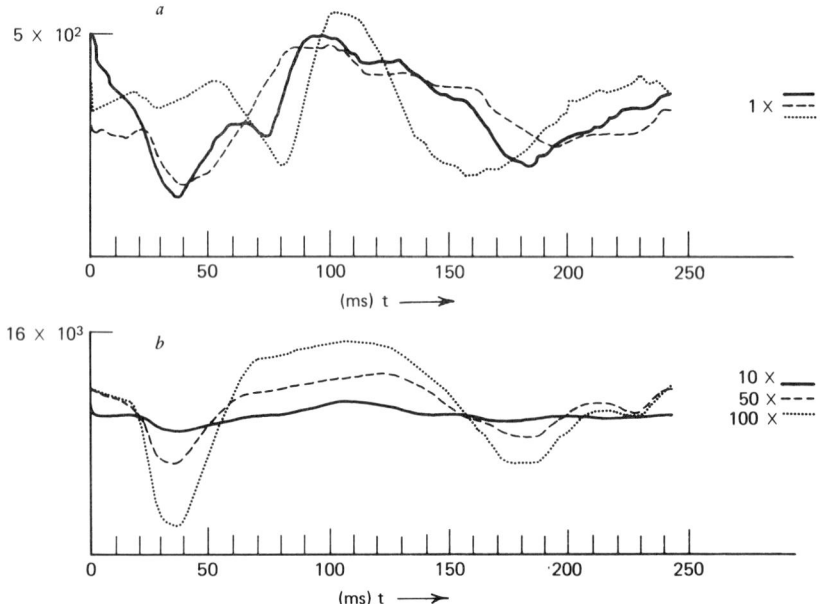

Fig. 1-12 (a) Individual, superimposed records of the EEG activity following the stimulus. (b) Emerging of the evoked response by summing the shown number of individual records.

statistics. This standard deviation plays important role in estimation of the validity of the evoked response evaluated simultaneously (biodata).

Auto and cross-correlograms and the corresponding spectral density functions are the most often used second-order statistics in connection with EEG processing.[18,19] Correlograms are evaluated by convolution, and the spectral density functions were formerly computed as their Fourier transform. Nowadays the spectral density is evaluated directly by digital filtering or by using the fast Fourier transforms.

Further interpretation of these statistics is rather difficult mainly because the assumptions under which these functions are evaluated do not hold and because the error due to the finite time of observation is neglected.[20] Both the correlation function and spectral density functions are equal insofar as the description of the stationary stochastic processes is concerned. Some examples of correlograms and spectrograms together with the EEG evaluated are presented in Figures 1.13 and 1.14.

The relationship between the power of the EEG activity in different frequency bands is in some cases very important from the clinical point of view, and thus the spectral density function may soon become an important clinical characteristic.

By means of the spectral analysis the different stages of the sleep or different levels of vigilance could be recognized.[21-23]

Modeling seems to be important in connection with the description of the EEG activity by their correlograms. Often it is possible to approximate the experimental correlation function by an analytical function and on that basis to define the normal process giving the same autocorrelation function.

The continuous signal could be reduced to the point process for some tasks, which enables a substantial simplification of the recording. Activity observed is reduced to the instants when the signal crosses zero (mean) level, as shown in Figure 1.15.[6,24]

Evaluation of the mutual dependence of the two activities of the different types such as the evaluation of the two-dimensional histogram of the amplitude of the observed EEG and of the instants of the occurrence of spikes from a single neuron must be mentioned in conclusion of this short review.

1.5 Behavioral or Communication Continuous Processes

A typical example of the continuous communication signal is the sound wave. Acoustical communication based on the sound wave modulation is used in human speech, in bird singing, and in terestial and aquatic animal communication. Insects are frequently using a pulse modulation of the

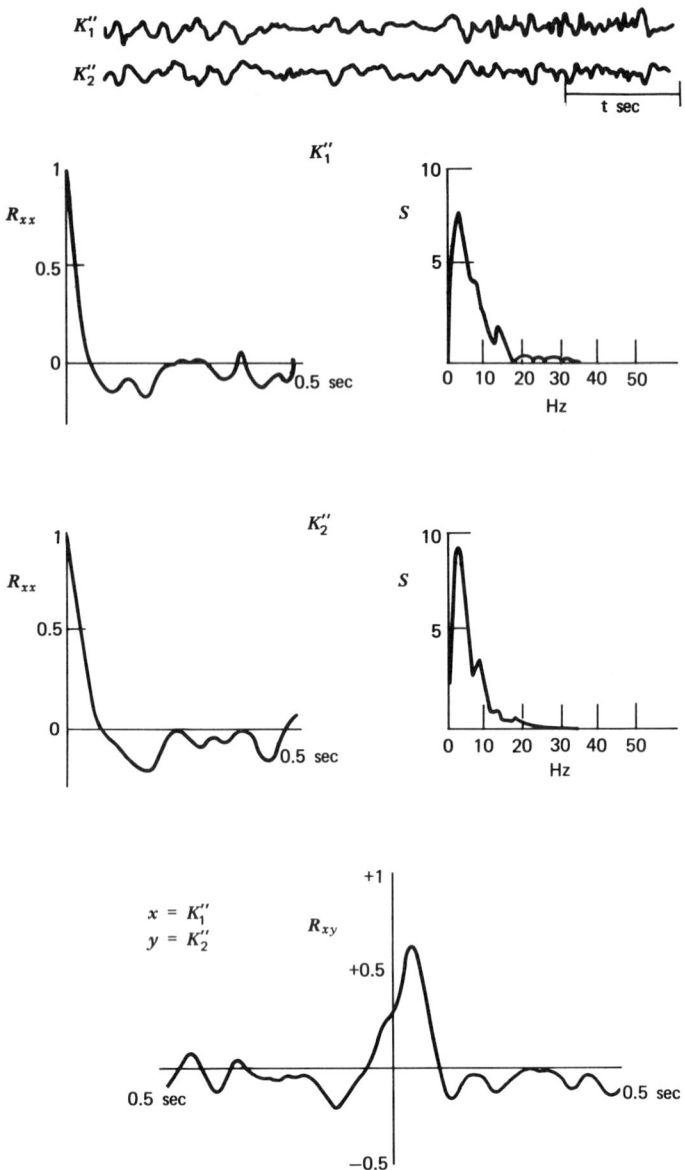

Fig. 1-13 An example of two original EEG records taken from the two different parts of the brain and their autocorrelograms (Rxx), spectrograms (S), and crosscorrelogram (Rxy).

Behavioral or Communication Continuous Processes

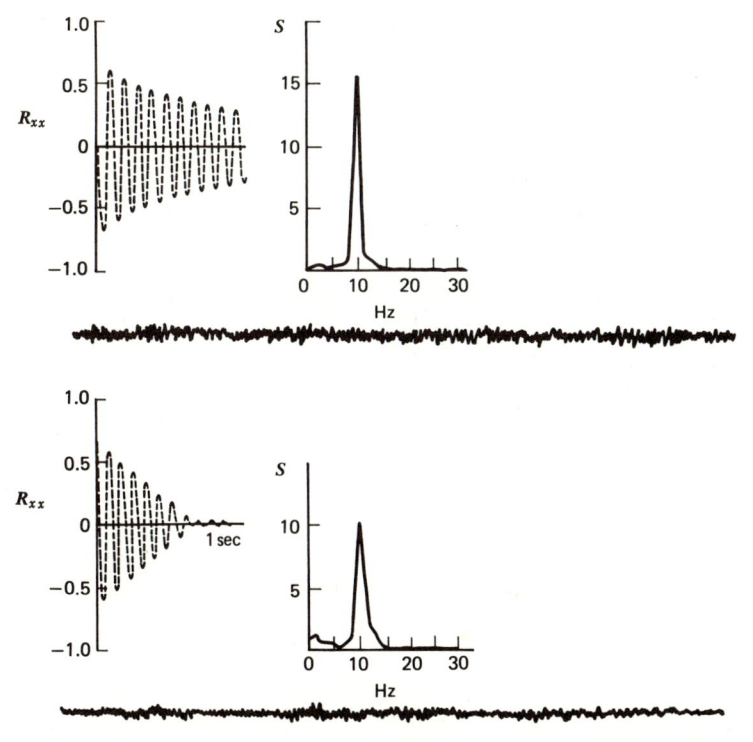

Fig. 1-14 Examples of periodical EEG activity and corresponding autocorrelograms (Rxx) and spectrograms (S).

sound wave producing short chirps. We have described these chirps as an example of the point process. On the other side, mammals and birds use amplitude, frequency, and phase modulation of the sound wave producing continuous communication signal. Computers present a major tool in studying continuous communication processes.

An example of the continuous communication process is presented in Figure 1.16. Figure 1.16a presents a recording of the bird song. The amplitude of the sound is recorded as a function of time. It is very difficult and often impossible to provide any conclusion based on the direct observation of such a continuous signal. Different computing techniques are used to transform the signal in the more meaningful forms.

Figure 1.16b shows the power spectrum of the same signal as a function of time. Such sonagrams are produced through the procedure of measuring the frequency content of the signal at various time intervals and are identical to the mathematical procedure of Fourier transforms.

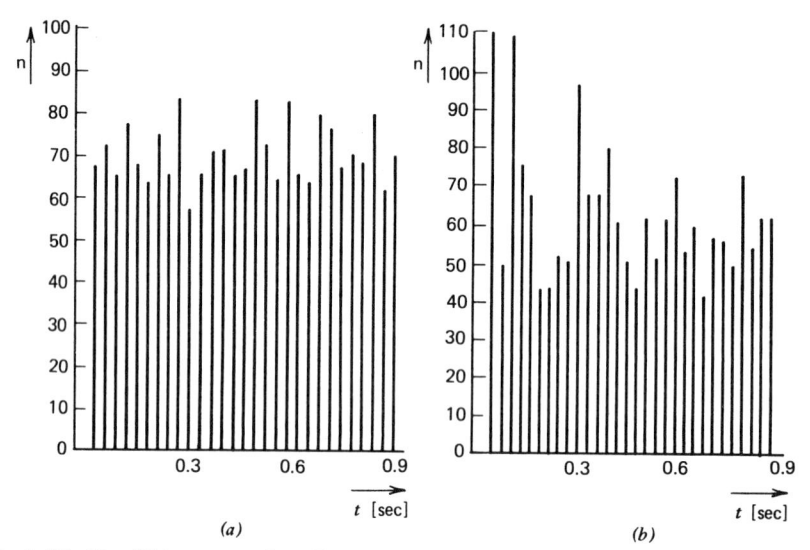

Fig. 1-15 Dwell histogram of the EEG zero level crossing instants relative to the stimulus. (*a*) Evoked response is absent. (*b*) Evoked response is present.

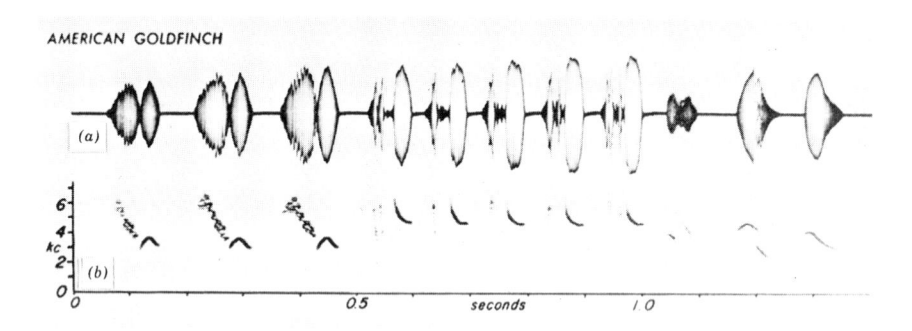

Fig. 1-16 Acoustical continuous communication process. (*a*) Amplitude of the sound as a function of time. (*b*) Power spectrum of the same signal as a function of time. (From C. H. Greenewalt: *Bird Song Acoustics and Physiology*, Smithsonian Institution Press, Washington, D.C., 1968.)

Digital machines could be used in many phases of continuous communication signal analysis. The most elementary phase is computer-controlled experiment, signal sampling and data acquisition. The most elaborate phase would be the application of computers for sonagram pattern recognition. By defining to the computer the message code for each sonagram, one could write the program to accept and analyze the sonagrams and to decode the message. Such techniques are especially of interest in studying man–machine communication through speech.

References

Neural Processes

1. Škvaril, J.: Short Review of the Basic Types of Biomedical Signals and Their Computer Processing, in *Data Processing* (B. Souček, Ed.), p. 111, Jurema, Zagreb, 1969.
2. Bures, J., Petran, M., and Zachar, J.: *Electrophysiological Methods in Biological Research*, Academia, Prague, 1967.
3. Nastuk, W. L.: *Physical Techniques in Biological Research*, Vol. 3, Academic Press, New York, 1964.
4. Ruch, T. C., and Fulton, J. F.: *Medical Physiology and Biophysics*, Saunders, Philadelphia and London, 1960.
5. Cox, D. R. and Lewis, P. A. W.: *The Statistical Analysis of Series of Events*, Methuen, London 1966.
6. Krekule, I.: Zero Crossing Detection of the Presence of Evoked Response, *Electroenceph. Clin. Neurophysiol.*, 25 (1968) 175–176.
7. Moore, G. P., Perkel, D. H., and Segundo, J. P.: Statistical Analysis and Functional Interpretation of Neuronal Spike Data, *Ann. Rev. Physiol.*, 28 (1966) 493–522.
8. Hendrix, C. E.: Theory of Transmission of Electric Fields in Cortical Tissue, Two Models for the Origin of the Alpha Rhythm. *UCLA Diss:-* 1964.
9. Saunders, M. G.: Amplitude Probability Density Studies on Alpha and Alpha-like Patterns, *Electroenceph. Clin. Neurophysiol.*, 15 (1963) 761–767.
10. Radil-Weiss, T., Krekule, I., and Chocholová, L.: Epileptoid EEG Activity during Chloralose Anaesthesia in Rats, *Physiol. Bohemosl.*, 16 (1967) 23–27.
11. Campbell, J., Bower, E., Dwyer, S. J., and Lago, G. V.: On the Sufficiency of Autocorrelation Functions as EEG Descriptors, *IEEE Trans. Bio-Med. Eng.*, BME, 14 (1967) 49.
12. John, E. R., Ruckin, D. S., and Villegas, J.: Experimental Background: Signal Analysis and Behavioral Correlates of Evoked Potential Configuration in Cats, *Ann. N. Y. Acad. Sci.*, 112 (1964) 362–420.
13. Giblin, D. R.: Somatosensory Evoked Potentials in Health Subjects and in Patients with Lesions of the Nervous System, *Ann. N. Y. Acad. Sci.*, 112 (1964) 92–142.
14. Rapin, I., and Graziani, L. J.: Auditory Evoked Responses in Normal, Brain-Damage, and Deaf Infants, *Neurology*, 17 (1967) 881–894.
15. Bendat, J.: Mathematical Analysis of Average Response Values for Nonstationary Data, *IEEE Trans. Bio-Med Eng.*, BME, 11 (1964) 72.
16. Sayers, B. McA.: *Signal and System Analysis*, Imperial College Press, University of London, 1968.
17. Ruchkin, D. S.: An Analysis of Average Response Computations Based upon Aperiodic Stimuli, *IEEE Trans. Bio-Med, Eng.*, BME, 12 (1965) 87.
18. Brazier, M. A. B., and Casby, J. U.: Crosscorrelation and Autocorrelation Studies of Electroencephalography Potentials, *Electroenceph., Clin. Neurophysiol.*, 4 (1952) 201–211.

19. Barlow, J. S., Brazier, M. A. B., and Rosenblith, W. A.: The Application of Autocorrelation Analysis to Electroencephalography, in *Proc. National Biophys. Conf.* (1967), Yale University Press, New Haven, 1959, 622–662.
20. Daniel, R. S.: Electroencephalographic Correlograms Ratios and Their Stability, *Science*, **145** (1964) No. 3633.
21. Rhodes, J. S., Brown, D., Reite, M., and Adey, W. R.: Computer Analysis of Chimpanzee Sleep Records, in *Symposium Sleep Mechanisms*, Lyon, France, 1963.
22. Rusinov, V. S., et al.: *Gagra Symposium*, 1965.
23. Radil-Weiss, T., Walter, D. O., Brown, D., and Adey, W. R.: EEG Manifestations of Sleep and Wakefulness in Rat Studied by Spectral Analysis, *Electroenceph. Clin. Neurophysiol.*, **18** (1965) 516.
24. Ertl, J. P.: Detection of Evoked Potentials by Zero Crossing Analysis, *Electroenceph. Clin Neurophysiol.*, **18** (1965) 630–631.
25. *Mathematics and Computer Sciences in Biology and Medicine*, Her Majesty's Stationary Office, London, 1965.

Behavioral or Communication Processes

26. Sebcok, T. A., Hayes, A. S., and Bateson, M. C., Eds., *Approaches to Semiotics*, Mouton, The Hague, 1964.
27. Sebcok, T. A., *Language*, **39** (1963) 465; *Am. Anthropol.*, **66**(1964) 954; *Science*, **147** (1965) 492.
28. Tinbergen, N.: *Study of Instinct*, Oxford Univ. Press, London, 1951; *Z. Tierpsychol.*, **20** (1963) 411. For a recent survey of ethology, see also E. H. Hess, in *New Directions in Phychology*, Holt, New York, 1963.
29. Alexander R. D.: *Science*, **144** (1964) 713.
30. Darwin, C.: *The Expression of the Emotions in Man and Animals*, Murray, London 1872.
31. Kainz, F.: *Die "Sprache" der Tiere. Tatsachen—Problemschau—Theorie*, Enke, Stuttgart, 1961.
32. Marler, P.: in *Darwin's Biological Works: Some Aspects Reconsidered* (P. R. Bell, Ed.) (Cambridge Univ. Press, London, 1959, pp. 150–206. Thorpe, W. H.: *Ann. Rev. Psychol.*, **12** (1960) 27. Lindauer, M.: *ibid*, **13** (1962) 35. Wood-Gush, D. G. M.: **14** (1963) 175. Mason, W. A., and Riopelle, A. J.: *ibid.*, **15** (1964) 143. Scbcok, T. A.: *Language*, **39** (1963) 448. Jinde, R. A.: in *Disorders of Communication: Proceedings of the Association for Research in Nervous and Mental Disease*, **1962** (D. McK. Rioch and E. A. Weinstein, Eds. (Williams and Wilkins, Baltimore, 1964, pp. 62–84. *Animal Communication*, Blaisdell, New York, 1964.
33. Among linguists, see especially C. F. Hockett, whose pertinent writings are listed in *Current Anthropol.*, **5** (1964) 166.
34. Marler, P.: *J. Theoret. Biol.*, **1** (1961) 295. Altmann, S. A.: *Ann. N. Y. Acad. Sci.*, **102** (1962) (Art. 2), 338. Scbcok, T. A.: *Behavioral Sci.*, **7** (1962) 430. Zinkin, N. I.: in *Acoustic Behavior of Animals* (R.-G. Busnel, Ed.) Elsevier, Amsterdam, 1963, pp. 132–180.
35. Moles, A.: in *Acoustic Behavior of Animals* (R.-G. Busnel, Ed.) Elsevier, Amsterdam, 1963, pp. 112–131.
36. Bühler, K.: *Sprachtheorie*, Fischer, Jena, 1934. Jakobson, R.: in *Style in Language, A Conference, Indiana University*, **1958** (T. A. Sebock, Ed.) M.I.T. Press and Wiley, New York, 1960, pp. 350–377.

References

37. Trager, G. L., *Stud. Linguist.*, **13** (1958) 1.
38. Wilson, E. O., and Bossert, W. H.: *Recent Progr. Hormone Res.*, **19** (1963) 792. Bossert, W. H., and Wilson, E. O.: *J. Theoret. Biol.*, **5** (1963) 443.
39. R.-G. Busnel, Ed., *Acoustic Behavior of Animals*, Elsevier, Amsterdam, 1963.
40. Thorpe, W. H.: *Bird-Song: The Biology of Vocal Communication and Expression in Birds*, Cambridge Univ. Press, London, 1961. Frings, M. R., and Frings, H. W.: *Sound Production and Sound Reception by Insects; A Bibliography*, Pennsylvania State Univ. Press, University Park, 1960.
41. W. N. Tavolga, Ed., *Marine Bio-Acoustics*, Macmillan, New York, 1964.
42. Wenner, A. M.: *Science*, **138** (1962) 446.
43. ———: *Sci. Am.*, **210** (Apr., 1964) 117.
44. Baerends, G. P.: *Arch. Necrl. Zool.*, **13** (1958) 401. Marler, P.: in *Vertebrate Speciation: A Conference, University of Texas,* **1958**, Univ. of Texas Press, Austin, 1961, pp. 96–121.
45. McElroy W. D., and Seliger, H. H.: *Sci. Am.*, **207** (Dec. 1962) 76.
46. Möhres, F. P.: *Naturwissenschaften*, **44** (1957) 431. Lissmann, H. W.: *J. Exptl. Biol.*, **35** (1958) 146.
47. Lindauer, M.: *Communication Among Social Bees*, Harvard Univ. Press, Cambridge, 1961.
48. Linsdale, J. M., and Tomich, P. Q.: *A Herd of Mule Deer; A Record of Observation, made on the Hastings Natural History Reservation*, Univ. of California Press, Berkeley 1953.
49. Thorpe, W. H.: *Nature*, **197** (1963) 774. Mowrer, O. H.: *Learning Theory and Personality Dynamics*, Ronald, New York, 1950, chap. 24.
50. Griffin, D. R.: *Listening in the Dark: The Acoustic Orientation of Bats and men*, Yale Univ. Press, New Haven, 1958. Kellogg, W. N.: *Porpoises and Sonar*, Univ. of Chicago Press, Chicago, 1961.
51. Mayr, E.: *Proc. Natl. Acad. Sci. U.S.*, **51** (1964) 939. The distinction was first mentioned in application to communication events by W. J. Smith, in preparation.
52. Marler, P.: in *Acoustic Behavior in Animals* (R.-G. Busnek, Ed.) Elsevier, Amsterdam, 1963, pp. 228–243, 794–797.
53. Lenneberg, E. H.: in *The Structure of Language* (J. S. Fodor and J. J. Katz, Eds.) Prentice-Hall, Englewood Cliffs, N. J., 1964, pp. 579–603;and in *New Directions in the Study of Language* (E. H. Lenneberg, Ed.) M.I.T. Press, Cambridge, 1964, pp. 65–88.
54. Stankiewicz, E.: in *Approaches to Semiotics* (T. A. Sebcok, A. S. Hayes, and M. C. Bateson, Eds.) Mouton, The Hague, 1964, pp. 239–264.
55. Thorpe, W. H.: *Learning and Instinct in Animals*, Harvard Univ. Press, Cambridge, 1963.
56. Barnett, S. A.: *The Rat: A Study in Behaviour*, Aldine, Chicago, 1963, p. 98.
57. Wynne-Edwards, V. C.: *Animal Dispersion in Relation to Social Behaviour*, Hafner, New York, 1962, p. 16.
58. MacKay, D. M.: *Cybernetics: Transactions of the 8th Conference* (H. von Foerster, Ed.) Josiah Mach, Jr. Foundation, New York, 1952, p. 224.
59. Schaller, G. B.: *The Mountain Gorilla: Ecology and Behavior*, Chicago Univ. Press, Chicago, 1963, p. 272.

60. Sebcok, T. A.: in *Natural Language and the Computer* (P. L. Garvin, Ed.) McGraw-Hill, New York, 1963, pp. 47–64.
61. Smith, W. J.: *Am. Naturalist*, **97** (1963) 122. Compare T. C. Schneirla, in *The Nebraska Symposium on Motivation*, Vol. 7 (M. R. Jones, (Ed.) Univ. of Nebraska Press, Lincoln, 1959, pp. 1–42.
62. Ruesch, J.: in *Toward a Unified Theory of Human Behavior* (R. R. Grinker, Ed.) Basic Books, New York, 1956, p. 37.
63. Goodall, J.: *Natl. Geograph. Mag.*, **124** (1963) 293.
64. Altmann, S. A.: *J. Theoret. Biol.*, in press. Lorenz, K.: in *Group Processes: Transactions of the First Conference* (B. Schaffner, Ed.) Josiah Macy, Jr. Foundation, New York, 1955, p. 179.
65. Köhler, F.: *Z. Bienenforsch.*, **3** (1953) 57.
66. Altmann, S. A.: in *Roots of Behavior* (E. L. Bliss, Ed.) Harper, New York, 1962, pp. 277–285.
69. Jakobson, R.: "Shifters, Verbal Categories, and the Russian Verb" (mimeograph, Department of Slavic Languages and Literatures, Harvard University, 1957.
68. Shannon, C. E., and Weaver, W.: *The Mathematical Theory of Communication*, Univ of Illinois Press, Urbana, 1949, p. 117.
69. Slama-Cazacu, T.: *Language et Contexte*, Mouton, The Hague, 1961.
70. Collias, N. E.: in *Animal Sounds and Communication* (W. E. Lanyon and W. N. Tavolga, Eds.) American Institute of Biological Sciences, Washington, D. C., 1960, p. 387.
71. Armstrong, E. A.: *A Study of Bird Song*, Oxford Univ. Press, London, 1963, p. 6.
72. Bartholomew, G. A., and Collias, N. E.: *Animal Behav.*, **10** (1962) 7.
73. Jesperson, O.: *Language: Its Nature, Development, and Origin*, Norton, New York, 1964, p. 123.
74. Tinbergen, N.: *The Herring Gull's World*, Basic Books, New York, rev. ed., 1961, p. 112.

Chapter 2

INFORMATION CODING

Introduction and Survey

Quantitative data about the information exchange between and within living organisms present a breakthrough in biological studies. Digital computers are new tools for data acquisition, signal and data processing, and system simulation.

This chapter deals with information coding. Information coding is important both as a computer technique and as a guide to understand biological communication systems.

Decimal and binary codes are compared, and definitions of bit, byte, and word are given. Basic logical operations are explained, and comparison between digital circuits and logical neurons is presented. Basic laws of information theory are derived, with emphasis on information content in biological systems. The chapter ends with the discussion of optimal and redundant codes. There is a strong evidence that different methods to achieve redundancy are used in biological communication systems.

2.1 Decimal, Binary, and Octal Number Systems

Decimal Number System The value of each position in a number is known as its position coefficient or weight. For example:

$$\begin{aligned} 346 = 3 \times 100 &= 300 \\ 4 \times 10 &= 40 \\ 6 \times 1 &= \underline{6} \\ & 346 \end{aligned}$$

A simple decimal weighting table is

$$\cdots 10^3\ 10^2\ 10^1\ 10^0\ \cdot\ 10^{-1}\ 10^{-2}$$

In general, in a base b system the successive digit positions, left to right, have the weights

$$\cdots b^3\ b^2\ b^1\ b^0\ \cdot\ b^{-1}\ b^{-2} \cdots$$

The symbol "$\cdot$" is called the radix point. The portion of the number to the right of the radix point is called the fractional part of the number, and the portion to the left is called the integral part. In the decimal system the radix point is called the decimal point.

Computing machines can be built on the basis of any number systems. However, all modern digital computers are based on the binary (base 2) system. Why has this new number system been brought into use? The reason is that it is easier to distinguish between two entities than between ten entities. Most physical quantities have only two states: a light bulb is on or off; switches are on or off; material is magnetized or demagnetized; current is positive or negative; holes in paper tape or cards are punched or not punched; and so on. It is easier and more reliable to design circuits which must differentiate between only two conditions (binary 0 and binary 1) than between ten conditions (decimal 0 through 9).

Binary Number System The base of the binary system is 2. The radix point can be called the binary point. The possible digits are 0 and 1. The successive digit position, left to right, have the weights:

$$\cdots 2^3\ 2^2\ 2^1\ 2^0\ \cdot\ 2^{-1}\ 2^{-2}\ 2^{-3} \cdots$$

This weighting table is used to convert binary numbers to the more familiar decimal system. For example, let us find the decimal equivalent of the binary number 10101 (Table 2.1).

TABLE 2.1

2^4	2^3	2^2	2^1	2^0	(Weight table)		
						Position	
1	0	1	0	1	(Binary number)	coefficient	
					1 X	1 =	1
					0 X	2 =	0
					1 X	4 =	4
					0 X	8 =	0
					1 X	16 =	16
						Decimal number =	21

Decimal, Binary and Octal Number Systems

The process of counting can be used to point out differences and similarities between decimal and binary systems. In the decimal system counting consists of increasing the digit in a particular position in the order 0, 1, 2, ..., 8. 9. When we reach 0 (10) in this position, we carry 1 to the immediate-left position. Since the binary number system utilizes only two digits, particular positions of the number can go only through two changes, and then we carry 1 to the immediate-left position. Thus the numbers used in the binary number system to count up to a decimal value 10 are as shown in Table 2.2.

TABLE 2.2

Decimal	Binary
0	0
1	1
2	10
3	11
4	100
5	101
6	110
7	111
8	1000
9	1001
10	1010

There are some special names widely used in computer technology, such as bit, byte and word.

Bit A *binary digit* is usually refered to as a *bit*. Thus a number such as 1010 is referred to as a 4-bit binary number, and 101 as a 3-bit number, and so on. The bit at the left end of the number is called the most significant bit (it has the largest weight). The bit at the right end of the number is called the least significant bit (it has the smallest weight). Figure 2.1 presents a binary number composed of 16 bits.

Byte The evolution of the computer and data equipment has brought about an 8-bit unit for information exchange between devices. Such an 8-bit unit is referred to as a byte. Many new types of digital computers and controllers thus express the numbers of 8, 16, 24, or 32 bits (1, 2, 3, or 4 bytes). Figure 2.1 presents a binary number composed of 2 bytes.

Word The computer is composed of a large number of cells or registers for storing binary information. Most of the registers in a given machine are

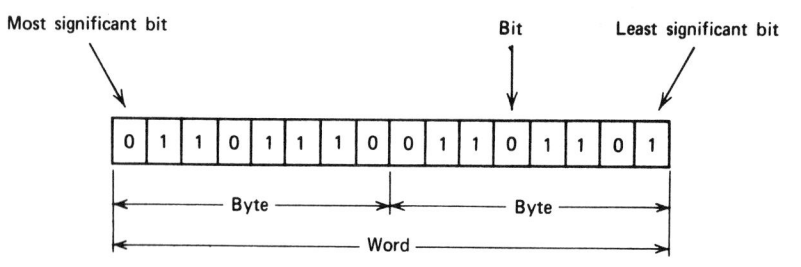

Fig. 2-1 Definitions of bit, byte, and word.

of the same length, n. Each register can be used to store n bits of binary information. Information stored in one register is also called a *word*. Figure 2.1 presents a word of a 16-bit computer.

Binary Addition Binary addition follows the same pattern as decimal addition except that a carry to the next position is not generated after the sum reaches 10, but is generated when the sum reaches 2 (1 + 1). For example:

$$101 = 5_{10}$$
$$+010 = 2_{10}$$
$$\overline{111} = 7_{10}$$

$$11 \leftarrow \text{carries}$$
$$111 = 7_{10}$$
$$+101 = 5_{10}$$
$$\overline{1100} = 12_{10}$$

Let us follow the second example (add 111 to 101). First, $1 + 1 = 0$ plus a carry of 1. In the second column, 1 plus the carry $1 = 0$ plus another carry. The third column is $1 + 1 = 0$ with a carry plus the previous carry, or $1 + 1 + 1 = 11$. Our answer is 1100 (decimal 12), which is the correct solution for $7 + 5$.

The addition of binary numbers in digital machines is performed by a special unit called a *functional adder*.

Octal Number System To express a number in the binary system, it is necessary to use substantially more digits than in the decimal system. For example, $(35)_{10} = (100011)_2$. It is very easy for humans to make errors in reading and writing large binary numbers. For easy notation of binary numbers, the octal number system can be used. The base of the octal system is 8, and the digits are 0 through 7. Thus the numbers used in the octal system to count up to a decimal value 10, are as shown in Table 2.3.

Basic Logical Operations

TABLE 2.3

Decimal	Binary	Octal
0	0	0
1	1	1
2	10	2
3	11	3
4	100	4
5	101	5
6	110	6
7	111	7
8	1000	10
9	1001	11
10	1010	12

Since the base of the octal system is $8 = 2^3$, to convert binary numbers into octal numbers one has to separate binary bits into 3-bit groups. These 3-bit groups can be represented by one octal digit using the table of equivalents in Table 2.3. For example:

```
110101111001           Binary number
110 101 111 001        3-bit groups
 6   5   7   1         Octal equivalent for each group
```

Hence $(110\ 101\ 111\ 001)_2 = (6571)_8$.

Conversion of a decimal number into an octal equivalent, and arithmetic operations with octal numbers, follows the same philosophy as for binary numbers.

It is important to notice that computers do not operate on the basis of the octal number systems. Computers operate in binary number systems. Octal notation is used sometimes as a help for humans to avoid reading and writing large binary numbers.

2.2 Basic Logical Operations

Numerical data and instructions for operations are represented in the computer in the binary number system. A binary signal can be either 1 or 0, and it may be considered as a special kind of variable, Fig. 2.2a. Binary variables can be represented by a letter. Let us suppose that a binary variable is denoted by A. To distinguish between possible states 1 and 0, we can use notation A and $\bar{A}$, respectively. $\bar{A}$ is called the complement of the variable A.

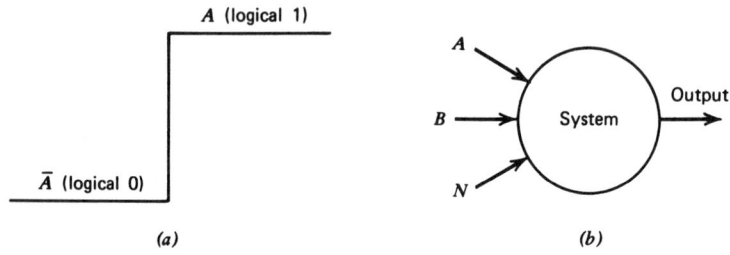

Fig. 2-2 (*a*) Binary variable, (*b*) Digital system.

To solve different problems, computers use logical operations between binary variables. Logical operations between binary variables are treated by a special branch of mathematics which is called Boolean algebra.

Figure 2.2*b* presents a general logical system with binary variables A, $B, \ldots, N$ as inputs. Boolean algebra can be used to find out the output of the system.

Input-output relationships of a logical system can also be expressed through the truth table. The truth table consists of two parts. The first part of the truth table is related to the inputs and presents a complete listing of all possible combinations of inputs $A, B, \ldots, N$.

The second part of the truth table is the output state as a function of combinations of inputs.

All logical systems can be deduced to a few basic logical circuits which will now be defined.

INVERTOR The invertor is one input circuit which produces the output, the state of which is equal to the complement of the input. Thus, if the input is a 0, the output is a 1; if the input is 1, the output is 0. Figure 2.3 shows the truth table of the invertor and a standard symbol which is used to indicate the invertor in block diagrams.

AND The AND circuit has two or more inputs. Figure 2.4*a* presents a simple circuit with two switches which illustrate the AND operation. If

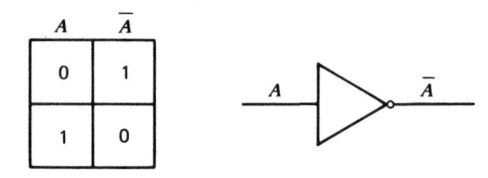

Fig. 2-3 Invertor.

Basic Logical Operations

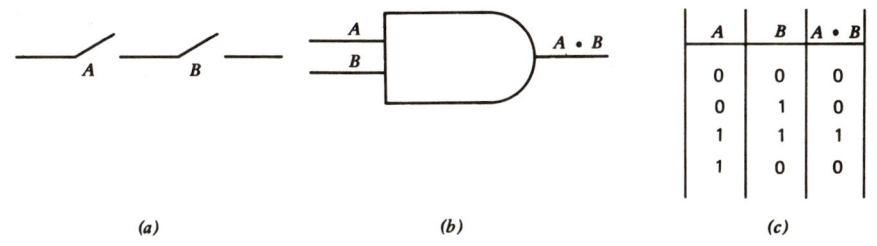

Fig. 2-4 AND circuit.

(a)   (b)   (c)

A	B	A · B
0	0	0
0	1	0
1	1	1
1	0	0

switch A is closed, the variable A has a value of 1. If the switch A is open, the variable A has a value of 0. The current can flow through the circuits only if both switches are closed (the output is 1, only if both variables A and B are 1). The AND operation is often indicated as $A \cdot B$. Thus the AND operation of three variables A, B, and C is indicates as $A \cdot B \cdot C$.

Figure 2.4b presents a standard symbol for the AND circuit. Figure 2.4c presents the truth table for the AND circuit.

OR The OR circuit has two or more inputs. Figure 2.5a presents a simple circuit with two switches which illustrates the OR operation. The current will flow if either of the switches A or B is closed. This operation is indicated by a plus sign between the variables. The truth table and block diagram are shown in Fig. 2.5 for two variables.

The three basic circuits, INVERTOR, AND, and OR, are sufficient to design a digital computer. These circuits can be combined to produce elaborate functions. Some of those functions, which are very frequently used, will be listed next.

NOR The NOR circuit is a combination of the OR and INVERTOR. Figure 2.6 shows the truth table of the NOR circuit and two block diagrams

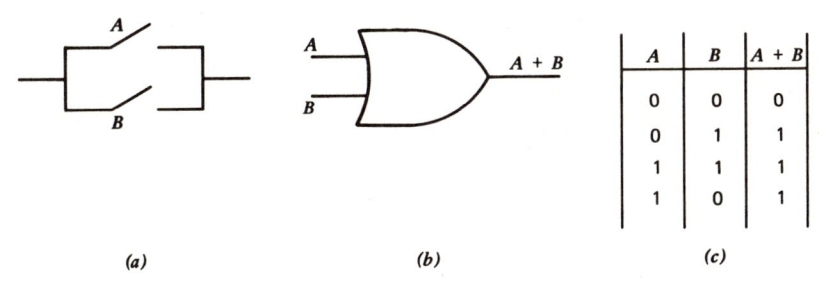

A	B	A + B
0	0	0
0	1	1
1	1	1
1	0	1

(a)   (b)   (c)

Fig. 2-5 OR circuit.

32   Information Coding

(a)

			OR →
A	B	A + B	$\overline{A+B}$
0	0	0	1
0	1	1	0
1	1	1	0
1	0	1	0
			← INVERTOR

(b)

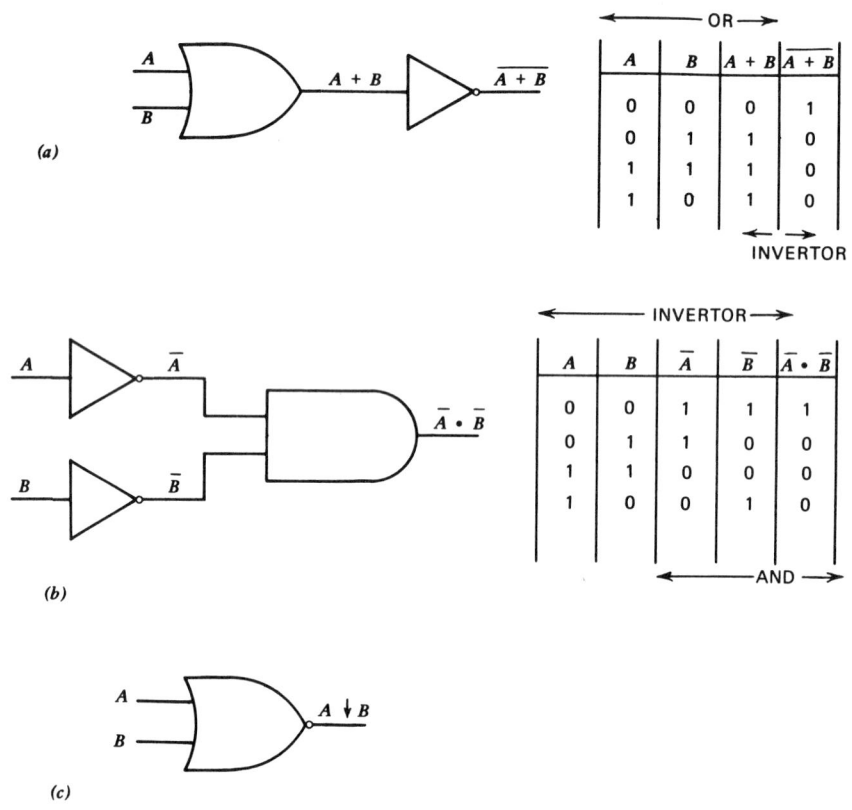

		← INVERTOR →		
A	B	$\overline{A}$	$\overline{B}$	$\overline{A} \cdot \overline{B}$
0	0	1	1	1
0	1	1	0	0
1	1	0	0	0
1	0	0	1	0
				← AND →

(c)

Fig. 2-6 NOR circuit.

for the NOR operation. The output is expressed as $\overline{A + B}$. This is simply the result of first performing the OR function $A + B$, and then the INVERT function. Because of the inversion, the output of the truth table for NOR circuit is exactly opposite of the output of the OR circuit. The truth table is composed of the truth tables for OR and INVERTOR as indicated in Figure 2.6a.

One can also perform NOR operation by different combinations of circuits (Fig. 2.6b). This circuit is composed of invertors and an AND gate, but one can see that its truth table gives the same output as the one in Figure 2.6a. In the second circuit, the output is expressed as $\overline{A} \cdot \overline{B}$. This is the result of first performing the INVERT functions $\overline{A}, \overline{B}$, and then the AND function. We see that

$$\overline{A + B} = \overline{A} \cdot \overline{B} \tag{2.1}$$

Basic Logical Operations 33

Equation 2.1 can be written in a more general form which is known as de Morgan's theorem and presents a fundamental theorem of Boolean algebra. This theorem states that the complement of a function is obtained by complementing each of the variables and interchanging the operations of OR and AND. The general form of de Morgan's theorem is

$$\overline{f(A_1, A_2, \ldots, +, \cdot)} = f(\bar{A}_1, \bar{A}_2, \bar{A}_3, \ldots, \cdot, +) \tag{2.2}$$

The bottom of Figure 2.6 presents a symbol for NOR operation.

NAND The NAND circuit is a combination of the AND and INVERTOR. Figure 2.7 shows the truth table and two block diagrams for NAND operations. In the first diagram, NAND operation has been expressed as $A \cdot B$. In the second diagram, NAND operation has been expressed as $\bar{A} + \bar{B}$. This is another example where de Morgan's theorem, Eq. 2.2, can be used

$$\overline{A \cdot B} = \bar{A} + \bar{B} \tag{2.3}$$

The bottom of Figure 2.7 shows a symbol for NAND operation.

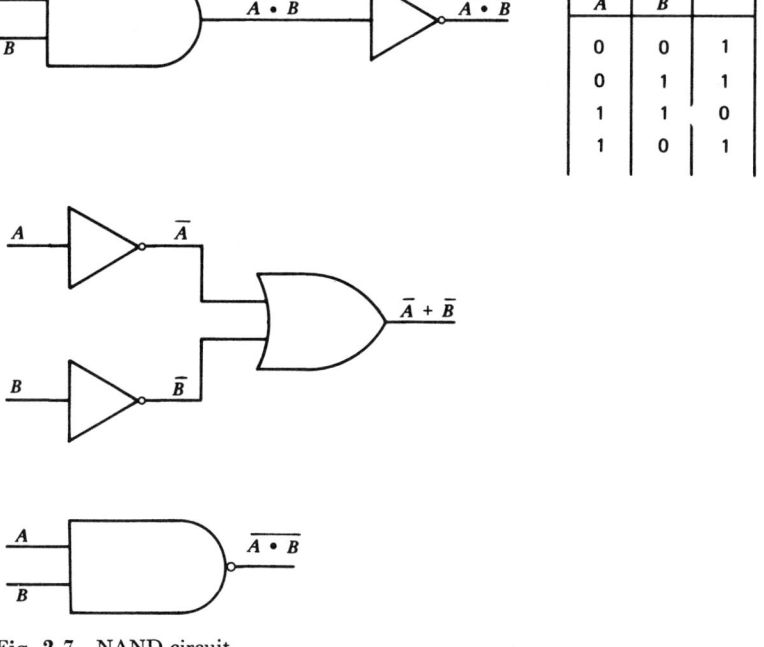

Fig. 2-7 NAND circuit.

EOR The EOR circuit presents the exclusive OR operation. The exclusive OR is similar to OR, with the exception that one set of conditions for A and B are excluded; if both A and B are in the state 1, the output will not go into the state 1 but will take the state 0.

Figure 2.8 presents a truth table, block diagram, and symbol of the EOR circuit.

The output of EOR circuit, according to the block diagram, is given by

$$Q = (A + B) \cdot \overline{(A \cdot B)} \qquad (2.4)$$

The operations, INVERTOR, OR, AND, NOR, NAND, and EOR, can be applied to binary numbers.

A binary number in the computer is composed of a number of bits. Each bit can be treated as one binary variable. Hence logical operations can be performed between two binary numbers, treating each pair of bits as two binary variables. For example:

$$\begin{array}{l} A_{12} A_2\ A_1 \\ A = 1\ 0\ 1\ 0\ 0\ 0\ 1\ 1\ 0\ 0\ 1\ 0 \quad \text{First number} \\ B = 0\ 1\ 1\ 1\ 0\ 1\ 0\ 1\ 1\ 0\ 1\ 1 \quad \text{Second number} \\ B_{12} B_2\ B_1 \end{array}$$

results in

$$\begin{array}{rl} A \cdot B = 0\ 0\ 1\ 0\ 0\ 0\ 0\ 1\ 0\ 0\ 1\ 0 & \text{AND} \\ A + B = 1\ 1\ 1\ 1\ 0\ 1\ 1\ 1\ 0\ 1\ 1 & \text{OR} \\ \overline{A \cdot B} = 1\ 1\ 0\ 1\ 1\ 1\ 1\ 0\ 1\ 1\ 0\ 1 & \text{NAND} \\ \overline{A + B} = 0\ 0\ 0\ 0\ 1\ 0\ 0\ 0\ 0\ 1\ 0\ 0 & \text{NOR} \\ (A + B) \cdot \overline{(A \cdot B)} = 1\ 1\ 0\ 1\ 0\ 1\ 1\ 0\ 1\ 0\ 0\ 1 & \text{EOR} \end{array}$$

Timing The truth tables of logical circuits show a steady state of input-output relationships. However, when electrical pulses are applied as inputs, it takes some time for the output to reach the steady-state level because of the internal delays in the circuits. The delay of one circuit is usually small, in the order of a few nanoseconds (10^{-9} sec). In the computer, binary signals of the pulse pass through many circuits, and delays might become substantial. In this situation, it is crucial to allot a specific amount of time for each step of an operation. If the operation is completed before this time has elapsed, the machine waits. In this way the speed of operation is somewhat slowed down, but high reliability is achieved because the synchronism between many parallel-going operations is guaranteed.

This basic synchronism for a computer is derived from a clock. This is normally a free-running oscillator designed to produce pulses with stable constant frequency (Fig. 2.9). Two levels of pulses correspond to two logical

Flip-Flop 35

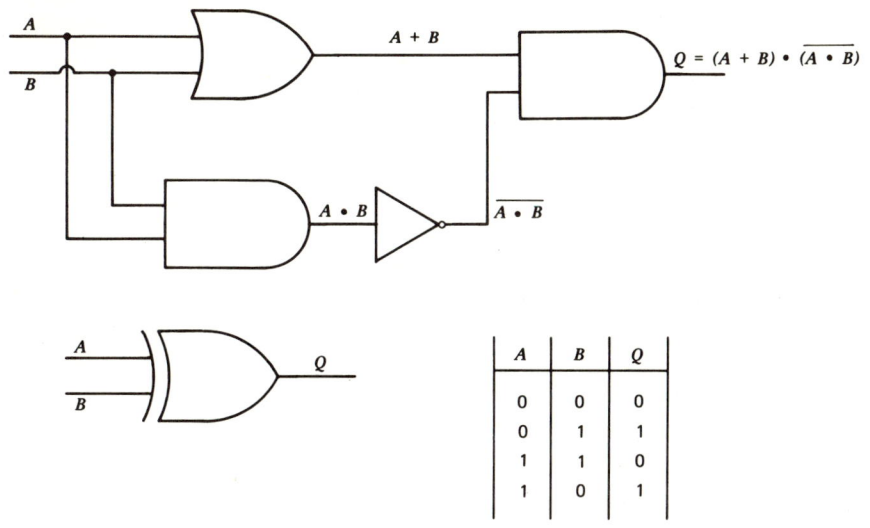

Fig. 2-8 Exclusive OR circuit (EOR).

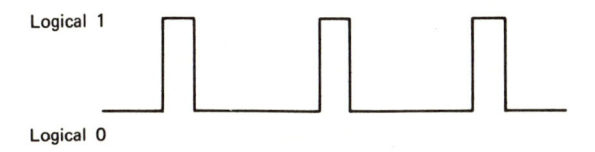

A	B	Q
0	0	0
0	1	1
1	1	0
1	0	1

Fig. 2-9 Clock.

states of binary variables, 0 and 1. Almost all operations in the machine are gated with the clock pulses.

However, input-output transfers between computer and peripherals devices are usually asynchronous.

2.3 Flip-Flop

The main active elements in the computer are the flip-flops, which are binary storage devices, each capable of storing a single bit of information. The flip-flop is also called a binary trigger, latch, bistable multivibrator, or, simply, bistable.

Figure 2.10 explains a basic principle of a device having two stable states.

The circuit has two inputs, the SET and RESET. Normally, both the set and reset inputs are at 0. If the SET input is brought to 1, the circuit output

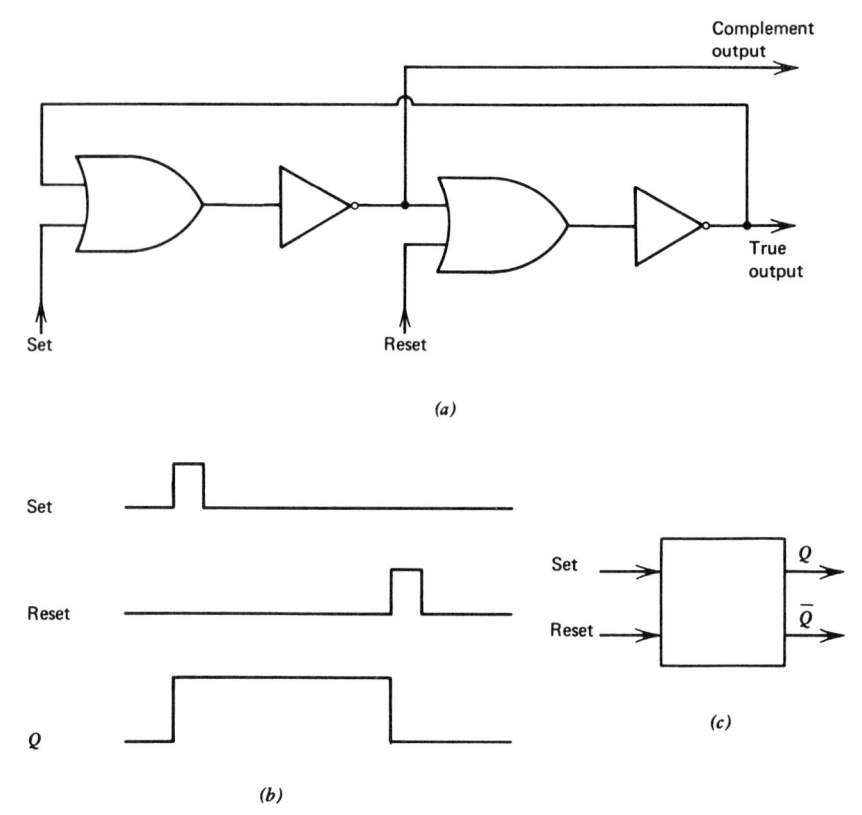

Fig. 2-10 Flip-flop.

goes to 1. Even when the input is removed, the circuit stays in the state 1, due to the internal feedback.

Figure 2.10a shows the flip-flop composed of OR circuits and invertors, and it explains the feedback action which keeps the circuit in the stable state. If the RESET input is brought to 1, the circuit goes to the opposite stable state and the true output goes to 0. The same feedback from the output will keep the circuit in the state 0 when the reset input is removed.

Figure 2.10b shows the input-output relationship for the flip-flop. Figure 2.10c shows a symbolic presentation of the flip-flop.

A flip-flop can be easily designed from INVERTORS and OR gates, as shown in Figure 2.10. However, since the flip-flops are very frequently used, they are produced as standard integrated units. Flip-flops are produced in few versions, which differ in triggering arrangements.

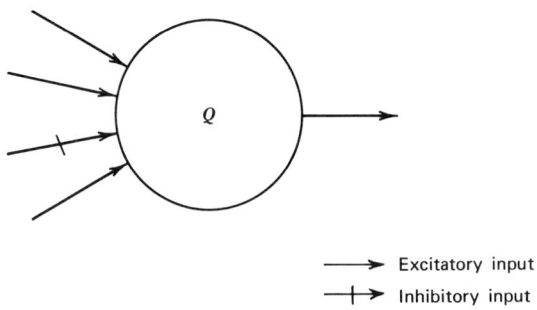

Fig. 2-11 Logical neuron.

2.4 Logical Neurons

McCulloch and Pitts[1] have developed the idea of logical or mathematical neuron. Its representation is shown in Figure 2.11. Logical neuron has a large number of inputs and one output. It will produce an output only if at least Q inputs are active, where Q is a threshold of a logical neuron. Logical neuron has many realistic features such as threshold, excitability, spatial summation, and all-or-none output.

A logical neuron can exist in one of two states, which may be called active and inactive. Hence logical neuron is a binary device, because it has two possible states

McCulloch and Pitts[1] had realized great similarity between input-output relations of their idealized neurons and the truth functions of symbolic logic.

2.5 Information Content of Memory and of Communication Channel

One-Symbol Messages Each communication operation transmits some quantity of information. It is important to define the unit to measure the quantity of information.

The simplest communication system is one transmitting only 1 bit. Only two messages, "yes" (1) or "no" (0) could be expressed with 1 bit.

Communication system transmitting 2 bits could convey four messages: 00, 01, 10, and 11. The general, communication system transmitting n bits could convey 2^n alternative messages:

n bits correspond to a choice between $a = 2^n$ alternatives

It is important to notice that similar relationship has been used by mathematicians to define the "base 2 logarithm":

If the number $a = 2^n$, than the number $n = \log_2 a$

From the above two relationships it follows that information content or capacity I could be defined as

$$I = \log_2 a \tag{2.5}$$

Thus defined information content I is equivalent to the base 2 logarithm of the number of possible alternations, and its unit is the bit.

Example 1 The English alphabet is composed of 26 symbols or alternatives. Hence the information content of the symbol is

$$I = \log_2 26 = 4.7 \text{ bits}$$

Example 2 Arabic numerals present 10 decimal digits or symbols 0, 1 ... 9. Hence the information content of the symbol is

$$I \log_2 10 = 3.32 \text{ bits}$$

Example 3 The digital memory has been designed, having 40 bits (say 40 flip-flops). Hence the information content of this memory is $I = 40$. The number of alternative messages that could be expressed with 40 bits is

$$a = 2^n = 2^{40} = 1099511627776$$

Note that even a fairly small information capacity, say 40 bits, can correspond to the store having a quite astronomical number of different distinct states.

Example 4 Nobody knows what the information capacity of the human memory is. Based on physiological considerations such as the visual input of information, it has been estimated that human memory may receive 10^8–10^9 bits/sec. If all this information were permanently stored, one would receive through the years say $n = 10^{20}$ bits. The number of different messages stored would be then

$$a = 2^n = 2^{10^{20}}$$

It is beyond human capacity to envisage the magnitude of the number a.

Content of Memory and Communication Channel 39

Multisymbol Messages One letter contains 4.7 bits of information because there are 25 other letters that can represent other messages. The next question is how to calculate information content of a message that consists of two or three letters.

The number of possible combinations or different messages expressed with three letters is

$$a = 26^3$$

Using equation 2.5, the information content of the three-letter message is

$$I = \log_2 (26)^3 = 3 \cdot \log_2 (26) = 3 \cdot 4{,}7 = 14{,}1 \text{ bits}$$

In general, if the message is composed of Q symbols, and if the number of symbol types is m, then the number of possible combinations is m^Q. The information content in this case is

$$I = \log_2 (m^Q) = Q \cdot \log_2 m \tag{2.6}$$

The Statistical Theory of Information Equation 2.6 shows that the information content depends on the number of different symbols available, m. If all the symbols are equally frequent, the relative frequency of each is equal to the reciprocal value of the number of symbol types m

$$p = \frac{1}{m} \tag{2.7}$$

Hence for equally frequent symbols, equation 2.6 could be rewritten in the form

$$I = Q \log_2 \frac{1}{p} = -Q \log_2 p \tag{2.8}$$

Equation 2.8 can be also used in the case when different symbols have different frequencies of occurrences. In this case p is the probability of occurrence of a given symbol.

Example 5 Information content of the letters in the language. Table 2.4 shows relative frequency of occurrence or probabilities of the various letters of the English alphabet. The right-hand column shows the information content I, obtained from equation 2.8, for one-letter messages ($Q = 1$).

Note that the most frequent letter E carries only 2.98 bits of information. On the other side, the least frequent letter Z carries 10.71 bits of information.

TABLE 2.4

Letter	Relative frequency %	Probability	Bits	Letter	Relative frequency %	Probability	Bits
E	12.68	0.1268	2.98	F	2.56	0.0256	5.29
T	9.78	0.0978	3.35	M	2.44	0.0244	5.36
A	7.88	0.0788	3.67	W	2.14	0.0214	5.51
O	7.76	0.0776	4.69	Y	2.02	0.0202	5.64
I	7.07	0.0707	3.82	G	1.87	0.0187	5.74
N	7.06	0.0706	3.82	P	1.86	0.0186	5.75
S	6.31	0.0631	3.99	B	1.56	0.0156	6.00
R	5.94	0.0594	4.08	V	1.02	0.0102	6.62
H	5.73	0.0573	4.13	K	0.60	0.0060	7.38
L	3.94	0.0394	4.67	X	0.16	0.0016	9.29
D	3.89	0.0389	4.68	J	0.10	0.0010	9.97
U	2.80	0.0280	5.16	Q	0.09	0.0009	10.12
C	2.68	0.0268	5.22	Z	0.06	0.0006	10.71

Equation 2.8 could be further modified. In the statistical theory we define a quantity H, which gives a total information content, or the degree of uncertainty of the system.

For a system with m possible outcomes having respective probabilities $p_1, p_2, \ldots, p_m$, H is equal to

$$H = -p_1 \log_2 p_1 - p_2 \log_2 p_2 \cdots = -\sum_{i=1}^{m} p_i \log_2 p_i \qquad (2.9)$$

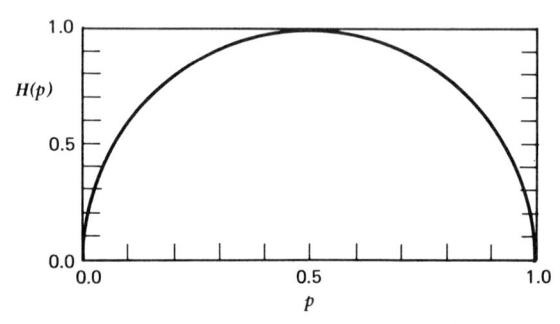

Fig. 2-12 Information content of the system with two possible outcomes, success, or failure. The probability of the success is p. (From J. S. Griffith: *Mathematical Neurobiology*, Academic Press, London-New York 1971.)

Redundancy and Message Distances

If the number of possible outcomes $m = 2$, with probabilities $p_1 = p$ and $p_2 = 1 - p$, we have

$$H = -p \log (p) - (1 - p) \cdot \log (1 - p) \qquad (2.10)$$

Equation 2.10 is plotted in Figure 2.12.

For $p = 0$, we are certain that the experiment will fail; hence the experiment acquires no information, $H = 0$. Also, for $p = 1$, we know that experiment produces always the same result, hence another experiment acquires no information, $H = 0$. However for the case $p = 0.5$, the outcome of the experiment is the most uncertain, and the new experiment acquires maximum information content $H = 1$.

2.6 Redundancy and Message Distances

Distance Figure 2.13 presents a digital system composed of 3 bits. Three bits could form $2^3 = 8$ alternatives or messages, 000, 001 $\cdots$ 111.

DISTANCE

Messages i and j are at the distance d_{ij}.

Distance d_{ij} is equal to the number of bits which are different in the two messages.

$d_{02} = 1$

$d_{34} = 3$

$d_{47} = 2$

Cube presentation: each dimension presents one bit.

Distance d_{ij} between two points i and j is equal to the number of coordinates for which X and Y differ.

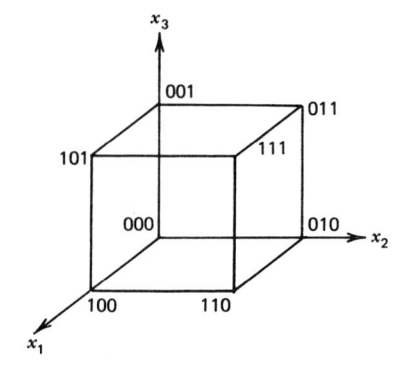

Fig. 2-13 Distance between messages.

The distance between the messages is defined as the number of bits that are different in the two messages. For example, message 000, presenting decimal 0, and message 010, presenting decimal 2, have a distance $d_{02} = 1$, because they are different in only 1 bit. Messages 100, presenting decimal 4, and 111, presenting decimal 7, have a distance $d_{47} = 2$, because they are different in two bits.

For a 3-bit system a cube presentation could be used to visualize the distances between the messages. In cube presentation each dimension presents 1 bit, as shown in Figure 2.13. If, say, bit X_3 is zero, then the X_3 coordinate is zero. If bit X_3 is equal one, the X_3 coordinate is one.

One could directly read the distances between messages from the cube: the distance d_{ij} between two points i and j is equal to the number of coordinates for which X and Y differ. For example, messages 000 and 110 are on the distance $d = 2$, because there are two edges of the cube between them.

Redundancy The information quantity is a function of the number of alternative messages used. It is implied that each message is different from all other messages. However, sometimes two messages carry in part the same information. This is known as repeated information or redundancy.

The redundancy is a function of a code used for communication. Concrete explanation of the redundancy is presented on a simple example in Figure 2.14.

REDUNDANCY

n = Number of used binary digits

m = Minimum number of bits necessary to convey the same message

REDUNDANCY $R = \frac{n}{m}$

Optimal code $d \geq 1$

```
0 0 0
0 0 1
0 1 0
0 1 1
1 0 0
1 0 1
1 1 0
1 1 1
```

$R = \frac{3}{3} = 1$

Redundant $d = 2$

```
0 0 0
0 1 1
1 0 1
1 1 0
```

Single error detection

$R = \frac{3}{2}$

Codes $d = 3$

```
000
111
```

Single error correction

$R = \frac{3}{1}$

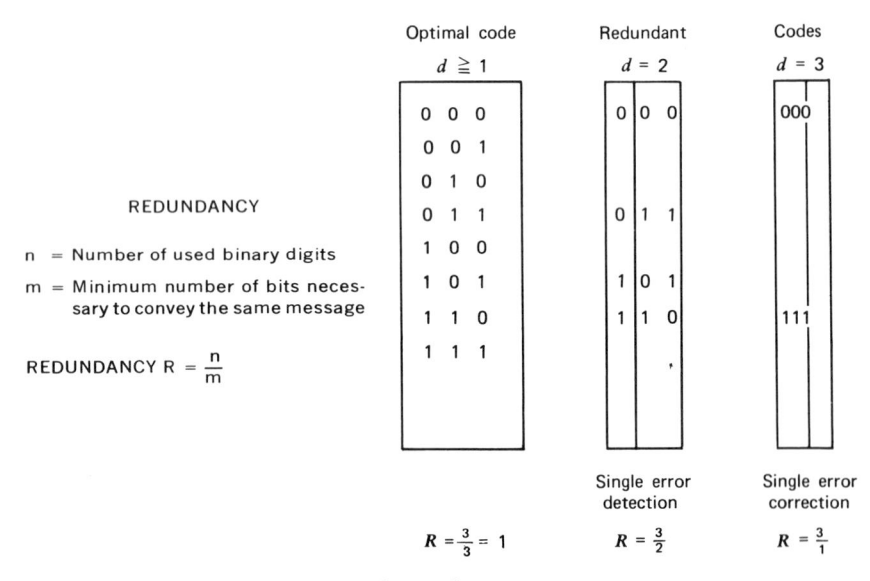

Fig. 2-14 Optimal code and redundant codes.

Redundancy and Message Distances

Figure 2.14 presents digital system comprised of 3 bits. Three bits could produce 8 different messages 000, ..., 111. These eight messages are presented in the left-hand side of Figure 2.14, and they together form optimal code.

Optimal code is a code that gives the maximal number of different messages. The major characteristic of optimal cose is that some of the messages have the distance $d = 1$. For example, messages 010 and 011 have a distance $d = 1$ because they differ only in 1 bit.

During the transmission of the message from sender to the receiver, the communication channel noise is added to the message, and it may cause a change in 1 or more bits of the message. The receiver would interpret the changed message as another message of the same optimal code. For example, if the message 010 is transmitted and the channel noise changes the right-hand side bit 0 into 1, a new message is formed, 011. However this is also legal message of the optimal code, and the receiver has no way to find out that the received message is produced under the influence of noise.

To distinguish the optimal code from other codes, the redundancy R is defined

$$R = \frac{n}{m}$$

where n = number of available binary bits in the system
m = minimum number of binary bits necessary to convey the same message.

For the example of Figure 2.14, 3 bits are used, hence $n = 3$. Also, eight messages are generated, hence $m = 3$, and it follows for optimal code

$$R = \frac{n}{m} = \frac{3}{3} = 1$$

The middle part of Figure 2.14 presents a redundant code with $d = 2$. Again a 3-bit digital system is used ($n = 3$). However, out of eight possible messages, only four are used. The four messages are selected in such a way that the distance between any of them is always $d = 2$. Only the selected four messages are legal messages. If the legal message 011 is transmitted and the channel noise changes 1 bit, say producing a new message 010, this message does not belong to the class of legal messages, and the receiver will immediately notice an error.

The basic property of redundant code with $d = 2$ is the single-bit error-detection feature.

In this example the minimal number of bits necessary to convey four messages is $m = 2$; hence

$$R = \frac{3}{2}$$

The right-hand side of Figure 2.14 presents a redundant code with $d = 3$. Again, a 3-bit digital system is used ($n = 3$). However, out of eight possible messages, only two are used. The two messages are selected in such a way that the distance between them is $d = 3$.

If the legal message 000 is transmitted and the channel noise changes 1 bit, say producing a new message 001, the transmitter will recognize that this message is illegal, and it will look for the closest legal message. The receiver will find out that the closest legal message is 000. Hence the basic property of redundant code with $d = 3$ is the single-bit error-correction feature.

In the example, the minimal number of bits necessary to convey two messages is $m = 1$; hence

$$R = \frac{3}{1}$$

It is clear that as the redundancy and distance are increased, the utilization of available bits become poorer, but the messages are better protected against the noise.

Another obvious way to achieve redundancy and to protect the message is to repeat the message a few times. In this way the receiver could compare received messages and introduce a correction if needed.

There is a strong evidence that different methods to achieve redundancy are used in biological communication, as is shown later.

References

1. McCulloch, W. S., and Pitts, W., Bulletin of Mathematical Biophysics, 5 (1943), 115.
2. Souček, B.: *Minicomputers in Data Processing and Simulation*, Wiley, New York, 1972.
3. Griffith, J. S.: *Mathematical Neurobiology*, Academic Press, London and New York, 1971.
4. Hassenstein, B.: *Information and Control in Living Organisms*, Chapman and Hall, London, 1971.

Chapter 3

INTERFACING THE EXPERIMENT TO THE COMPUTER

Introduction and Survey

The introduction of integrated circuits and tens of thousands of small computers known as minicomputers and microprocessors is rapidly changing the profile of laboratory research and of measurement and control techniques. Cost of microprocessors and minicomputers, from few hundred to few thousand dollars, is now comparable with the cost of typical laboratory instruments. As a result, minicomputers and microprocessors are becoming standard laboratory tools.

Direct connection of the minicomputer into measuring chain results in a new, highly sophisticated kind of experiment, with significantly increased accuracy and large volume of data.

This chapter gives an overview of laboratory minicomputers and microprocessors. Interfacing the minicomputer to experiment and to the process control systems is explained. Programmed input output transfer of data is described. Interrupt mode of operation and high-speed direct memory access are shown.

Recently, catheters have been implanted into a human body. Under the control of the stored program microprocessor, such a system processes the biological signals. For example, a surgically inserted catheter can monitor the flow rate and the blood pressure in the various chambers of the heart. The attached microprocessor immediately analyzes the data.

3.1 Laboratory Mini- and Microcomputers

General Trends Application of digital computers started a quarter of a century ago, and today computers are found everywhere. Most of those applications have been in the area of numerical analysis, business data processing, and statistics.

Recently, a new trend has started in the computer world. The main characteristic of this new trend is introduction of new kind of computers, called minicomputers and microprocessors. These machines are small in size, yet powerful as computational or control tools.

Figure 3.1 shows a cross section of the computer world. On the right-hand side are large, classical computers. They provide high accuracy of 32 to 64 bits, necessary for mathematical applications. These machines have large memories, capable of storing millions of words of information, and they are equipped with all kinds of sophisticated input-output devices. The cost of those machines is in hundredths of thousands of dollars. Typically one finds such machines in computer centers.

The middle part of Figure 3.1 deals with microprocessors and minicomputers. They provide accuracy of 8 to 16 bits, which is usually best matched for general computation, process control, and automated data acquisition. This machine could have memories with storage capacity starting with only a few hundred words, going up all the way to 1 million words. Typically they are equipped with a moderate, yet quite powerful set of input-output devices.

The cost of microprocessors and minicomputers runs from a few hundred to a few thousand dollars, which has become comparable with the cost of typical laboratory instruments, such as a display scope or a counter. As a result, minicomputers and microprocessors have found their way in scientific and industrial laboratories. Each year thousands of those machines enter the area of experiment and process control and system simulation.

The main merits of microprocessor and minicomputer application in scientific or industrial laboratory are

- Money and Manpower saving.
- Speedup of experiment or production time.
- Highly improved quality of results.
- Highly increased volume of data.
- New kinds of experiments, which are feasible only through computer-controlled instrumentation.
- New kinds of production processes, which are feasible only through computer control of the process.

Early process control computers asked for highly skilled professionals for program development in assembly language who were also to design electrical interfaces between machines and instruments. Modern minicomputers and microprocessors have come much closer to the end uses. Many of those machines could be programmed by the end-user in easy-to-learn high-level

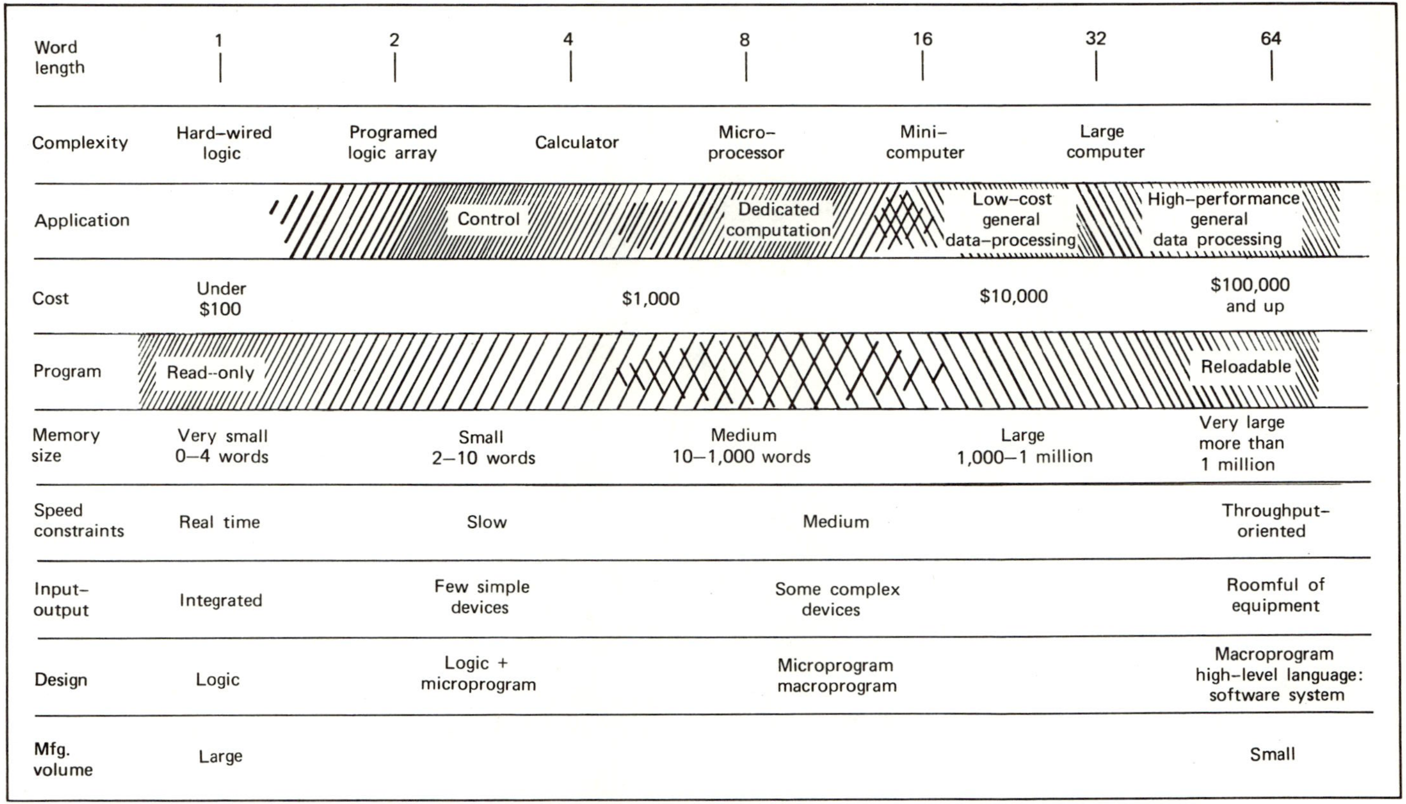

	1	2	4	8	16	32	64
Word length							
Complexity	Hard-wired logic	Programed logic array	Calculator	Micro-processor	Mini-computer	Large computer	
Application	Control			Dedicated computation	Low-cost general data-processing		High-performance general data processing
Cost	Under $100		$1,000		$10,000		$100,000 and up
Program	Read-only						Reloadable
Memory size	Very small 0–4 words		Small 2–10 words	Medium 10–1,000 words	Large 1,000–1 million		Very large more than 1 million
Speed constraints	Real time		Slow	Medium			Throughput-oriented
Input-output	Integrated		Few simple devices	Some complex devices			Roomful of equipment
Design	Logic		Logic + microprogram	Microprogram	macroprogram		Macroprogram high-level language: software system
Mfg. volume	Large						Small

Fig. 3-1 Computer world. Comparison between large computers, minicomputers, and microprocessors.

Fig. 3-2 Typical laboratory microcomputer. Such a system could be used for data acquisition, process control, data processing, and simulation. (Reprinted with permission, copyright 1975 by Digital Equipment Corporation. All rights reserved.)

language. The most frequently used languages are real-time FORTRAN, real-time BASIC, and PL/M. Also, interface design has moved from the level of digital circuit design to the level of plug-in interfacing cards.

Figure 3.2 shows a typical microcomputer, and Figure 3.3 shows one of recently developed microprocessors. Such machines are becoming standard tools in measurement and in control environment.

Structure of Minicomputer or Microprocessor Figure 3.4 presents simplified block diagram of a typical minicomputer or microcomputer: memory, major registers and data ways between part of the machine.

Memory is composed of large number of locations or words. Each location has an unique address and can store one piece of information. This informa-

Laboratory Mini- and Microcomputers 49

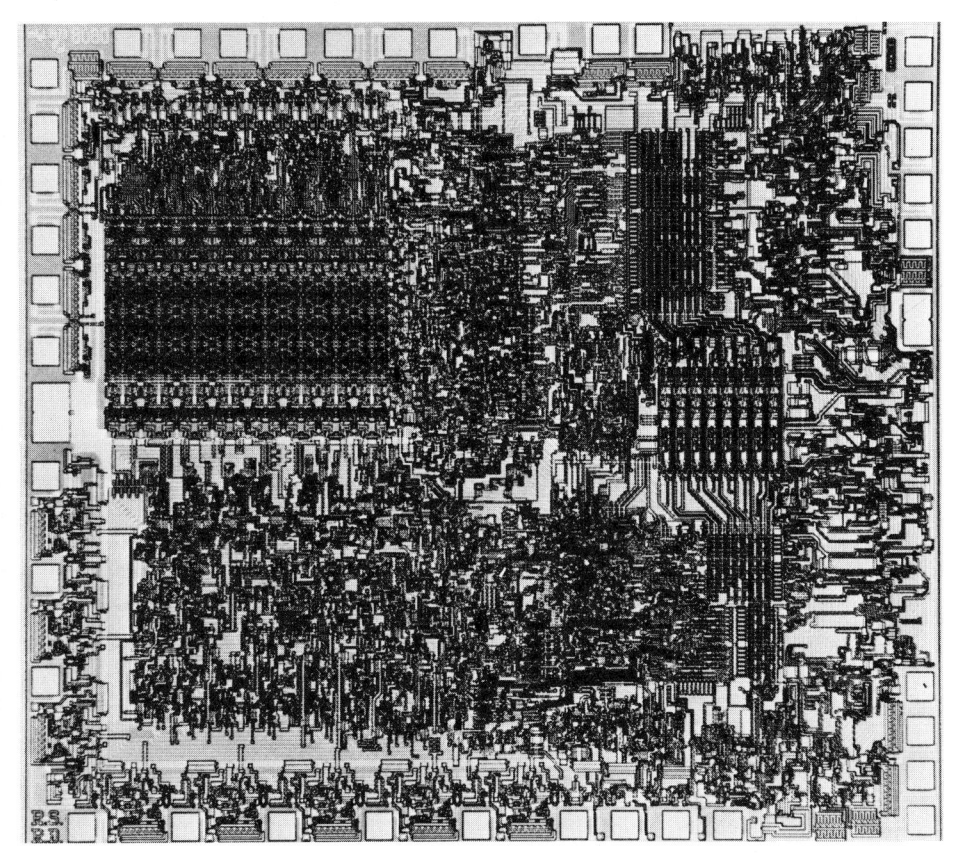

Fig. 3-3 Microprocessor, new tool in instrumentation and in process control. Actual size less than 1 cm^2. (Reprinted with permission, copyright 1973 by Intel Corporation. All rights reserved.)

tion can be data, but it can also be an instruction to tell the computer what to do.

Accumulator (AC) is prime register of the computer. It is used to keep the data and the results during arithmetic or logical operations. It is also used as the end-point for input-output transfer of data between the computer and peripheral world.

Memory data (MD) register is used for reading the data out of the memory or for writing the data into memory.

Instruction register (IR) is used to keep the operation code of the instruction currently being preformed by the computer. The IR decodes the operation

50 Interfacing the Experiment to the Computer

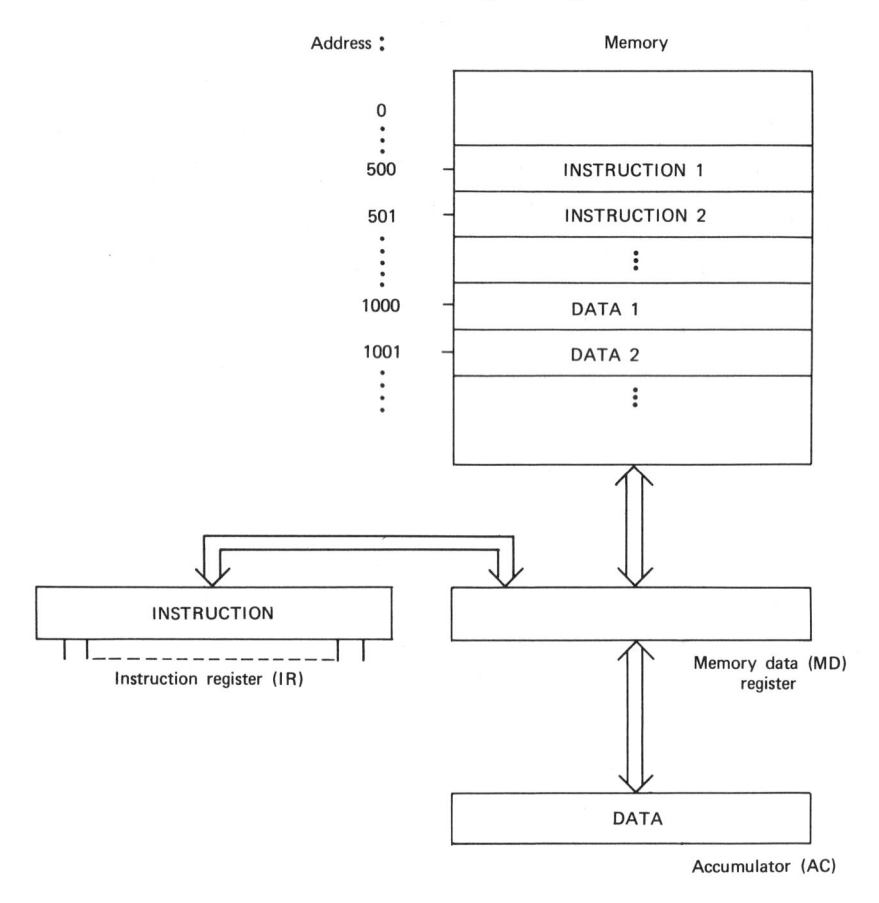

Fig. 3-4 Simplified block diagram of a small computer.

code of the instruction and initiates the steps necessary for the execution of the instruction.

During the computer operation, instructions are read from the memory and routed to instruction register to tell the computer what to do. The data are routed to the accumulator. The results are formed on the accumulator, then routed back into different memory locations. During input-output transfer, data are routed on one side between the accumulator and the peripheral devices and on the other side between the accumulator and the memory.

3.2 Programmed Input-Output Transfer

One of the major applications of small computers is in the domain of process control, data collection, and measurement. For those applications it is of importance that the computer be capable of communicating with the devices of the measuring and control chains. The exchange of information between the peripheral device and the computer is controlled either by the computer program or by specially designed elements of the peripheral device. Input-output transfers controlled by the computer program are called programmed data transfers. The transfer controlled by the peripheral device is performed without program intervention through special data channels which steal time slices from the central processor whenever necessary. Hence such a transfer is called the cycle stealing transfer. In some machines this transfer is called by other names, such as data break, data channel, and direct memory access (DMA). In this section we discuss only programmed data transfers.

A programmed data transfer is performed by the I/O transfer instruction. A small computer has at least one IOT instruction. This instruction can be used for the following tasks:

- To send the command to the peripheral device, instructing the device what to do. For example, a magnetic tape unit can be instructed to backspace the tape by one record.
- To receive and test the information describing the status of the peripheral device. An example is a test to determine if a magnetic tape transport is rewinding or if it is ready for recording.
- To output the data from the computer to the peripheral device. An example is the output of X and Y data to be used as coordinates of the point to be displayed on CRT.
- to input the data from the peripheral device to the computer. An example is the input of the digitized data from the measured process.

The above tasks can be performed in one of the three ways called unconditional transfer, conditional transfer, and program interrupt.

Unconditional Transfer This transfer is rarely used, only for processes whose timing is fixed and known. The peripheral device must be ready for communication. The program for unconditional transfer is simple and straightforward, as shown schematically in Figure 3.5. The IOT instruction is inserted in the program between other instructions at the place where the transfer is needed.

Conditional Transfer This transfer is used very often. It is performed under program control, but only if the peripheral device is ready for com-

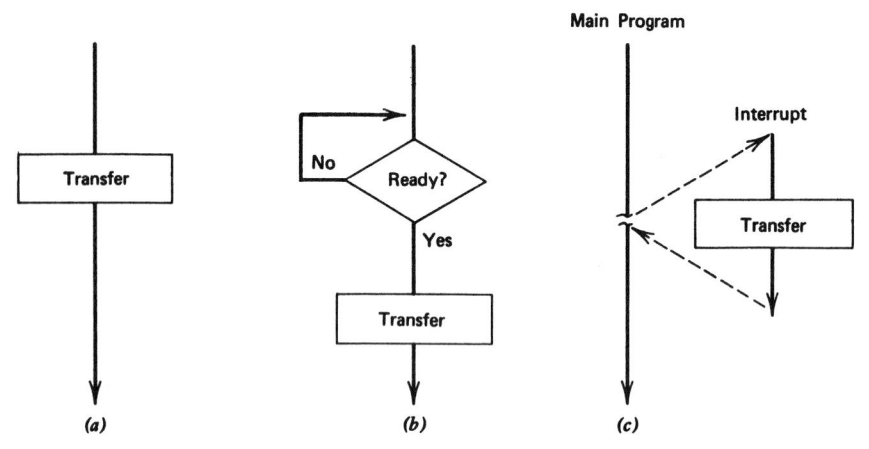

Fig. 3-5 Programmed input-output transfer. (*a*) Unconditional, (*b*) Conditional. (*c*) Program interrupt.

munication. The computer program is shown schematically in Figure 3.5. Two IOT instructions are usually used to perform the transfer. The first instruction is used to bring into the computer the information describing the status of the peripheral device. The program then tests the status and makes the decision. If the device is not ready, the program can perform the loop and check the status repeatedly. When the device becomes ready, the program executes the second IOT instruction, which performs the desired action. The main advantage of the conditional transfer is that it enables the synchronization between the computer operation and the peripheral-device operation. The main disadvantage is the waste of computer time if one has to wait for the device to become ready.

Program Interrupt Program-interrupt transfer makes more efficient use of the computer time possible. The transfer will be performed under the program control, but the computer does not have to check repeatedly if the device is ready for transfer. The computer program can perform a calculation that is independent of the transfer. Let us call this program the background job. When the peripheral device is ready, it will interrupt the computer and cause it to leave its background program for a moment and to perform a special interrupt subroutine for transfer. This operation is shown schematically in Figure 3.5. The IOT command is part of the interrupt subroutine. When this subroutine is completed, the control is returned to the background program.

Party Line for Programmed Input-Output

Simple Example Let us compare the unconditional, conditional, and program-interrupt transfer in the following way.

One member of the family is "programmed" to watch over the milk and to take it off the stove when cooked.

Unconditional operation: go in the kitchen at, say, 8:25 and take the milk off the stove, without caring whether it is cooked.

Conditional operation: look at the milk once every minute. When cooked, take it off the stove.

Interrupt operation: set the alarm to ring when the milk starts boiling. Perform the background job of writing homework. When the alarm rings, interrupt the background job for a moment and take the milk from the stove. Resume the background job from the point where interrupted.

3.3 Party Line for Programmed Input-Output

Normally, peripheral devices are slower than the computer. Hence a computer can communicate, if necessary, with a number of peripheral devices.

The I/O lines of a computer form a bus to which all the peripheral devices are connected.

Figure 3.6 shows the computer bus with a number of peripheral devices. This mode of operation is called the party line operation. The input/output lines from the computer are bused to all devices on the party line, so the devices appear as a single device to the computer. Each peripheral device must have its own controller. The controller must have a device selection address, so that the computer can call a specific device for transfer.

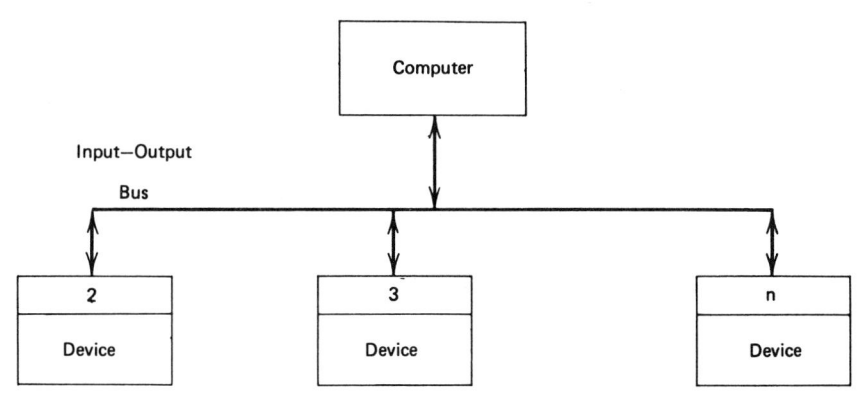

Fig. 3-6 Party line input-output transfer. A single computer controls a number of measuring and control points.

The controller must take care of the following:

- To decode the device selection code received from the computer and to respond only if the code is identical with its address.
- To decode the command code received from the computer and to initiate operations.
- To send the computer the information describing the status of the peripheral device.
- To perform the gating for data transfer between the computer and the device.

IOT Instruction The computer communicates with the controller through the IOT instruction. An example of the IOT instruction is shown in Figure 3.7. The instruction has three parts: operation code, device selection address, and command code.

Operation Code In the example, 3 bits are used for operation code and one out of eight possible combinations of those bits is the code of the IOT instruction. These bits are loaded in the computer instruction register. The specific action for the I/O instruction is initiated.

Device Selection Address In the example, 6 bits are used for device selection address. Hence this computer can identify up to 64 different devices. The decoding is performed in the peripheral device.

Command Code In the example, 3 bits are used for the command code. Hence this computer can send up to eight different commands to the peripheral device. The decoding is performed in the peripheral device.

Figure 3.8 shows the information flow which affects a programmed data transfer with the peripheral device. The IOT instruction, like any other instruction of the program, is read from the main memory into the MD buffer register. The operation code (first 3 bits) are transferred in the IR.

IO instruction code	Device selection	Command
←— 3 —→	←——— 6 ———→	←— 3 —→

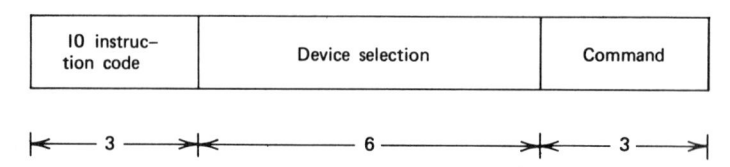

Fig. 3-7 Instruction format for input-output transfer. Twelve-bit computer word is presented. A similar structure is also in machines with 8 or 16 bits.

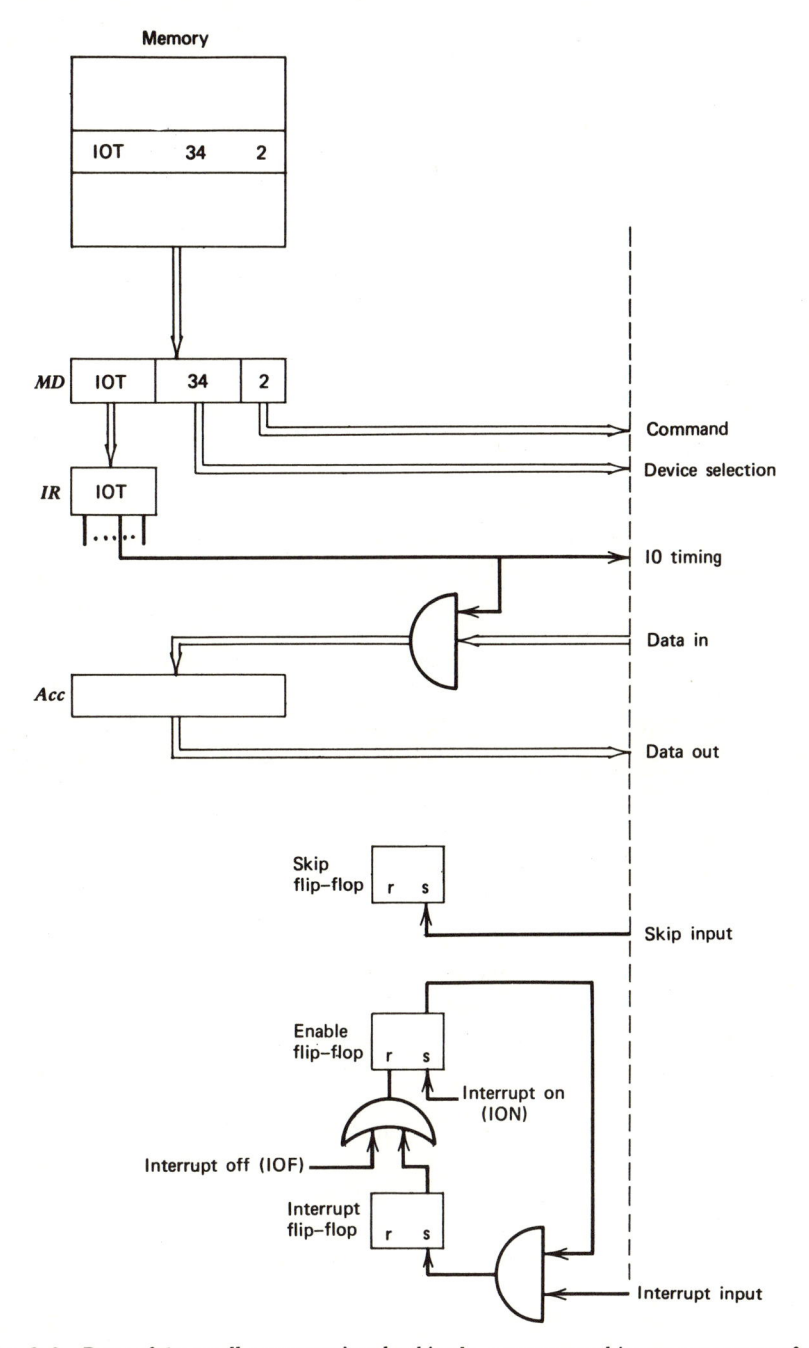

Fig. 3-8 Parts of the small computer involved in the programmed input-output transfer.

The instruction decoder activates the IOT timing generator. As a result, the computer enters the I/O state.

I/O State and Timing Signals The I/O state is characteristic of the I/O instruction. During the I/O state, the computer generates a number of timing signals, which are used to perform the operations needed for communication with the peripheral device, for strobing the data lines and gating the command lines.

Device Selection Lines The device selection bits of the I/O instruction sit in the memory data register MD. Using those bits, the I/O instruction selects the peripheral device. These bits are transmitted over the party line bus to all peripheral devices. The device selection code presents the key to the peripheral device. Each peripheral device has its specific lock, called the device selector. The device selection code will match with only one device selector; hence only one device will be coupled for communication with the computer. In the example, 6 bits are used for device selection; hence six device selection lines go out of the computer.

Command Lines The command code of the I/O instruction sits in the memory data register. Using those bits, the I/O instruction tells the selected peripheral device what to do.

The command lines are connected to all peripheral devices. However, the device address code will couple one device to the bus. Hence the command will be received only by the peripheral device selected by the device selection part of the I/O instruction. In the example 3 bits are used for command code, hence three command lines got out of the computer.

Data Lines The data transfer is performed between the peripheral device and the accumulator. Twelve input and twelve output lines present the data way. Data lines are connected to all peripheral devices. Each peripheral device has the gating logic for data lines. Only the device selected by the device selection part of the I/O instruction will open the gates for data lines.

The output lines are provided with driving amplifiers. These lines present the contents of the accumulator throughout the operation. The selected peripheral device should strobe these lines into its data buffer register.

The input lines bring the data from the peripheral device. These lines are gated inside the computer, so that they do not disturb the contents of the accumulator. Only during the execution of an I/O transfer instruction will the computer strobe the input lines into the accumulator.

Skip Line The computer has one skip input. The computer senses the flag of the peripheral device through this input. The flag is a 1-bit register

Party Line for Programmed Input-Output

(flip-flop) in the peripheral device. If the device is ready for communication, it will set its flag. If the device is not ready, it will keep its flag reset.

If the flag shows that the device is ready, the computer may issue an I/O instruction and it can perform the transfer. On the other hand, if the flag shows that the device is not ready, the computer will have to delay the I/O transfer for some later time. The flags of all peripheral devices are connected to the same one skip line. Each device has a gating logic between the flag and the skip line, controlled by the device selector.

The computer examines the flag by issuing an I/O command with the address of the device. The selected device connects its flag to the skip line.

The states of the flag will be copied through the skip line into the skip flip-flop in the computer. The computer is so wired that each I/O instruction checks the skip flip-flop.

If the skip flip-flop is in the state zero, after the I/O instruction is performed, the program counter will point to the next instruction in the program.

If the skip flip-flop is in the state one, after the I/O instruction is performed, the program counter will be incremented by two and the next instruction will be skipped.

In this way the program can be split in two branches; the first branch, if the device is busy, and the second one, if the device is ready for transfer.

Interrupt Line The computer has one interrupt line. The computer receives requests from peripheral devices through this line. Upon receiving a request, the computer stops the background program and performs the interrupt routine for the peripheral device. The interrupt line is connected directly without gating, to the flags of the peripheral devices. As a result, if any of the peripheral devices requests the interrupt service, the computer will receive the request.

A simplified interrupt logic is composed of the input gate, enable flip-flop, and interrupt flip-flop, Figure 3.8.

The input gate is controlled by the enable flip-flop, and is closed if the flip-flop is not set.

The enable flip-flop is under program control. The computer has the instruction interrupt on (ION) which will set the enable flip-flop. Another instruction, interrupt off (IOF) will reset the enable flip-flop. Hence by the program control it is possible to disconnect (disable) or to connect (enable) the interrupt line from the computer. (The two interrupt instructions, IOF and ION, belong to the class of I/O instructions with a special code.)

If the programmer wants to use the interrupt feature, the ION instruction must be issued. The input gate will be open and will wait for a signal on the interrupt line.

The signal on the interrupt line will set the interrupt flip-flop. Each computer instruction examines the interrupt flip-flop. When the interrupt flip-flop is set, the following takes place:

- The computer will complete the current instruction of the program.
- The contents of the program counter will be automatically deposited in a specific memory location, for example in address 0.
- The specific address, for example, 1, will be set in the program counter.

The two characteristic interrupt addresses are fixed for a given computer. In our example we use addresses 0 and 1.

The features described above can be used to transfer the program control from the background program to the specific interrupt routine whenever the device sets the request. Upon the execution of the interrupt routine, the background job is resumed.

3.4 Example of Programmed Input-Output Transfer

Here we show a simple example of the data transfer between the computer and the peripheral device. The computer collects the data from a number of measuring points. Each measuring point is treated as an independent peripheral device. A selection address must be assigned to each device. Figure 3.9 shows one of the measuring points, for which the designer has decided to use selection address 34. The interface is composed of flag/flip-flop, selector, and data register.

Flag flip-flop will notify the computer when the process has generated new data. The computer examines the flag by issuing a skip instruction, which connects the flag flip-flop to the skip input. If the computer finds out that the flag is set, the computer will start service routine for this device. As a first step, the computer will issue instruction to clear this flag.

Selector presents the specific lock, in this example lock with selection address 34. The computer will use the code 34 to distinguish this peripheral device from other devices.

Data register is a one-word storage device, into which the measured data is stored. The computer will issue a read instruction, to open the gates, and transfer the datum over party-lines into accumulator.

3.5 Direct Memory Access

Direct memory access (DMA) provides the possibility for I/O transfer without program intervention. The transfer is performed through special channels

Direct Memory Access

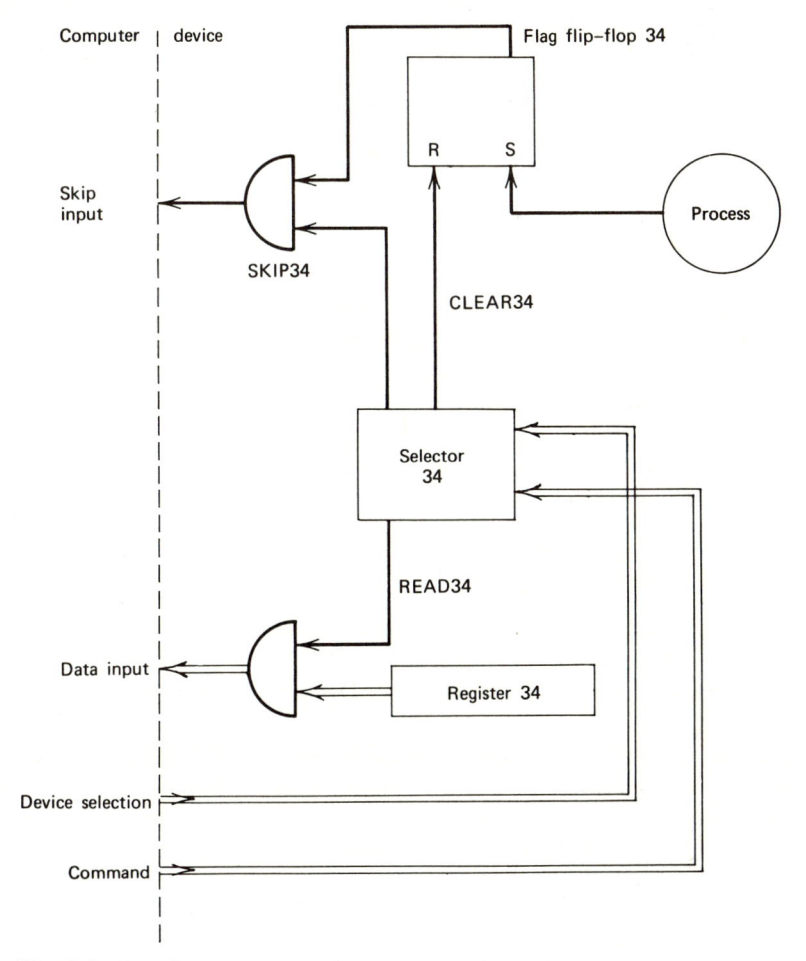

Fig. 3-9 Interface components for programmed, conditional transfer.

which steal time slices from the central processor whenever necessary. During each stolen time slice one transfer is performed. This kind of transfer is also known by other names, such as data channel, data break, and cycle stealing transfer.

The computer logic performing the DMA is basically independent of the logic involved in the programmed transfer. The main point is that the DMA does not perform the transfer with the accumulator. Rather the transfer is performed via the memory data register directly with the computer memory. Since the program execution is not involved in the DMA transfer, the computer working registers are not disturbed.

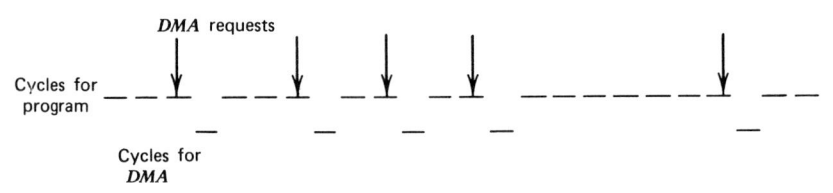

```
              DMA requests
                │    │    │    │                          │
 Cycles for ── ─▼── ─▼─ ─▼─ ─▼─ ──────────────▼──── ──
  program
     ───       ───  ───  ───  ───                  ───
          Cycles for
            DMA
```
Fig. 3-10 Principle of cycle stealing for direct memory access.

Each basic step of DMA is under the control of the peripheral device which sets the request for transfer. For inputting the data in the computer, the device sets the request when the data are prepared; for outputting the data from the computer, the device sets the request when it needs the data and is ready to receive them.

Having the DMA facility, the computer can perform in parallel two jobs which might be entirely independent:

- Programmed job.
- DMA transfer.

The computer examines the DMA requests during the execution of every instruction of the programmed job. If the DMA request is received, the next time slice is given to the DMA transfer. The programmed job is delayed for this time slice.

Figure 3.10 shows the principle of cycle stealing. The upper line presents the time slices during which the computer performs the programmed job. The arrows indicate the DMA requests, received from the peripheral device. The lower line presents the time slices stolen from the programmed job and given for execution of the DMA transfer.

The DMA transfer is particularly useful for devices with high speed and a large amount of data in block form. An example is the high-speed magnetic tape system or high-speed drum memories.

References

1. Souček, B.: *Minicomputers in Data Processing and Simulation*, Wiley, New York, 1972.
2. Korn, G.: *Minicomputers for Engineers and Scientists*, McGraw-Hill, New York, 1973.
3. Weitzman, C.: *Minicomputer Systems*, Prentice-Hall, Englewood Cliffs, N. J., 1974.
4. Souček, B.: *Microprocessors and Microcomputers*, Wiley, New York, 1976.

Chapter 4

LABORATORY COMPUTER SYSTEM

Introduction and Survey

Minicomputers and microprocessors have found their way into tens of thousands of scientific and industrial applications. As a result, special units have been designed for laboratory application. Also, large number of different "turn key" computer-oriented laboratory systems are available.

It is important to notice that laboratory computers could be easily reprogrammed and in this way modified for new experiments. Today many laboratory plug-in units are available. Also, programming languages are simplified, and they are user oriented.

This chapter describes typical laboratory data acquisition and processing chain. Conversion of analog signals into digital data is explained. Time digitizers and digital-to-analog decoders are reviewed. The differences between computer driven, event driven, and real-time clock-driven experiments are emphasized. Direct connection of the cathode ray tube display to the laboratory computer is explained.

The chapter starts with an overall description of computerized laboratory system. It then concentrates on basic principles, which will help one to choose the right laboratory devices and to operate computerized experiments.

4.1 Computerized Experiments

Typical System Computer-oriented measurement and control systems could be divided into two families: data acquisition systems and direct digital process control systems. In both cases the transducers take a physical parameter such as amplitude, interval, temperature, strain, or position and convert it into electrical voltage or current. Once in electrical form, all further processing of the signal is done by electronic circuits. The signal first passes through analog shaping and filtering. After that signal is converted into digital form and then fed to the computer for digital processing. Computer-produced results are then used as feedback controls or for display. If control elements and display units are analog devices, digital computer output must pass through digital-to-analog converters, producing analog-driving signals.

A complete representative data acquisition, conversion and data processing system is illustrated in Figure 4.1. The system is composed of a number of different units, connected to the computer party line bus.

The party line bus of the computer is divided into three subbuses: device selection and command bus, data-in bus, and data-out bus.

Device selection and command bus is used by the computer to select one device at a time for communication, and to specify the operations performed by the selected device (such as read, write, input, output, reset).

The data-in bus carries the data from the selected peripheral device data register into the computer. The data-out bus carries the data from the computer into selected peripheral device data register.

Each device interface is composed of device data register and input/output gates, device selector and command decoder, and flag flip-flop. Flag flip-flops are connected to the computer skip and interrupt inputs. Flag flip-flop of a particular device informs the computer if the device is busy or ready for communication.

The system in Figure 4.1 is composed of the following units: transducer, amplifier, filter, analog multiplexer, sample and hold circuit, analog-to-digital convertor, digital-to-analog convertor, cathode ray tube display, real-time clock, trigger circuit, buffered digital input, and buffered digital output.

Amplifiers and Filters The first part of the data acquisition and processing system is concerned with extracting the signal that is to be measured. The initial signal processing is done with an amplifier and possibly a filter. The purpose of the amplifier is to perform one or more of the following functions: boost the amplitude of the signal, buffer the signal, convert a signal current into a voltage, or separate differential signal from common noise. For most analog multiplexers, sample and holds and analog-to-digital converters, the desired voltage level at the amplifier output is 5 to 10 V full scale.

Following the amplifiers in the system, it may be necessary to use a filter. Filters are usually used for two reasons: (1) to reduce the noise and to improve signal to noise ratio and (2) to limit the bandwidth of the signal and to avoid high-frequency components in the signal, if those components are not needed. In this way the signal could be sampled at a moderate sampling rate, and the data processing is less expensive.

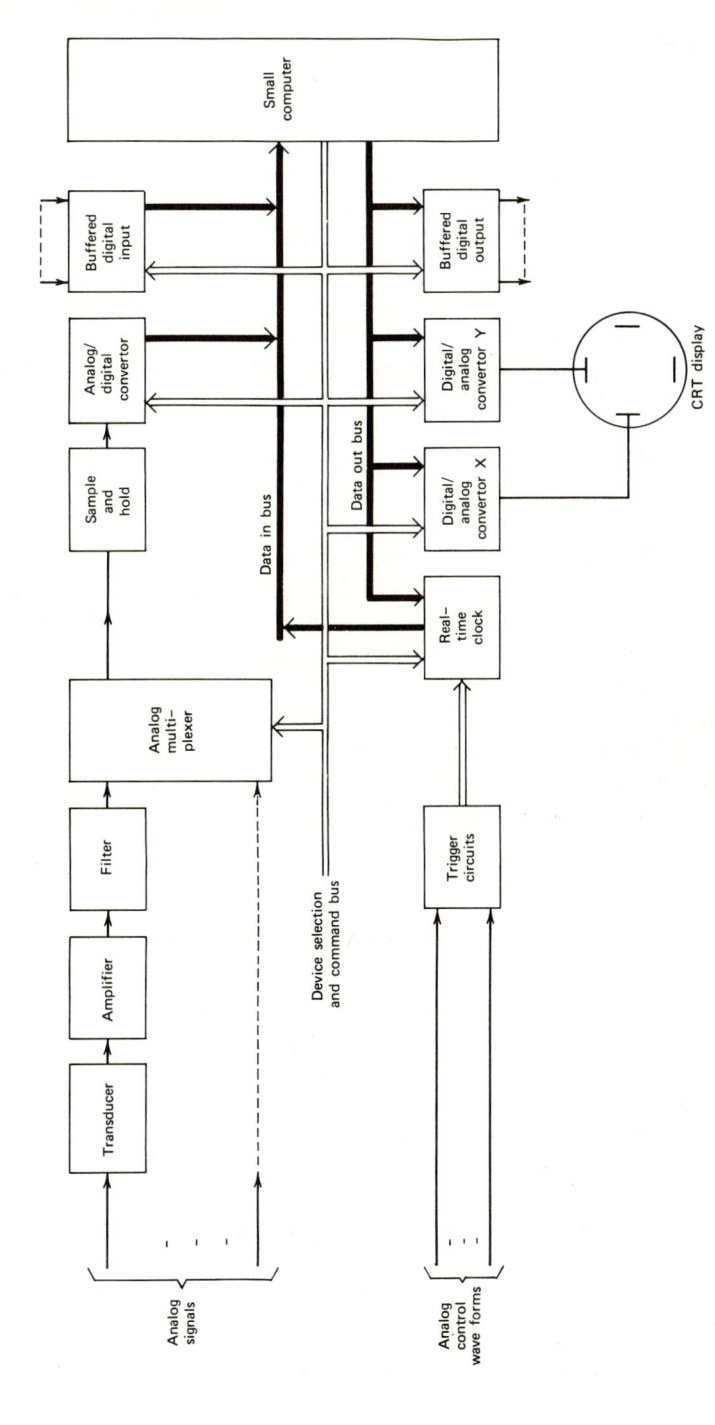

Fig. 4-1 Typical data acquisition, conversion, and data processing system.

Analog Multiplexers and Sample Hold Circuits Analog multiplexers are used for time sharing of analog-to-digital convertors between a number of different analog channels. Analog multiplexer has many analog inputs and only one output. The multiplexer is composed of a number of analog switches. Each switch connects one analog input to the common output. The switches can be addressed by a digital input code through the computer bus. Only one input is connected to the output at any one time. Usually the input channels are sequentially connected to the output of the multiplexer.

The output of the analog multiplexer goes into a sample and hold circuit which samples the output of the multiplexer at a specified time and then holds the voltage level at its output until the analog-to-digital convertor performs its conversion operation.

Buffered Digital Input and Output In some cases the peripheral unit is a digital device. In such circumstances the data are already in a digital form, and direct communication with the computer is possible. To eliminate the need for interface design, many laboratory systems provide buffered digital input unit and buffered digital output unit, Figure 4.1. Interface components, such as device selector, flag flip-flop, data registers, and gates, are already built into the unit. Digital input or output lines are directly available to the user.

If the user has an instrument that generates digital data, the output lines of the instrument could be directly connected to the digital input lines.

If the user has a control or display device that operates on digital data, the digital output lines could be directly connected to the input lines of such a device.

4.2 Digital-to-Analog Decoder

Digital-to-analog conversion is an operation opposite to quantization; information presented as a digital number is converted into proportional analog form (voltage, current, and so on). Digital information is presented on a number of lines. There are as many lines as there are bits, each at one of two voltage levels (0, 1), making a binarily weighted digital word. Analog information is presented on a single line as a signal whose magnitude gives the value of the information.

Several techniques are available to convert a piece of digital information into its analog form. We show the basic, most frequently used technique.

Figure 4.2 represents the basic circuit of a digital-to-analog convertor (DAC). The digital number X_D is expressed by n digits, D_0, D_1, D_{n-1}, and is

Digital-to-Analog Decoder

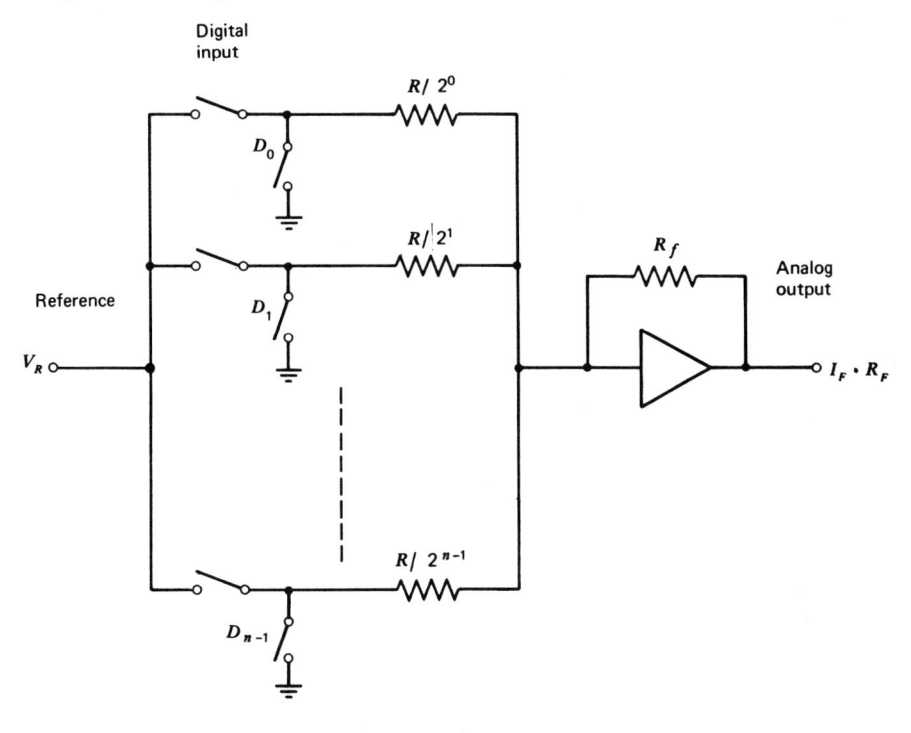

Fig. 4-2 Digital-to-analog convertor (DAC).

loaded into an *n*-stage register. The digital number can be expressed as

$$X_D = D_0 2^0 + D_1 2^1 + \cdots D_{n-1} \cdot 2^{n-1} \tag{4.1}$$

where $D_k = 0$ or $1, k = 0, 1, 2, \cdots (n-1)$.

Each flip-flop of the digital data register is connected through an electronic switch to a network of resistors. The values of resistors are chosen to be $R/2^0, R/2^1 \cdots R/2^{(n-1)}$.

One side of each resistor is connected through the switch to the constant reference voltage V_R. The other side is connected to an operation amplifier summing input, which has approximately zero voltage level.

Each flip-flop of the register controls its switch. If the flip-flop k is in the state zero ($D_k = 0$), the switch k connects its resistor to the ground. If the flip-flop k is in the state one ($D_k = 1$), the switch k connects its resistor to the reference voltage V_R and produces the current $(V_R/R) \cdot 2^k$.

Each bit of the decoder in the state one acts as a current source, and the network output current is a summation of individual currents from each bit. The total network output current is equal to the feedback current

$$I_F = I_0 D_0 + I_1 D_1 + I_2 D_2 + \cdots I_{n-1} D_{n-1}$$
$$= \frac{V_R}{R} D_0 + 2 \cdot \frac{V_R}{R} \cdot D_1 + \cdots 2^{n-1} \cdot \frac{V_R}{R} \cdot D_{n-1}$$
(4.2)

The output voltage of the operational amplifier is $I_F \cdot R_F$; hence, according to equation 4.2, it is proportional to the digital input number X_D.

The summing network branches must be composed of accurate resistances, especially for the most significant bits. Stable and accurate networks can be made in the thin-film technique.

Digital-to-analog convertors are available as modules, with a basic convertor and sometimes also with a buffer register to store the digital input information.

4.3 Analog-to-Digital Convertor

A method similar to that described previously can be used to convert a piece of analog information (say voltage) into its digital form. The principle is shown in Figure 4.3a and is based on the use of the successive approximation technique. An analog-to-digital convertor (ADC) is shown in Figure 4.3b.

A voltage V (approximately 9.2 volts in the figure) is to be digitized to the nearest integral voltage between 0–16 V. The digitization is accomplished by a series of test steps. The first step tests whether V is greater or less than 8 V (that is, the voltage range is divided in half). It is greater, so the most significant output binary digit is set equal to one. Next the upper half of the range is itself divided in half, and a test made to see which of these quarters contains V. In this example V is located in the range 8 to 12 volts, so a binary zero is entered as the next significant digit. The procedure continues through steps three and four, yielding a binary zero and one as the two succeeding digits. Thus, after four steps, an unknown voltage is placed in channel 9 of 16 possible channels. In general, it takes only n steps to quantize an unknown voltage into one of 2^n channels.

The block diagram of ADC, with successive approximations is shown in Figure 4.3b. It uses a digital-to-analog convertor which reconverts the digital output X_D to a voltage $\alpha E X_D$, which is compared with the analog input X_A. Assuming $|X_A| \leq E$ and $|\alpha X_D| \leq 1$ we vary X_D during each conversion

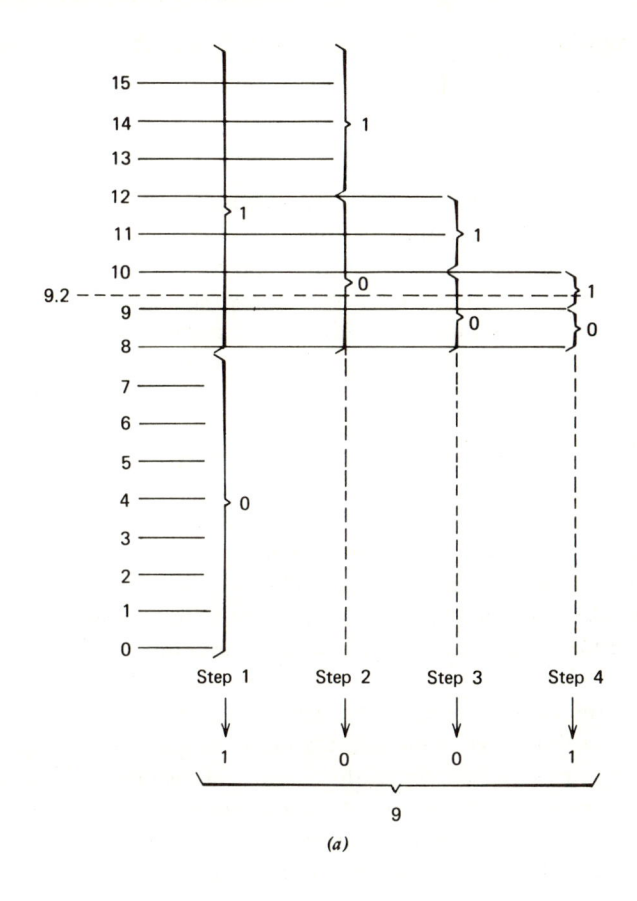

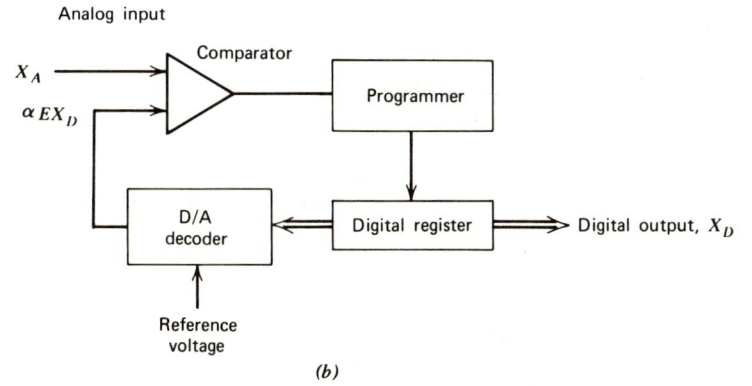

Fig. 4-3 Analog-to-digital convertor (ADC). (*a*) Successive approximation conversion principle. (*b*) ADC circuit diagram.

cycle until the magnitude of the analog error $e_A = \alpha E X_D - X_A$ is sufficiently small, that is,

$$X_D = \frac{X_A + e_A}{\alpha E} \qquad |e_A| < \frac{2E}{2^n}$$

This ADC requires n steps for encoding to an n-bit word. As each bit is tried, it can be used for serial output. At the end of conversion the whole piece of information is available in the counter register for parallel output.

The main advantage of this kind of conversion is high speed. This technique is rapid, but it requires special care to achieve a constant quantization step from channel to channel.

4.4 Time-to-Digital Conversion

Continuous information to be quantized must not necessarily be an amplitude of the signal. Various parameters might be quantized. Now we shall show the technique for quantization of time intervals, Figure 4.4.

The time interval to be quantized is presented as a pulse controlling an AND gate. Its digital equivalent will be stored in the counter. This is done by simply opening the AND gate and counting clock pulses. The number of pulses received by the counter will be proportional to the time interval T.

It should be noted that direct time digitization is simple, very accurate, and linear. The accuracy is a function of the stability of the clock frequency.

To achieve the accuracy, the cycle time of clock pulses must be substantially shorter than the time interval T to be digitized. For this reason this kind of conversion is limited to long intervals (well over one μsec).

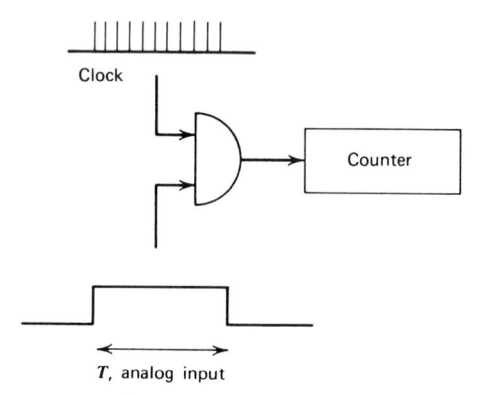

Fig. 4-4 Time-to-digital conversion.

4.5 Real-Time Clock

Programmable Clock Computer-oriented experiment and control systems could be divided in three classes:

Event-driven systems.
Computer-driven systems.
Clock-driven systems.

Event-driven systems are typical in the data acquisition area. The events, in the form of external stimuli, are usually physical or biological processes that must be responded to in a time-critical manner. The design of a real-time system is predicated upon being able to respond to events not only at a rapid rate but also in terms of servicing these events under peak load conditions.

Computer-driven systems are typical in the process or experiment control area. The computer program is the actual master of the system. The program determines both the timing of operations and the scheduling. The computer program also determines the sampling rates and the data display rates. This mode of operation is usually used for relatively slow processes, because it is difficult to achieve precise timing using computer programmes. It is especially difficult to estimate the time needed to execute a part of the program written in a high-level language.

Clock-driven systems are found in all real-time applications. A special unit, called real-time clock, is actual master of the timing and scheduling operations. The real-time clock is an oscillator, producing pulses at a constant rate. The clock pulse rate is under program control. Usually computer program can control the clock rate in the range from 1MHz to 100Hz.

Real-time clock is usually connected to the computer interrupt input. Interrupt requests produced by the clock could force the computer to perform different operations with precise timing. Typical clock driven operations are

- Sampling or display at precise rate.
- Switching from one task, or experimental mode, to another task at precise time intervals.
- Digitizing the time intervals, such as interspike intervals, by counting the number of clock pulses in between two spikes.

Parts of the Real-Time Clock Figure 4.5 shows the basic parts of the real-time clock: oscillator, clock counter, buffer-preset register, and overflow flag.

Typical interactions between the computer and the clock are timing mode and digitizing mode.

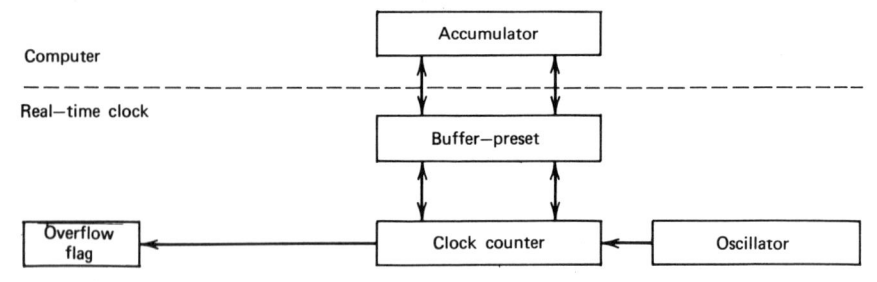

Fig. 4-5 Basic components of the real-time clock.

Timing mode The computer program determines the pulse rate of the oscillator and presets the count into clock count register. The program then connects the oscillator to the clock counter. The computer now switches its attention to the background program, which might be an entirely independent operation. Simultaneously, the clock counter is incremented by the oscillator. Eventually, the clock counter will produce an overflow, and it will set the overflow flag. The overflow flag is used to interrupt the computer. In this way the computer is switched from one task to another at precise time, as dictated by the real-time clock.

The buffer-preset register is usually set through the accumulator to the negative (2's complement) value of the number of counts desired before overflow. Note that the computer programmer has not only determined the rate of counting but also the number of counts before overflow, thus allowing him two dimensions in selecting the time intervals between overflow.

Digitizing mode This mode is useful for determining the total elapsed time between some initial event (e.g., a stimulus) and subsequent events that might be caused by the initial event (e.g., muscle reaction).

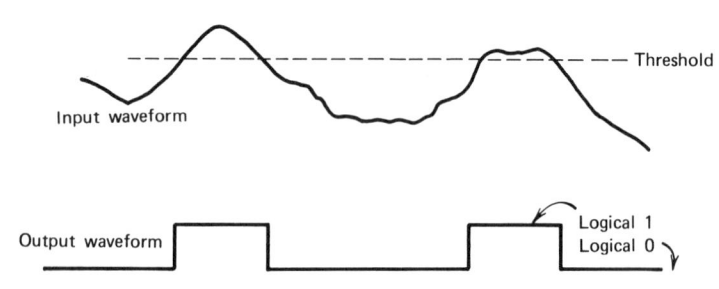

Fig. 4-6 Schmitt trigger, input, and corresponding output waveforms.

Real-Time Clock

Fig. 4-7 Turn key laboratory computer system, built around the minicomputer PDP-11. (Courtesy of Digital Equipment Corporation.)

Real-time clock is connected to a set of trigger circuits or control inputs (known also by the name Schmitt trigger, Figures 4.1 and 4.6). The Schmitt trigger firing is governed by setting the threshold control. Each time the input waveform crosses the preset threshold voltage, the Schmitt trigger fires, causing a pulse to be generated. The Schmitt trigger is in fact a 1-bit analog-to-digital convertor: if the input waveform is more positive than the threshold, the Schmitt trigger produces the output equal to logical 1. If the input waveform is less positive than the threshold, the Schmitt trigger produces the output equal to logical 0. The positive pulse produced by the Schmitt trigger is used to start or stop the real-time clock.

A poststimulus histogram can be generated as follows. First, a stimulus is issued to the subject. It is also applied to the Schmitt trigger, and it starts the real-time clock. A subsequent subject reaction event is applied to another Schmitt trigger, and it transfers the clock counter content into the buffer-preset register. In this way the elapsed time between the stimulus and the subsequent event can be determined, and it is proportional to the saved time count. The computer program reads the saved count from the buffer-preset register and treats it as the digital equivalent of the measured time interval.

All the units presented in Figure 4.1 are available as modules or as plug-in cards. Also, many "turn key" computer oriented laboratory systems are available. One such laboratory system is shown in Figure 4.7. New miniature low-cost microprocessors are finding their way into the laboratory as integral parts of the new instruments, or as experiment controlling and data acquisition systems.

References

1. Souček, B.: *Minicomputers in Data Processing and Simulation*, Wiley, New York, 1972.
2. Korn, G.: *Minicomputers for Engineers and Scientists*, McGraw-Hill, New York, 1973.
3. Weitzman, C.: *Minicomputer Systems*, Prentice- Hall, Englewood Cliffs, N. J., 1974.
4. Finkel, J.: *Computer-Aided Experimentation*, Wiley, New York, 1975.
5. Souček, B.: *Microprocessors and Microcomputers*, Wiley, New York, 1976.

Chapter 5

BASIC PROGRAMMING

Introduction and Survey

Many new computer languages have been developed recently. These languages are user oriented, and some of them are easy to learn. The simplest, yet very effective, is the BASIC language. The BASIC language is available on minicomputers as well as on large machines. BASIC is interpreter language that operates directly on a source program in memory. The interpreter translates the instructions of the source program one by one and executes them immediately. Hence it is especially suitable for laboratory operations, because it provides an easy way to write, modify, debug, and run the program.

This chapter presents a short review of BASIC programming language. Control operations, loops, subroutines, arithmetic, and input-output operations are explained. A number of short programs are discussed. Presented statements should suffice of writing many programs. However, for more elaborate functions and especially for the features that are specific for a given computer, one should consult manufacturers manuals.

5.1 Basic Language

In many applications advanced programming languages can be used, that is, languages that are problem oriented and allow the programmer to use more meaningful commands than elementary computer instructions. The following languages are of particular importance:

COBOL for common business-oriented language.
ALGOL for algebraic language.

FORTRAN for formula translation language.
BASIC for Beginners All-purpose symbolic instruction code.
PL/M for programming language for microcomputers.

Because of its wide use, especially in scientific and technical programming, we have chosen BASIC to be presented here.

BASIC (beginners all-purpose symbolic instruction code) is a common and universal language used on computers throughout the world. This language is used by the computer to solve the various problems directed at it. The user defines how to solve the problem by describing the solution with a series of statements called a program. These statements are composed of a series of statement numbers, verbs, and variables, and so on, all of which are discussed in this section.

BASIC is especially designed for programming problems that can be expressed by simple mathematical formulas. The basic operations of addition, subtraction, multiplication, division, and exponentiation are written like algebra, with a special sign between the variables concerned. The signs used are

$$+, -, *, /, \uparrow$$

Hence the arithmetic expression in BASIC may look like

$$Y = (X1 + X2) * (X3 - X4)/(X5 \uparrow 3)$$

The computer, like a high school student, will first perform the calculation within the parentheses and then the operations between the parentheses. An additional fact should be remembered for the BASIC: some operations will take precedence over the others. The priority will be

Exponentiation ($\uparrow$)
Multiplication and division ($*,/$)
Addition and subtraction ($+,-$)

Hence the equation

$$Y = 3.0 * X1 + X2$$

will be treated by the BASIC as

$$Y = (3 * X1) + X2$$

Although we did not indicate the parentheses, multiplication takes precedence over addition.

The equation

$$Y = X1 + X2/X3$$

Basic Language

will be treated by the BASIC as if it were written

$$Y = X1 + (X2/X3)$$

The BASIC will first perform division and then it will add X1 to the quotient.

Character Set, Variables, Numbers The *characters* allowed within a BASIC statement are

A through Z
0 through 9

A *variable* is defined as a letter (A to Z), or a letter immediately followed by a number (0 to 9). Variables are used to represent numeric values. For example:

$$M5 = 96.7$$

M5 is a variable; 96.7 becomes the value of the variable M5. *Numbers* within the BASIC system can range between a maximum of 10^{38} (2^{127}) to a minimum of 10^{-38} (2^{-127}). Zero is included in this range. BASIC will always attempt to display six significant digits but will display fewer if significance is not changed. Thus although one may input the number as shown in the left-column of Table 5.1, the number will be output and stored in the programs as shown on the right in Table 5.1. Note that in the last example, 6.000596 has been rounded off to 6.0006 with the trailing zeros suppressed. E Notation (or E-format) is used to express numbers having more than six digits in the form of a decimal number raised to some power of 10. BASIC will also

TABLE 5.1

Input to BASIC	Displayed
1	1
1.0	1
1.	1
0.0	0
0.00	0
0.	0
.0	0
.0010	.001
.50	.5
0.50	.5
3.14159	3.14159
6.66666	6.66666
6.000054	6.00005
6.000594	6.00059
6.000596	6.0006

accept a number having more than six digits but will immediately convert the number to E notation. All six digits are always shown in E notation. For example:

1000000	is equal to	1.00000E+06
1.02000E−04	is equal to	.000102
1.02000E+04	is equal to	10200

When using E-formats, leading and trailing zeros may be omitted. The plus sign and leading zeros may be omitted from the exponent. The precision of all numbers is six to seven decimal digits.

5.2 Control Operations

One of the most important features by which to distinguish a digital computer from a desk calculator is the ability to program the computer to make decisions. It is quite possible to have a useful program in which there are no arithmetic operations or a program without input-output transfer, and so on. On the other hand practically every program has to make at least a few decisions.

Without decision making the program would be composed of a number of commands, which would be executed sequentially in the order in which they are stored in the memory. The decision-making feature enables one to write the program composed of a number of branches, each branch to be executed if a given condition is fulfilled. The program then looks like a tree with many branches and loops, rather than like a straight sequence of commands.

Let us suppose that the computer controls an experiment in which the temperature is the critical variable. The computer can read the temperature T from the digital voltmeter. The program can be written to compare the temperature T with the prescribed value T_0. Three possible outcomes may result $T < T_0$; $T = T_0$; $T > T_0$. The computer program can have three different branches, and it can be a general control program which is the thread binding together separate parts and instructions.

In writing the control program, one has to solve three problems; labeling, testing, and directing.

Labeling In BASIC any statement may be identified by a number written to the left of the statement. For example

            71   Z = X + Y

The number 71 presents the label of this statement. In this program no other statement is allowed to be labeled with the same number. The statement in BASIC does not mean the memory address; its sole purpose is to be used to

Control Operations

identify the statement when one wishes to transfer the control to that statement.

As a programming aid, the computer will sort the statements by their numbers before executing them. Thus, if one inputs statements in the sequence 30, 20, 10, the computer will arrange them in the order 10, 20, 30. If the user desires to go back and change or add a statement in a series of statements, all that is required is to type a new statement with either the same statement number for a replacement or a new number for an addition; the computer will place the new statement on top of or between the appropriate prior statement. In order to save room for possible future insertions, it is usually desirable to number the statements by fives or tens.

Testing: IF – THEN Statement In the BASIC language there is one basic conditional control statement that can be used for testing and at the same time for directing the program control. This is the IF statement. This statement will test the specified condition, and it will provide a branching. For example, the statement

$$\text{IF } (T = 0) \text{ GO TO } 70$$

will test the variable T and will transfer the program control to the line labeled 70.

Since numbers are not always represented exactly in the computer, the = operator should be used carefully in IF . . . . THEN statements. Limits, such as $< = > =$, should be used in an IF expression, rather than $=$, whenever possible.

If the specified condition for transfer is not true, the program will continue executing in sequence.

Directing: GO TO Statement In the BASIC language there is one basic unconditional directing statement, GO TO. If at some point in the program we wish to have the control transferred to statement 71, we should write the statement

$$\text{GO TO } 71$$

Example The program compares the temperature T with the critical value T_0. If $T > T_0$, the program modifies T, such that $T = T_0$. If $T < T_0$, the program leaves the value T unchanged. The solution is shown in Table 5.2.

Note the following points concerning the example. Line 10 checks if $T > T_0$, and if yes, forces the program to go to line 40. If the statement $T > T_0$ is not true, the program continues into line 20.

TABLE 5.2

```
10  IF (T = > T0) GO TO 40
20  GO TO 80
40  T = T0
80  END
```

Line 20 forces the program to go to line 80, the end of the program, leaving the value of T unchanged. Line 40 with the statement $T = T_0$ states that the old value of the variable T is to be replaced by the present value of the variable T_0, as requested.

5.3 Loops

DIM Statement The program loop is a set of commands that is repeatedly executed. Looping is one of the most powerful techniques in programming the computer. Often the computer has to perform the same task but on different sets of data. In such situations the program can be repeated in the loop rather than writing the same set of commands for each set of data.

As a problem-oriented language, BASIC is very elegant in dealing with lists. List processing and looping can be performed with statements we have thus far learned. We still have to learn the statement for space allocation: DIM statement.

A DIM statement is an instruction to the compiler, not to the machine. It is used to reserve space for the list. Thus to reserve space for the X of 100 items, we write the statement

DIM X(100)

To reserve space for the list of X of 100 items, and for the list Y of 56 items, we write the statement

X(100), Y(56)

The statement can also be used for a variable with more than one subscript. Thus to reserve space for a two-dimensional array X_{ij} of 10 by 36 items, we write

DIM X(10,36)

Here we give a few example of programming lists and loops in BASIC language.

Example Add 100 numbers stored in the array D, and place the result into R. The solution is shown in Table 5.3.

Loops

TABLE 5.3

```
10  DIM D(100)
20  I = 1
30  R = 0
40  R = R+ D(I)
50  I = I + 1
60  IF (I < = 100) GO TO 40
70  END
```

Note the following:

- Statement 10 tells the computer to reserve space for 100 items of the array called D.
- Statement 20 sets the index of the list to one.
- Statement 30 sets the initial value of the result, R = 0.
- Statement 40 is the basic operation of the loop. In each iteration it adds one item from the list to the existing contents of R.
- Statement 50 increments the index I, in each iteration.
- Statement 60 checks the index I. If I < 100, the program is switched back to statement 40, to perform the next iteration of the loop.

Example Perform the operation $Z_i = X_i + Y_i$, where Z, X, and Y are three lists, each of 100 items. The program is presented in Table 5.4.

TABLE 5.4

```
10  DIM X(100), Y(100), Z(100)
20  I = 1
30  Z(I) = X(I) + Y(I)
40  I = I + 1
50  IF (I < = 100) GO TO 30
60  END
```

FOR ... NEXT Statements We have seen a few examples of BASIC programs of count-controlled loops. All these problems have been coded using the statements we have thus far learned. As count-control loops are very frequently encountered, the BASIC language has a special statement for their programming, the FOR statement. A FOR statement may look like

FOR I = 1 to 20

This statement takes care of a few basic operations needed in programming the loop.

1. It initializes the index I. In the above example, the first value of the index will be 1.
2. The FOR statement is outside the loop. Statements composing the loop are all statements following the FOR statement until the statement specified as NEXT I is reached.
3. Because of the FOR statement, the loop in this example will be executed 20 times. Each time the value of the index I will be increased by one and the Ith item from the list will be processed. After the number of passes specified in the FOR statement is finished, the program exists out of the loop.

Example Code the problem shown in Table 5.3 using the FOR statement. The program is shown in Table 5.5. Statement 50 will be executed 100 times, each time increasing the index I by one.

$$\text{First pass } R = R + D\ (1)$$
$$\text{Second pass } R = R + D\ (2)$$
$$\text{ith pass } R = R + D\ (i)$$
$$\text{100th pass } R = R + D\ (100)$$

TABLE 5.5

```
10  DIM D(100)
30  R = 0
40  FOR I = 1 TO 100
50  R = R + D(I)
60  NEXT I
70  END
```

In this example the index I has been incremented in steps of 1. However the step value need not be 1. For instance, the statement

$$80 \text{ FOR } N = 1 \text{ TO } 2 \text{ STEP } 0.1$$

produces 10 loop executions, incrementing N by 0.1 each time, Also, negative step size may be used.

Example Code the program shown in Table 5.4 using the FOR statement. The program is shown in Table 5.6.

Subroutines

TABLE 5.6

```
10  DIM X(100), Y(100), Z(100)
20  FOR I = 1 TO 100
30  Z(I) = X(I) + Y(I)
40  NEXT I
60  END
```

5.4 Subroutines

The ever-increasing use of computers has been accompanied by a number of different programs. Writing programs may take rather long. Hence it is of interest to have a means for taking advantage of already-written programs and to "steal" from them as much as possible.

Programs solve various problems with a substantially unlimited number of variations of operations and parameters. Hence there is little chance that one can copy somebody's whole program, except in the case of exactly the same application. Fortunately, modern programs are written in modular form. It is possible that some modules can be copied from one application to another.

Modularity is achieved by using subroutines in writing programs. A subroutine is the part of the program which by itself solves some basic operation. It can be composed of only a few instructions, but it can also be quite a long program.

The main purpose of writing a subroutine is to use it at different points in the same program, or to copy it from one program to another. It is of utmost importance that a subroutine should provide aid to the programmer who is not capable of developing the required routine himself.

Suppose we want to calculate the area function of the parameter at a few places in a program. The area function is a very basic function, but it is not easy to program it by using elementary machine instructions; we shall most probably have to calculate the area function in a power series.

Since the area is a function often required, we may be lucky to find somewhere an already-written program for calculating this function. Nowadays there are organized program libraries and organizations collecting programs from many sources and making them available to programmers.

The program for the area function will be written as an independent unit or subroutine. It will clearly define how to enter the parameter in the routine and how to get the result from the routine. In such a case the user can copy the subroutine and use it in his program, without even having to know how the subroutine is written.

If we buy a small computer for laboratory applications, we shall develop a number of programs in the course of time. It is very useful to split programs into subroutines from the very beginning. In this way we can build our own library of subroutines useful for particular applications. When it is necessary to write a new program, we can copy, whenever possible, the subroutines already existing in the library.

Since a subroutine can be used at a later time, or even by another person, it is very important to provide a clear description and definition of the subroutine and, if necessary, to make a flow chart of it.

GOSUB ... RETURN Statements GOSUB transfers control to the specified statement number. RETURN transfers control to the statement following the GOSUB statement. GOSUB ... RETURN eliminates the need to repeat frequently used groups of statements in a program. The format is

GOSUB statement number
.
.
.
RETURN

The portion of the program to which control is transferred must logically end with a RETURN statement. RETURN statements may be used at any desired exit point in a subroutine. At the end of the subroutine, RETURN statement returns control to the next statement in the calling program. The following example could be considered a typical situation. The arrows indicate the flow of the logic, Figure 5.1.

Example Take the variable X and replace it with the value

$$(2 \cdot X)^3/P$$

Repeat the same procedure for the variables Y and Z. Write the program using the subroutine. The solution is shown in Table 5.7.

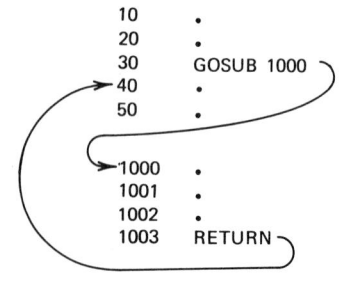

```
  10      .
  20      .
  30      GOSUB 1000
→ 40      .
  50      .

→1000     .
 1001     .
 1002     .
 1003     RETURN
```

Fig. 5-1 Transfer from the main program into the subroutine and back.

Input-Output Programming

TABLE 5.7

```
  5   P = 3.14
 10   X = 5
 20   Y = 10
 30   Z = 17
 40   R = X
 50   GOSUB 500
 60   X = R
 70   R = Y
 80   GOSUB 500
 90   Y = R
100   R = Z
110   GOSUB 500
120   Z = R
150   END
500   R = ((2*R) ↑3)/P
510   RETURN
```

Note the following.

- Statements 5 to 30 define the initial conditions.
- Statement 40 uses the variable R to prepare the first datum to enter the subroutine.
- Statement 50 forces the program to enter the subroutine at line 500. The subroutine performs desired operation and stores results into R. Statement 510, returns the control back to the major program, t.e. to line 60.
- Statement 60 replaces the variable X with the value of R.
- Statements 70, 80, and 90 use the subroutine at line 500 to replace the variable Y.
- Statements 100, 110, and 120 use the subroutine at line 500 to replace the variable Z. In this way the same subroutine has been used three times, called from different parts of the program.

5.5 Input-Output Programming

INPUT Statement When the INPUT statement is encountered, the program stops and a "?" is displayed on the input device. The "?" is the indication that an INPUT statement has caused the suspend. The INPUT statement may have any number of parameters. Data from the keyboard must be entered until all the parameters are satisfied, or the response will be another "?". Thus the statement

```
INPUT A,B,C,D
```

and the response 1, 2, 3, would cause another "?" to appear, indicating that the variable "D" had not yet been fulfilled. When satisfying the INPUT statement, items may be separated by either spaces or commas to indicate the beginning of a new number. These two delimiters are required between parameters in an INPUT statement. For example:

```
100 INPUT A,B,C,D
```

If, in satisfying the above statement one inputs too many numbers, for example, 1, 2, 3, 4, 5, 6, 7, 8, the excess (5, 6, 7, +8) will be ignored. The information will not be retained for a subsequent INPUT statement.

PRINT Statement   The PRINT command is a means of having a program write on an output device. Various options control horizontal spacing, output device, tabbing, and space allowed within printed material. Although the format of the PRINT statement is "automatic" to help beginning programmers, the experienced programmer may use several of the following features to control his output format.

When commas are used as separators, each line output to the terminal is divided into five print fields. The fields begin at print spaces 0, 15, 30, 45, and 60. The first four fields contain 15 spaces, and the last field contains 12. The comma signals the computer to move to the next print field, or if in the last field, to move to the next line.

If a PRINT statement does not have a comma or semicolon following it, a carriage return/line feed is generated following each output. For example, the following is a loop to print 30 numbers:

```
20 X = 1
30 PRINT X
40 X=X+1
50 IF X<30 GO TO 30
60 END
 .
 .
 .
```

would result in the 30 numbers being strung out down the output device.

```
1
2
.
.
.
30
```

Input-Output Programming

However, ending a PRINT statement with a comma causes output to fill all five fields on a line before moving to the next line. Note that trailing comma in statement 30 in the sequence:

```
20 X = 1
30 PRINT X,
40 X=X+1
50 GO TO 30
```

produces the output shown in the five fields

```
1 2 3 4 5
6 7 8 9 10
.
.
.
.
```

Text Printing Any characters, or text, placed within double quotes (" . . ") will be printed verbatim. For example,

```
PRINT "VOLTAGE =";V
```

in which the value of V is 20, the result will look like:

```
VOLTAGE = 20
```

Note that for readability a couple of spaces were left after the " = " sign in the example above. Also note that a semicolon was used to cause the following information to begin in the next available location.

Example Write the program to accept 20 data values from the teletype, to print the heading "SINE TABLE", and to print the input and its sine on the teletype. The program is shown at the top of Table 5.8, and the output produced by the program on the teletype is shown at the bottom of Table 5.8.

TAB The TAB command (part of a PRINT statement) controls horizontal spacing like the tab on a typewriter. The command can also be used, under the proper conditions, to modify the rules above.

The TAB command controls horizontal spacing in the framework, the leftmost margin being position "0" (not "1"), and 71 successive positions:

```
PRINT TAB (14), X, Y
```

If the number is negative or less than the current print head position, the function has no effect. If the number is greater than 71, the effect is to gen-

TABLE 5.8

```
60 PRINT "SINE TABLE"
100 FOR J=1 TO 20
110 INPUT A
120 B=SIN(A)
130 PRINT A,B
140 NEXT J
150 END

RUN
SINE TABLE
-.97            -.8248857
-.911           -.7901171
-.872           -.7656171
-.723           -.6616371
-.719           -.6586325
-.61            -.5728675
-.502           -.4811798
-.346           -.3391376
-.33            -.324043
-.283           -.2792376
-.175           -.1741081
-.155           -.1543801
-.02            -.01999867
 .03             .0299955
 .093            .092866
 .127            .1266589
 .13             .1296341
 .42             .4077605
 .529            .5046703
 .632            .5907596
READY.
```

erate a blank line and set the implied pointer at the beginning of the next successive line. The effect of commas or semicolons immediately preceding or following "TAB" is overridden by the "TAB" function.

Figure 5.2 shows an example of the sort of graph that can be drawn with BASIC using the TAB function.

5.6 Other Features

READ and DATA Statements   READ and DATA statements are used to input data into a program from a memory array. One statement is never

Other Features

```
30 FOR X=0 TO 15 STEP .5
40 PRINT TAB(30+15*SIN(X)*EXP(-.1*X));"*"
50 NEXT X
60 END
```

RUN

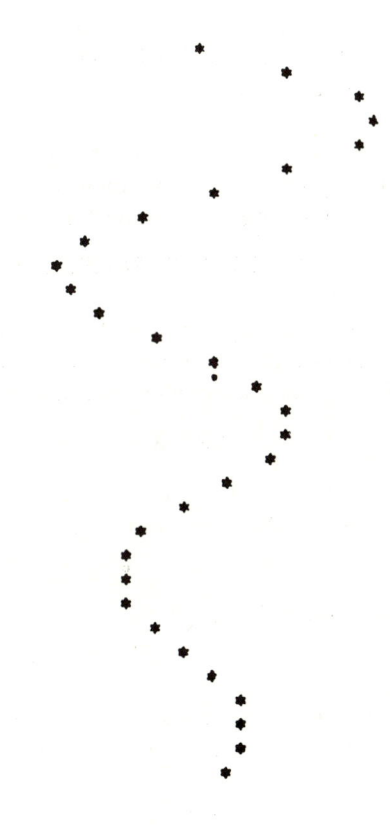

```
                              *
                                  *
                                     *
                                      *
                                     *
                                 *
                         *
                       *
                      *
                         *
                             *
                            *
                             *
                              *
                              *
                             *
                           *
                         *
                        *
                        *
                         *
                           *
                             *
                              *
                               *
                                *
                                 *
                                 *
                                  *
                                  *
                                   *
```

READY.

Fig. 5-2 TAB function. Program to calculate the sin function and the output produced by the program. (Courtesy of Digital Equipment Corporation.)

used without the other. For example,

```
10 READ A,B,C
 .
 .
 .
100 DATA 1,2,7,4,5
```

Statement 10 will assign the following values to A, B, and C:

$$A = 1, B = 2, C = 7.$$

Assuming that it is proper to execute the program more than once, it is desirable to have the implied "pointer" that went from number to number in the DATA statements "reset" to the beginning of the first DATA statement. RESTORE will do this by setting the implied pointer to the first number in the first data statement of the program.

REM The REM statement allows the programmer to insert comments or remarks into a program without these comments affecting execution. The BASIC compiler ignores everything following REM. For Example:

```
10 REM PROGRAMMERS NAME IS BOB
```

LET The LET statement allows the assignment of a value to a variable. This value may be a direct assignment (X = Y), or the result of a calculation (X = SIN (Y)). The assignment can also be across variables:

```
LET X = Y = Z = 0
```

It is possible to use a little bit of shorthand at the keyboard by omitting the word LET, since BASIC will add it for you. Hence, in typing:

```
X = 15
```

will be stored and listed as:

```
LET X = 15
```

END The END statement must be the last statement of the entire program. This statement acts as a signal that the entire program has been executed.

Functions BASIC performs several mathematical calculations for the programmer, eliminating the need for tables of trig functions, square roots, and logarithms. These functions have a three-letter call name, followed by an argument, x, which can be a number, variable, expression, or another function. Table 5.9 lists the functions available in 8K BASIC. Most are self-explanatory; those that are not are marked with asterisks, and are explained further on in the text.

Sample Program Statements are the programmer's tools for telling the computer what to do and how to do it. A few are used to define data or to reserve space for it. The following sample program shown in Figure 5.3 and Table 5.10 shows the general format of a series of statements and how they

Other Features

TABLE 5.9

Function	Meaning
SIN(x)	Sine of x (x is expressed in radians)
COS(x)	Cosine of x (x is expressed in radians)
TAN(x)	Tangent of x (x is expressed in radians)
ATN(x)	Arctangent of x (result is expressed in radians)
EXP(x)	e^x ($e = 2.718282$)
LOG(x)	Natural log of x ($\log x$)
*SGN(x)	Sign of x—assign a value of $+1$ if x is positive, 0 if x is zero, or -1 if x is negative
*INT(x)	Integer value of x
ABS(x)	Absolute value of x ($|x|$)
SQR(x)	Square root of x ($\sqrt{x}$)
*RND(x)	Random number
*TAB(x)	Print next character at space x
*GET(x)	Get a character from input device
*PUT(x)	Put a character on output device
*FNA(x)	User-defined function
*UUF(x)	User-coded function (machine language code)

are arranged to provide a desired result. This example is given to show the overall structure of BASIC.

Lines 10 through 65 represent remarks or comments that are present in the program listing to permit the programmer to annotate the program for anyone who will read the listing at a later time.

Line 70 initializes key variables to zero.

Line 80 causes the program to ask for input from the operator console by displaying a "?". After the operator enters a value and a carriage return, the program will continue.

Line 90 "looks for" a number (999999) that flags the end of a series of numbers to be averaged.

Line 100 looks for a number (-999999) used as a flag to indicate when all work is done and the operating system is to be called again.

Line 110 is a remark telling the reader what is going to happen in the following three lines.

Line 120 adds 1 to the number of inputs that have been entered.

Line 130 accumulates the total of the numbers being averaged.

Line 140 sends the logical flow of the program back to that point in the program which again reads in a new number. This is a beginning of a loop or a section of coding done time-after-time with the same procedure.

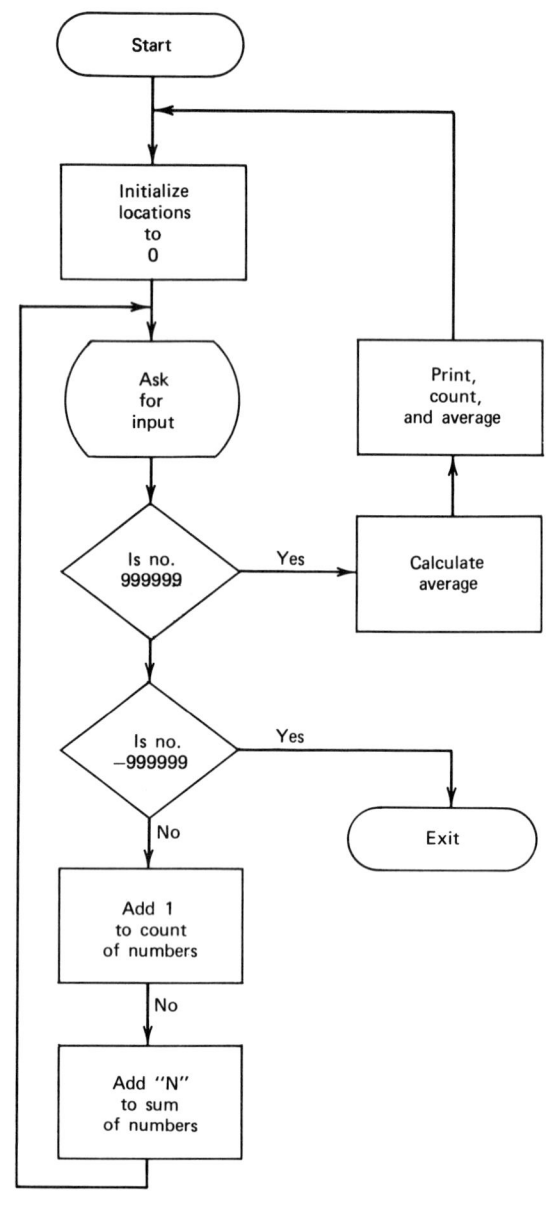

Fig. 5-3 Flowchart of the sample program to calculate the average value. (Courtesy of Hewlett-Packard Co.)

TABLE 5-10

```
10        REM AVERAGING ROUTINE
20        REM TO FIND THE AVERAGE OF
25        REM A SERIES OF NUMBERS.
30        REM WHEN "?" APPEARS' INPUT
35        REM EACH NUMBER WITH A CARRIAGE
40        REM RETURN.
50        REM AFTER ALL NUMBERS HAVE
55        REM BEEN INPUT, INPUT 999999.
60        REM TO EXIT PROGRAM,
65        REM INPUT -999999
70        LET S=C=0
80        INPUT N
90        IF N=999999 GOTO 1000
100       IF N=-999999 GOTO 9999
110       REM ADD TO SUM AND COUNT
120       LET C=C+1
130       LET S=S+N
140       GOTO 80
150       REM CALCULATE RESULT,
160       REM PRINT, AND CLEAR
1000      LET R=S/C
1010      PRINT "AVG = ";R "INPUTS = ";C
1020      GOTO 70
9999      END

RUN
?2
?4
?6
?8
?9
?999999
AVG = 5.8 INPUTS = 5

?2.23234
?3.4567
?999999
AVG = 2.84452 INPUTS = 2
? -999999

READY
```

Line 150 and 160 are remarks telling the reader what actions will take place in the next three lines.

Line 1000 calculates the actual average from all of the prior information.

Line 1010 writes on the output device the results of the calculations done in the lines above.

Line 1020 causes the logical flow of the program to go back to the beginning to prepare the program for a new set of parameters.

Line 9999 is a standard way of stopping the program and telling the interpreter that there are no more statements to be processed.

References

1. Gateley, W. Y., and Bitter, G. G.: *BASIC for Beginners*, McGraw-Hill, New York, 1970.
2. *Programming Languages*, Digital Equipment Corporation, PDP-8 Computer Manual, 1971.
3. *Real-Time Executive BASIC Software System*, Hewlett-Packard Company, Manual, 1974.

Chapter 6

REAL-TIME PROGRAMMING

Introduction and Survey

In real-time systems the computer is directly electrically connected to the data source or to the process control element. The data must be processed and the results must be generated immediately upon receiving the request from external source or control element. Typical real-time applications include laboratory data acquisition, process control, machine monitoring, power system monitoring, laboratory automation, and communication.

Recently, new languages have been developed for a real-time applications. Real-time languages include commands for analog input sampling and quantizing, for display control, for control element positioning, and for communication and operation of external clock. Programming in a real-time language is by an order of magnitude more efficient than is programming the real-time problems in an assembly or a machine language.

In this chapter real-time BASIC for minicomputers is described, and a number of simple examples are shown. Real-time executive is described. Real-time executive is crucial for multitask and multiprogramming operations. Also, microprocessors programming is included in this chapter. Real-time PL/M language specifically designed for microprocessors is described.

Real-time programming is becoming one of the most important tools in a laboratory research, measurement, and process control.

6.1 Real-Time BASIC for Minicomputers

Most of the modern minicomputers are supported with real-time BASIC language. The main task of the real-time BASIC is to provide an efficient way to program input-output operations. As a result, the real-time BASIC language is adapted to serve input-output devices available in a particular minicomputer family, because the language is somewhat different from one minicomputer family to the next. However, the basic principles are the same, and by understanding the real-time BASIC for one family of minicomputers, the reader will have no difficulty to switch to some other family. In this section real-time BASIC for Digital Equipment Corporation minicomputers are described.

Digital Equipment Corporation (DEC) has two families of minicomputers the PDP-8 family and the PDP-11 family. Both families are supported with real-time BASIC. Real-time BASIC has all the functions of regular BASIC, plus a number of commands to control input-output operations through DEC's laboratory units. Those additional commands are divided into the following groups:

> Cathode ray tube scope commands.
> Real-time clock commands.
> Analog-digital convertor sampling commands.
> Test and pause commands.
> User-defined commands and functions.

The most frequently used commands are explained, and short application programs are described. The examples are written in real-time BASIC for PDP-8 family of minicomputers. For additional commands, for more detailed descriptions, and for differences between minicomputers, one should consult the minicomputer manuals.

The material* in this section has been adapted in part from a publication of Digital Equipment Corporation. The material so published herein is the full responsibility of the authors.

PLOT The display scope can be programmed to plot points on its screen. The scope commands provide complete control for graph location and size, display time, and number of points displayed.

When plotting on the scope, which is rectangular, BASIC uses these dimensions for its perimeter:

$$\emptyset < X < 1.1$$
$$\emptyset < Y < 1.\emptyset$$

Thus any plot must be within these limits, which can easily be accomplished by inserting a scaling factor.

---

*Programs and tables are courtesy and copyright © 1972 by Digital Equipment Corporation. All rights reserved.

Real Time BASIC for Minicomputers

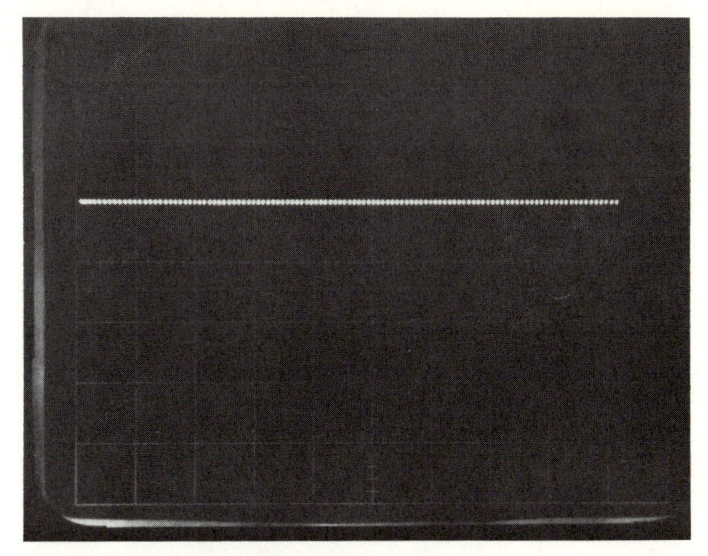

Fig. 6-1 Display of a straight line.

The PLOT command causes the appropriate point to be displayed on the scope. It is issued in the format:

PLOT x,y

where x and y are any expressions that equal the actual X and Y coordinates of the point to be plotted. An acceptable sequence for plotting a straight line across the middle of the scope is shown in Table 6.1. The produced display is shown in Figure 6.1. Remember that every X,Y set must be within the specified ranges. If it is not, that data set is simply not displayed.

BASIC displays points on the scope when it is not doing any internal calculations. Data are displayed, for example, while waiting for input or output. If a calculation is required during a plotting sequence, the data are not displayed until all the calculations are completed. Thus when plotting a decaying sine wave, the function is not displayed until all the points have

TABLE 6.1

```
 5  B=.5
10  FOR A=0 TO 1.1 STEP .01
20  PLOT A,B
30  NEXT A
```

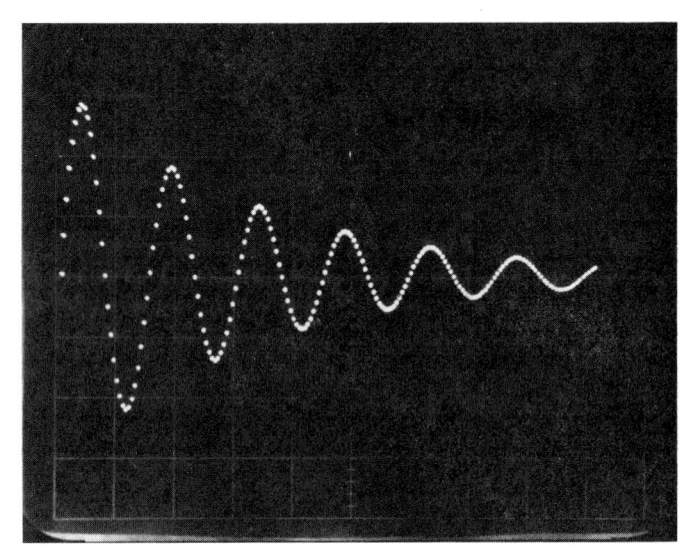

Fig. 6-2 Display of the sin function.

been calculated. Table 6.2 shows the program, and Figure 6.2 shows the produced display.

DELAY BASIC provides a command that refreshes the scope after each calculation so that the progress of the graph can be seen. This command, DELAY, causes BASIC to display all x,y points calculated up to this statement. Thus the decay of the sine wave above can be viewed after each point is calculated by adding to the above example the command:

215 DELAY

The DELAY command provides the additional time for BASIC to display the point before continuing to the next statement.

CLEAR BASIC also permits erasing the scope under program control. By inserting the statement:

CLEAR

all points currently displayed are removed from the screen. Thus if more than one plot is required by a user program and it is not necessary for them to overlap, a CLEAR command between calculations erases the scope for the second plot.

In the next example two compounded interest sums, $400 at 7% and $450 at 6.25% per annum compounded yearly for 30 years, are plotted. If the

Real Time BASIC for Minicomputers

TABLE 6.2

```
200  FOR S=0 TO 1.10 STEP .006
210  PLOT S, SIN(S*35)*EXP(-S*2.5)/3+.5
220  NEXT S
```

intercept point is to be noted, then line 120 can be omitted, and at completion the two curves will be displayed together. Lines 510 and 520 include scaling factors for the scope. Table 6.3 shows the program, and Figure 6.3 shows the display.

Display Buffer The technique for plotting points employed by BASIC includes creating a buffer in the user's area of core to store all the calculated points before they are displayed. (This buffer area is considered as a dimensioned variable; thus executing the commands SCRATCH, RUN, and END removes the buffer from core. The CLEAR command does not delete the buffer, it merely erases its contents.) When a PLOT command is encountered, BASIC checks to see if a display buffer has already been assigned, and if not, then space sufficient for about 500 points is allocated. If this amount of room is not available in core, the error message TOO BIG is printed.

USE The area created by the PLOT command is approximately equal to a 333-dimensional array. To conserve space, if less than 500 points are to be plotted, or to plot more than 500 points, a buffer dimensioning command is

TABLE 6.3

```
100  I=.07
102  P=400
104  T=30
110  GOSUB 500
120  CLEAR
130  I=.0625
134  P=450
140  GOSUB 500
150  STOP
500  FOR N=1 TO T
510  X=N/35
520  Y=(P*((I+1)↑N)/4000)
530  PLOT X,Y
540  DELAY
550  NEXT N
560  RETURN
```

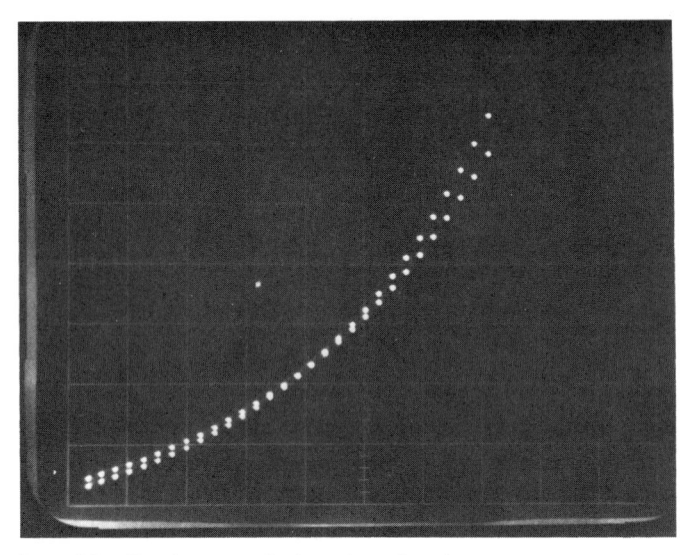

Fig. 6-3 Simultaneous display of two functions.

provided so that core can be allocated optimally. This command, USE x, is implemented as an array, as follows (x is always a variable):

```
20   DIM P(200)
30   USE P
```

Line 30 says: use P as a storage buffer for a future PLOT command; do not create an additional array at that time. Line 20 creates enough room for about 300 data points. If a user-assigned or BASIC generated buffer is not large enough and overflows during execution, the error message TOO BIG is printed.

Only one USE statement is effective for any plot sequence, and if it is to be used, it must be issued before the PLOT command. The variable associated with the USE command is active until one of the buffer removing statements (RUN, SCRATCH, END) is encountered, but it can be used for another purpose when not currently required for displaying.

SET RATE To set the clock to interrupt at a specific rate, use the command

SET RATE mode, time

where mode is the desired clock speed (0-7) and time is the number of clock "ticks" between interrupts, up to 4096 counts. The appropriate mode value is derived from Table 6.4.

TABLE 6.4

Mode	Rate
0	Stop
1	external input
2	10^{-2} sec
3	10^{-3} sec
4	10^{-4} sec
5	10^{-5} sec
6	10^{-6} sec
7	Stop

Thus for the clock to interrupt at 1-sec intervals, an acceptable command is

$$\text{SET RATE } 3,1000$$

which causes the clock to wait 1000 one-msec ticks. If the specified clock rate is too fast, so that the interrupt cannot be serviced in time, the error message RATE ERROR AT (line number) is printed and the clock stopped. The line number printed is that of the statement currently being executed.

Note that in the 100 to 200 μsec elapsed total time range BASIC is servicing the interrupt correctly, but is not executing any BASIC commands because of the high rate; in this case, the processing has been suspended.

SET CLOCK BASIC provides another command for setting the clock rate for the Schmitt triggers. Its format is the same as that for the SET RATE command, namely:

$$\text{SET CLOCK mode, time}$$

except that mode is a 12-bit decimal number that will be used to load the clock enable register, thus permitting the user to enable any function he chooses. The time parameter is specified in the same manner as with the SET RATE command.

Any time either of the SET statements is encountered, the time counter is zeroed. Then, any time a clock interrupt occurs, this counter is incremented. Up to approximately 16,000,000 counts can occur before this counter resets itself to zero.

TIM At any time in the program, the current count (number of elapsed interrupts) can be determined via the function TIM(n).

This function can be used in conjunction with any of the BASIC commands so that the value can be printed or the next action to be performed by the program can be dependent on the count. The format of the function call is TIM(n) where n is any argument (the argument is not checked by BASIC). In the following program the elapsed time for the plot determines the next action; prints the count and halt or, for 50 data points, prints the sines and the terminating count and then stops, Table 6.5.

In the first run, the time elapsed. By changing line 1∅, the sine table was generated.

TABLE 6.5

1∅ A=1∅	RUN
2∅ SET RATE 2,2∅	1∅
3∅ FOR M=∅ TO 1 STEP .∅1	READY.
4∅ PLOT M,M↑2	
5∅ DELAY	
6∅ NEXT M	1∅ A=12
7∅ IF A>TIM(∅) THEN 2∅∅	RUN
8∅ PRINT TIM(23)	∅
9∅ STOP	.∅1999867
2∅∅ FOR Z=∅ TO 1 STEP .∅2	.∅3998933
21∅ PRINT SIN(Z)	.
22∅ NEXT Z	.
23∅ PRINT TIM(C)	.
24∅ END	

WAITC Another application of the clock is to halt program execution until a clock interrupt occurs. The WAITC command performs this function, thereby permitting BASIC to display on the scope while waiting for the interrupt to signal continuation of program execution.

ADC Any A/D channel can be sampled at any time by using the single function ADC(n), where n is the channel number, ∅ to 15, to be sampled in a direct or indirect statement. The ADC function performs an immediate conversion; the clock, however, can be incorporated so that sampling occurs at an established clock rate. In the next program, BASIC waits for a clock tick and then prints the value of the clock, using the TIM function, and of A/D channels 3 and 4 for 50 samples, Table 6.6.

REAL-TIME The ADC function is restricted to nontime critical work, because the finite amount of time elapsed between clock ticks may not be sufficient to perform the tasks requested between ticks (e.g., printing three

TABLE 6.6

```
300  SET RATE 2,60
310  FOR P=1 TO 50
320  WAITC
321  A1=ADC(3)
322  A2=ADC(4)
323  T1=TIM(0)
330  PRINT T1,A1,A2
340  NEXT P
350  STOP
```

values in Table 6.6). Also, more than one channel cannot be sampled in the same time quanta (for example, sampling channels 3 and 4 above).

For time-critical operations, the REAL TIME command should be used because it provides a buffer to hold the sampled value prior to processing. Its format is

REAL TIME v,c1,n1,n2

where v is a subscripted variable to be used as the data buffer. The variable is assigned in a manner analogous to the USE command for the scope, namely, as a dimensioned array. Because only one value is to be taken per sample, three samples are stored per buffer word. Thus a dimension of 100 can store 300 data items. (Note: The data presented in BASIC is a floating-point mode. Three 12-bit computer words are used for each datum. On the other side, ADC produces a 12-bit data item, which can be stored in one computer word.) The array cannot be dimensioned larger than approximately 750, or a maximum of approximately 2200 points. The parameter C1 is the first channel to be sampled, n1 is the number of consecutive channels to be sampled, and n2 is the number of clock ticks for which to sample. To prepare to sample channels 1 and 2 once every millisecond for 150 msec, the suitable code is

```
SET RATE 4,10
DIM G(100)
REAL TIME G, 1,2,150
```

Operation of the REAL TIME command is independent of the BASIC statement processing speed; as long as there is sufficient buffer space, the REAL TIME statement will work.

ACCEPT AND REJECT The REAL TIME statement only creates the specified data buffer. To actually initiate sampling, the statement ACCEPT is required. Then sampling will start at the next clock tick. There must be an active REAL TIME statement, or the ACCEPT is ignored. A REAL TIME command becomes inactive when the clock count equals zero. To suspend sampling use the command REJECT. This command is also useful for executing subsequent REAL TIME statements.

In the next example, the ACCEPT and REJECT statements are used to be sure sampling occurs at the specified rate for only the designated number of counts. Statement 500 stops the clock, 530 prepares it so that the first sample is taken after processing statement 540. The REJECT at 560 assures that after 100 counts no extra samples are taken at the rate of 10 msec. The example is shown in Table 6.7.

TABLE 6.7

```
500    SET RATE 7,0
510    DIM G(100)
520    REAL TIME D,1,2,150
530    ACCEPT
540    SET RATE 2,10
550    IF TIM(4)<100 THEN 550
560    REJECT
570    SET RATE 3,10
580    ACCEPT
590    END
```

ADB To retrieve data collected by REAL TIME and ACCEPT sequences that are placed in a buffer, the ADB(n) function is required. Data is withdrawn from the buffer in the same order in which it was entered. Thus if four A/D channels, 1 to 4, are being sampled, the order of the data in the buffer is

$$1_i 2_i 3_i 4_i \quad 1_{i+1} 2_{i+1} 3_{i+1} 4_{i+1} \quad 1_{i+2} \ldots$$

The argument in the ADB function is ignored. The items will be removed consecutively. If there is no REAL TIME or ACCEPT statement or if there is no data remaining in the buffer (because the number of clock ticks, n2, in the REAL time statement had expired), the error message NO A-D and a line number are printed.

The example in Table 6.8 is an expansion of the one in Table 6.7 to incorporate the ADB function. Lines 590 through 610 have been added so that the sampled values can be printed.

WAIT If a pause time is required by a program, the WAIT command can be included in the program. BASIC processing will be halted until any interrupt occurs. Note that a clock interrupt is sufficient to reactivate BASIC.

TABLE 6.8

```
500  SET RATE 7,0
510  DIM G(100)
520  REAL TIME G,1,2,150
530  ACCEPT
540  SET RATE 2,10
550  IF TIM(4)<100 THEN 550
560  REJECT
570  SET RATE 3,10
580  ACCEPT
590  FOR A=1 TO 150
600  PRINT ADB(1),ADB(2)
610  NEXT A
630  END
```

6.2 Real-Time Executive BASIC System

Introduction Hewlett-Packard Company (HP) has developed Real-Time Executive BASIC Software System for her minicomputers.

The HP Real-Time Executive BASIC (RTE-B) Software System is a BASIC-programmable real-time system capable of providing a time- and event-scheduled operation of up to 16 different user tasks. It is designed for real-time measurement and control applications. Built-in time referencing and event interrupt handling relate RTE-B system operations to external processes occurring in real time. Priority levels from 1 to 99 provide flexible discrimination among the relative urgencies of tasks in the multitasking environment. The RTE-B system is specifically supported by device control routines for:

- Analog and digital input/output.
- Data logging and plotting.
- Punched tape input/output.
- Punched or mark-sense card input.
- Magnetic tape positioning and input/output.
- Single or multiple keyboard input/output terminals.

The RTE-B system support library includes bit manipulation routines as well as all the computational routines usually expected in a computer-automated measurement and control system.

The material* in this section has been adapted in part from a publication of Hewlett-Packard Company. The material so published herein is the full responsibility of the author.

*Programs, figures and tables are courtesy and copyright © 1974 by Hewlett-Packard Company. All rights reserved.

This section is a review of commanding, programming, and scheduling RTE-B system operations. It also provides instructions for configuring and generating the RTE-B system, and for starting system operation and initializing the system's real-time clock.

The RTE-B Real Time BASIC Software System operates in the following minimum hardware configuration:

1. HP 2100 series Computer with 12K memory, Extended Arithmetic Unit, Floating Point Hardware, Time Base Generator, and Teleprinter Communications Channel.
2. Punched Tape Reader Subsystem.
3. Teleprinter.

Task Definition and Priority The task is a series of program statement instructions with which the user specifies system operations to be performed, such as measurements, computations, digital and analog input and output, data logging, and data plotting. The task is a module of the total real-time BASIC program that is used to solve a real-time measurement and control application. The task may be oriented toward a given device, control a specific response to the occurrence of an external event, perform a given function, or satisfy a common requirement.

The RTE-B system provides for the assignment of priority levels from 1 to 99 to the various tasks operating in its multitasking environment. This enables the user to assure that system response to specific events is fast enough to maintain real-time relationship to external processes.

Time and Event Scheduling Up to 16 different tasks may be time and/or event scheduled by the RTE-B system. The tasks are all programmed as subroutines of a single HP Real-Time BASIC program, as shown in Figure 6.4.

The RTE-B scheduler checks for time and event interrupts after the execution of each program statement. As diagrammed in Figure 6.5, it examines task timing and priorities and either continues running the current task or starts running a newly scheduled task having higher priority.

Input–Output Processing An I/O scheduling and control monitor (IOC) in the RTE-B operating system coordinates the I/O transfers of all standard devices in the system. The I/O devices are referenced by a logical unit number rather than the actual physical I/O channel. Later, if it becomes necessary to change the configuration of the system, the I/O channels can be changed without changing the logical unit numbers.

Real-Time Executive BASIC System

```
Real-time BASIC
    program
┌─────────────────────────┐
│   Priority              │
│   and                   │
│   time & event          │
│   scheduling            │
│   statements            │
├─────────────────────────┤     Run task
│   250 REM TASK 1        │ ◄──────────────
│   255 START(250,900)    │     Reschedule task
│         *               │ ─ ─ ─ ─ ─ ─ ─►
│         *               │
│   390 RETURN            │     Task completed      ┌──────────┐
│                         │ ─────────────────►      │  RTE-B   │
├─────────────────────────┤     Run task            │scheduler │
│   500 REM TASK 2        │ ◄──────────────         │          │
│   505 START(500, 1200)  │     Reschedule task     │          │
│         *               │ ─ ─ ─ ─ ─ ─ ─►          │          │
│         *               │                         │          │
│   610 RETURN            │     Task completed      │          │
│                         │ ─────────────────►      │          │
├─────────────────────────┤     Run task            │          │
│   700 REM TASK 3        │ ◄──────────────         │          │
│   705 START(700, 500)   │     Reschedule task     │          │
│         *               │ ─ ─ ─ ─ ─ ─ ─►          │          │
│         *               │                         │          │
│   990 RETURN            │     Task completed      │          │
│                         │ ─────────────────►      └──────────┘
└─────────────────────────┘
```

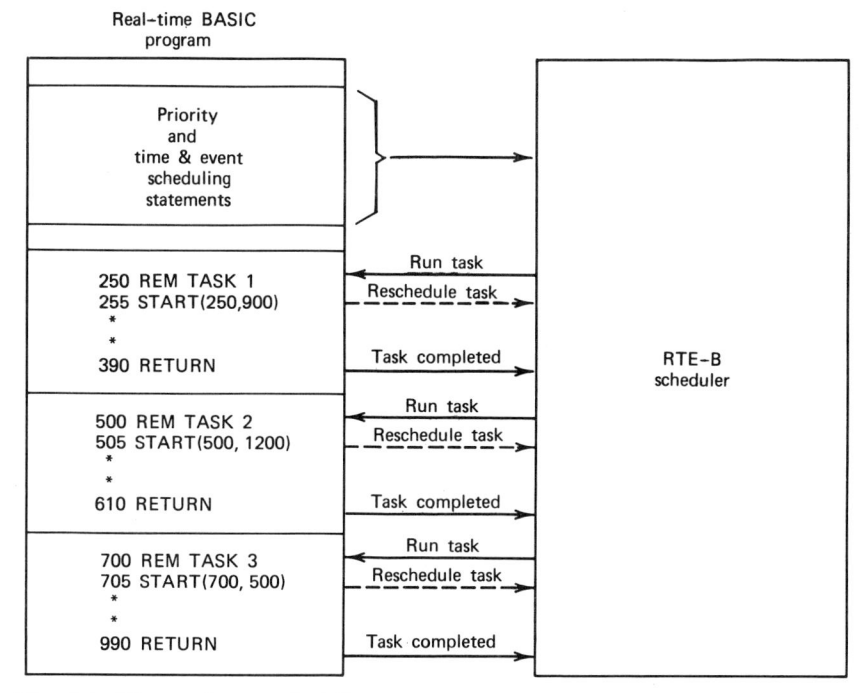

Fig. 6-4 Time and event scheduling.

The RTE-B system requires one teleprinter for system control, data logging, and tape punching. However, the system can also operate with several additional I/O or data-logging devices (teleprinters, CRT terminals, line printers, tape punches, card readers, or magnetic tape units). Multiterminal operations include task scheduling and input of information from the operator at any of the auxiliary keyboard input units in the system, when provided by the user's real-time BASIC program. In addition, instructions and data can be directed individually to any of the data logging units in the system. The RTE-B system is thus capable of concurrently exchanging information with several different remote locations.

IOC stacks output requests and provides for automatic memory buffering of output directed to low- or medium-speed peripherals. This speeds the processing and output of finished results because the system is not kept waiting while a low-speed device completes a printout.

System Communications The user exercises dynamic control of the system via system requests from his programs during real-time operations.

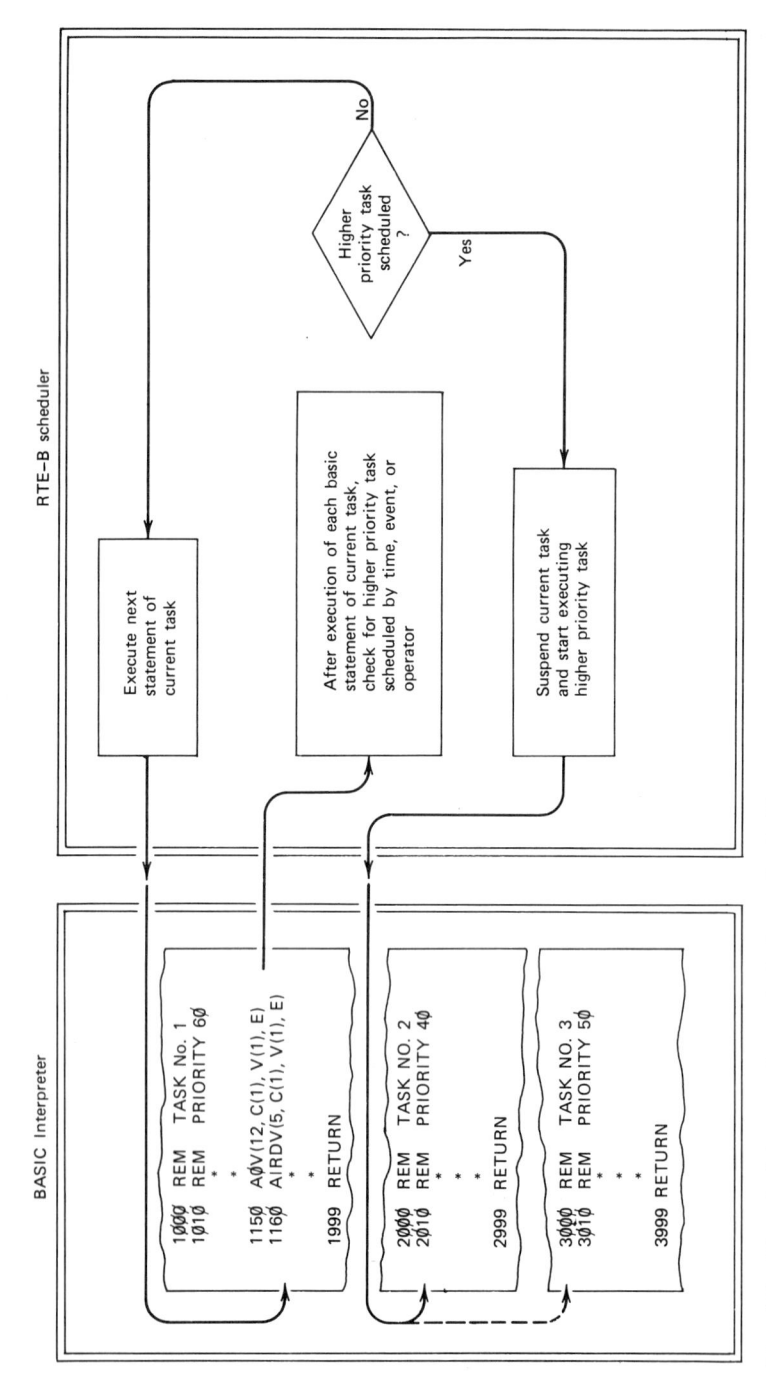

Fig. 6-5 The scheduler examines task timing and priorities.

Real-Time Executive BASIC System

System requests are made in the form of Real-Time BASIC program statements to perform any of the following executive functions:

- Read from any input device.
- Write to any output device.
- Control peripherals, such as magnetic tape unit.
- Schedule tasks to be run.
- Request current time of day.

The available program statements, functions, mathematical operators, and relators are summarized in Table 6.9.

Real-Time Task Scheduling The RTE-B system is called a real-time system because the order of processing is governed by time or by the occurrence of external events rather than by a strict sequence defined in the program itself. Because these events can occur in random order and require different amounts of processing, a real-time system must be capable of resolving conflicts between tasks.

A task is defined as a group of BASIC statements that are initiated by one of the RTE-B scheduling statements, such as START and TRAP, and terminated by a RETURN statement. It is an entity that can be scheduled for execution. Each task is uniquely identified by the line number of the first statement in the task. If a task is initiated by executing line 2000 of the BASIC program, that task will be represented as 2000 in all scheduling statements.

Methods of Initiating Tasks A task can be initiated in three different ways

- At a specified time of day.
- After a specified delay.
- Upon the occurrence of an external event.

To initiate a task at a given time of day, use the TRNON statement. The following example shows how TRNON will cause the task at 2000 to be executed at 5 sec after 1245.

$$100 \text{ CALL TRNON } (2000, 124505)$$

When the RTE-B system clock shows the time 124505, the BASIC interpreter, if it is the EXECUTION mode, will execute the task at 2000. Just like any other clock, the RTE-B system clock must be set if it is to reflect the correct time of day. To do this, use the SETIME operator command after loading the system into the computer.

TABLE 6.9

	Statements	Uses	Examples
General Program	COM	Allocates common program storage	100 COM P(8),T(10)
	DIM	Allocates program storage	105 DIM A(3),B(3),K(5)
	REM	Remarks in program testing	110 REM TASK 1
	LET	Assigns value ("LET" is optional)	115 R=R1*100+R2
	FOR	Starts repetitive operations	120 For I=1 TO 5
	NEXT	Terminates FOR operations	200 NEXT I
	READ#	Reads variables input by operator at unit Z	205 READ#Z;A(1),A(2),A(3)
	READ	Reads values from DATA statement	210 READ D(1),I(1),I(2)
	DATA	Provides vaues to READ statement	270 DATA 1,0,16,0
	RESTORE	Resets DATA statement for next use	275 RESTORE
	IF–	Conditional action	280 IF M1=24 Let M1=0
	GOTO	Unconditional transfer	285 GOTO 2480
	PRINT	Print on system keyboard input unit	290 PRINT "SHUTDOWN!!!"
	PRINT#	Print on data logging unit X	295 PRINT#X;T;TAB(13);R1
	GOSUB	Transfer to subprogram	300 GOSUB 2450
	RETURN	Return from subprogram	305 RETURN
Event/Time Control	TRAP	Links event interrupt to task	310 TRAP1 GOSUB 250
	TIME	Gets time of day for program	315 TIME(T)
	SETP	Sets task priority	320 SETP(250,1)
	TRNON	Turns on task at specific time	325 TRNON(250,1200)
	START	Starts task after delay	330 START(250,95)
	DSABL	Disables task	335 DSABL(250)
	ENABL	Enables task	340 ENABL(250)
	TTYS	Schedules trap N task from teleprinter	350 TTYS(U,N)
Analog/Digital Input/Output	AOV	Analog Output	390 AOV(6,C(1),V(1),E)
	AISQV	Sequential mode analog input	400 AISQV(5,1,V(1),E)
	AIRDV	Random-scan analog input	405 AIRDV(5,C(1),V(1),E)
	PACER	Sets pacer interval	410 PACER (1,5,2)
	SGAIN	Sets gain of analog channel(s)	415 SGAIN(1,1000)
	RGAIN	Reads gain of analog channel(s)	420 RGAIN(1,G)
	DAC	Analog output	425 DAC(101,C4)
	RDBIT	Reads digital input bit	430 RDBIT(18,B,V)
	RDWRD	Reads digital input word	435 RDWRD(18,W)
	WRBIT	Writes digital output bit	440 WRBIT(23,B,V)
	WRWRD	Writes digital output word	445 WRWRD(23,W)
	SENSE	Links event to trap number T	450 SENSE(5,B,1,T)

Category	Code	Mnemonic	Uses	Examples
Magnetic Tape Input/Output	600	MTTRD	Reads from magnetic tape	MTTRD(U,V(1),64,F,N2)
	605	MTTRT	Writes onto magnetic tape	MTTRT(U,V(1),64,F,N2)
	610	MTTPT	Positions magnetic tape	MTTPT(U,5,-3)
	615	MTTFS	Controls magnetic tape unit function	MTTFS(U,0)
Bit Manipulation	700	IOR	Adds M and N, bit-by-bit	IOR(M,N,R)
	705	INOT	Returns complement of M	INOT(M,R)
	710	IEOR	Adds M and N exclusively	IEOR(M,N,R)
	715	IAND	Logically multiples M and N	IAND(M,N,R)
	720	ISHFT	Shifts M by ±N bit positions	ISHFT(M,N,R)
	725	IBTST	Returns state S of bit B in V	IBTST(V,B,S)
	730	IBSET	Sets bit B in value V	IBSET(V,B,R)
	735	IBCLR	Clears bit B in value V	IBCLR(V,B,R)
	740	ISETC	Sets K equal to octal constant	ISETC("177077',K)
Plotter Output	800	PLOT	Plots line	PLOT(X,Y,Z)
	805	SYMB	Plots characters	SYMB(X,Y,H,"VOLTS",90,1)
	810	URITE	Moves pen to upper right for paper change	URITE
	815	LLEFT	Sets origin to 0,0	LLEFT
	820	SFACT	Sets paper size	SFACT(15,10)
	825	FACT	Sets scale factor	FACT(.6667,1)
	830	WHERE	Returns current position	WHERE(X,Y)

Functions	Uses	Examples
ABS	Absolute value	ABS(X+Y-Z)
EXP	Base e exponential value	EXP(X)
INT	Integer part	INT(X)
LN	Natural (base e) logarithm	LN(X)
RND	Random numbers (0 through 1)	RND(X)
SGN	Sign of variable	SGN(X)
SQR	Square root	SQR(89)
SWR	Status of switch register bit	SWR(6)
TAB	Lab set for PRINT statement	TAB(13)
SIN	Sine	SIN(X)
COS	Cosine	COS(X)
TAN	Tangent	TAN(X)
ATN	Arc Tangent	ATN(X)
OCT	Printing of Octal variable	OCT K
LOG	COMMON (base 10) logarithm	LOG(X)

Operators		Uses	Examples
+		Addition	X+Y
−		Subtraction	X−Y
*		Multiplication	X*Y
/		Division	X/Y
		Exponentiation	X'2
AND		Logical AND	X AND Y
OR		Logical OR	X OR Y
NOT		Logical NOT	NOT X

Relators	Uses	Examples
#	Not equal to	X#Y
<	Less than	X<Y
>	Greater than	X>Y
<=	Less than or equal to	X<=Y
>=	Greater than or equal to	X>=Y
=	Equal to	X=Y

To initiate a task after a delay, use the START statement. The following example shows how START will cause the task at 2000 to be executed 15 sec after this statement (100) is itself executed.

100 CALL START (2000,15)

The real power in this form of scheduling lies in the ability to schedule a task repetitively. Consider the following example, Table 6.10.

TABLE 6.10

```
100  CALL START(2000,15)
999  GOTO 999
2000 CALL START(2000,10)
2010 PRINT "TASK 2000 AT 10 SEC INTERVALS"
2020 RETURN
```

Fifteen seconds after executing line 100, the task at 2000 will begin execution. Its first action is to schedule itself for execution 10 sec later. As a result, it will thereafter execute every 10 sec.

Line 999 is called an IDLE LOOP and keeps the program in the execution mode indefinitely. When no task is executing, the program continuously cycles, waiting for something to happen. All real-time programs written for the RTE-B system should have some form of idle loop. To terminate this example, press any key on the system console. The RTE-B system will respond by printing a colon on the system console. Type "AB" and press the RETURN key. The program will be aborted and the system will return to the command mode.

One common error is to ask that a task schedule itself every few seconds, for example, every 2 sec. If the statement that asks for the rescheduling is at the end of the line of coding, then the delay will not be 2 sec but 2 sec plus the amount of time necessary to execute all statements up to the point of the initiating statement, plus all interim processing by the operating system. The amount of time in some routines can be considerable. Having the delay introduced once probably will not make much difference, but if the error is within a loop, the cumulative effect will soon become apparent.

Another error is to try to turn on each task with a delay of zero (immediate start) by giving a series of consecutive task initiation commands, each one turning on a task immediately. The programmer often forgets that the first task encountered with an immediate turn on must run to completion before another statement in the program can be executed. This does not allow the computer enough time to execute all the task-initiating statements, which

Real-Time Executive BASIC System

makes the future tasks late in starting. The usual technique is to have all the tasks initiated 1 sec after the call for initiation, allowing enough time for all of the initiating statements to be completed.

To initiate a task in response to an external event, the task must first be associated with a trap number. The following example shows how the TRAP statement will associate the task at 2000 with trap number 5.

> 100 TRAP 5 GOSUB 2000

Once this is done, various events can be associated with trap number 5 using SENSE calls and/or a TTYS call (for an auxiliary teletype event). TRAP is not used if there is no HP SENSE Subsystem or auxiliary teletype.

Priorities Since a program may contain up to 16 tasks, it is possible that more than one task will be trying to be executed at the same time. To resolve these conflicts, tasks may be assigned a priority number by the user writing in BASIC. Priorities are established to provide some delineation between actions that must occur immediately and actions that have a more relaxed requirement. Thus a higher priority task can take over the computer from a lower priority task. A priority can be assigned to a task with 1 being the highest and 99 being the lowest possible. A task of higher priority can interrupt the processing of a lower-priority task should the interrupt occur during the processing of the lower-priority task.

The following statement will set the priority of the task at 2000 to 50.

> 110 CALL SETP (2000,50)

Any task whose priority has not been explicitly specified will be given a priority of 99 by the system.

Time and Event Scheduling Using the information presented in the previous paragraphs, it can be seen that a structure is designed into the RTE-B system that will support different types of task scheduling. Some of the schedulings take place because of the time of day as determined by the computer's internal clock (e.g., CALL TRNON (150,1000000)). In addition, there are tasks that are scheduled after some delay. This might be the delay after a valve is opened and the computer is waiting for conditions to stabilize (e.g., CALL START (150,100)). Another situation that event scheduling would be expected to process would be the scheduling of an operator's panel task as a result of the pressing of a button on the panel. The pushing of a button is usually a random asynchronous event, and the resultant interrupt from the computer must cause the scheduling of a processing task.

The RTE-B system is designed to sense when the external event interrupt has occurred (e.g., pushbutton), and signals that such an event has occurred by recording significant information about the event. The recording of an event will take place concurrently with the execution of a statement in BASIC language.

At the completion of the BASIC statement being executed, the RTE-B scheduler checks to determine whether a task has been scheduled by time or external event. If such is the case and if that task is a higher priority than the task currently being executed the latter task will be suspended and the just-scheduled task will begin execution.

The RTE-B scheduler keeps track of tasks and tells the BASIC interpreter when to initiate execution of each task. The task scheduling statements pass information to the scheduler which maintains that information and uses it to make decisions concerning the various tasks.

A series of subroutines are used to cause the subsystem to perform various operations. These subroutines are "called" into play by executing the special statement:

CALL name (parameters as required)

Example This subroutine will read analog inputs in a random manner.

CALL AIRDV (numb, achan, volt, error)

Where:

numb = The number of channels to be read. If numb is negative, the readings will be paced by the system pacer.

achan = An array whose contents are channel numbers and whose positional relationship is the same as the voltage in volt. Hence, designating a channel number in achan will cause the corresponding voltage to appear in the corresponding position in volt.

volt = The first location of an array where the voltages to be read will be placed.

error = This variable is set to one of the following values upon return from the subroutine:

0 = No error.

1 = Overload has occurred. The voltage available at the sensor multiplied by the gain specified by the programmer, exceeds the voltage range of the analog-to-digital converter.

2 = Pace error. The subsystem was not ready when a pace pulse occurred. Normally this is caused by too rapid a pace rate or failing to turn off the pacer after a paced operation.

The following example will read 75 values into "V" (assuming the channel number array "A" has been initialized to the channel numbers). The values in array "V" will correspond on a one-to-one basis to the channel numbers in array "A" (i.e., the first value corresponds to the first channel number, etc.).

```
DIM V(75),A(75)
LET N = 75
CALL AIRDV (N,A(1),V(1),E)
```

6.3 Real-Time PL/M Language for Microprocessors

General Features A PL/M program is arranged as a sequence of *declarations* and *statements* separated by semicolons. The declarations allow the programmer to control allocation of storage, define simple macros, and define procedures. Procedures are subroutines that are invoked through certain statements in PL/M. These procedures may contain further declarations that control storage allocation and define nested procedures. The procedure definition capabilities of PL/M allow modular programming; that is, a particular program can be divided into a number of subtasks, such as processing teletype input, converting from binary to decimal forms, and printing output messages. Each of these subtasks is written as a procedure in PL/M. These procedures are conceptually simple, are easy to formulate and debug, are easily incorporated into a large program, and form a basis for library subroutine facilities when writing a number of similar programs.

In addition to the procedure declaration facilities, PL/M allows a number of data types to be declared and used in a program. The two basic data types are *byte* and *address*. A byte variable or constant is one that can be represented in an 8-bit word, while an address variable or constant requires 16 bits (double byte).

Input-Output The following PL/M sample program reads data from input ports 0 and 1 and writes the larger of these two values at output port 0. Note that the two pseudovariables INPUT(0), and INPUT(1) act like PL/M single-byte variables, but have the effect of reading the values latched into input ports 0 and 1, respectively. Similarly, the pseudovariable OUTPUT(0) can be used in an assignment statement in order to write values to output port 0.

The complete PL/M program for performing this simple function is shown in Table 6.11 (Intel 8080 micro system).

The symbol EOF (end-of-file) is required in PL/M to indicate the end of the program. Note also that the GO TO statement causes program control to restart at the point labeled 'LOOP:' where input values are read again.

TABLE 6.11

```
DECLARE (I,J,MAX) BYTE;
/* READ INPUT PORT 0 AND SAVE IN VARIABLE I */
LOOP:
    I = INPUT(0);
/* NOW READ INPUT PORT 1 AND SAVE IN VARIABLE J */
    J = INPUT (1);
/* SET MAX TO THE LARGER OF THESE TWO VALUES */
    IF I > J THEN MAX = I; ELSE MAX = J;
/* WRITE THE VALUE OF MAX AT OUTPUT PORT 0 */
    OUTPUT(0) = MAX;
/* GO BACK AND READ THE INPUT PORTS AGAIN */
GO TO LOOP;
EOF
```

(Courtesy of Intel Corporation.)

The built-in procedure INPUT and build-in variable OUTPUT were introduced earlier. In general, the input call takes the form

INPUT(constant)

where the constant is in the range 0 to 7. The effect of the call is to read the input port designated by the constant. The result of the call is the byte value latched into the port. The call to INPUT can appear as a part of any valid PL/M expression.

The pseudovariable OUTPUT can only be used as the destination of an assignment. The form is

OUTPUT(constant) = expression;

where the constant is in the range 0 to 23. The value of the expression is latched into the output port designated by the constant.

The TIME Procedure A built-in procedure, called TIME, is provided in PL/M for waiting a fixed amount of time at a particular point in the program. The form of the call is

CALL TIME(expression);

where the expression evaluates to a byte quantity n between 1 and 255. The wait time is measured in increments of 100 μsec; hence, the total timeout for a value n is

n(100 μsec).

Thus the call to TIME shown below results in a 4500 μsec (4.5 μsec) timeout

CALL TIME(45);

Since the maximum timeout is 255*100 μsec = 25500 μsec = 25.5 msec, longer wait periods are affected by enclosing the call in a loop. The following loop, for example, takes 1 sec to execute

```
DO I = 1 TO 40;
CALL TIME(250);
END;
```

References

1. Souček, B.: *Minicomputers in Data Processing and Simulation*, Wiley, New York, 1972.
2. *Lab 8/E Software User's Manual*, Digital Equipment Corporation, 1972.
3. *Real-Time Executive BASIC Software System Manual*, Hewlett-Packard Company, 1974.
4. *PL/M Programming Manual*, Intel Corporation, 1975.

Chapter 7

SIMULATION

Introduction and Survey

The first step toward the experiment is the design of the experiment. This plan can be done analytically if the system is linear, deterministic, and involving a small number of parameters. For more realistic, nonlinear, and nondeterministic systems and parameters, the experiment design can be done through simulation and modeling. The objective of the experiment design is to predict how a system under study will perform by conducting studies on the model of the system. We define a model as the body of information about a system gathered for the purpose of studying the system. Different models of the same system are possible, depending on the nature of the information that is gathered. One physical system can be modeled with another physical system, or it can be represented as a mathematical model in which the activities are described by mathematical functions and interrelations of variables. Mathematical models can be simulated by electronic computers.

System simulation is the technique of solving problems by following the changes over a period of time of a dynamic model of a system. The simulation does not attempt to solve the equations of a model analytically; instead, it observes the way in which all variables of the model change with time.

Since the advent of electronic analog computers about two decades ago, the analog computer has been widely used for the simulation. Recently, many modeling techniques utilizing digital computers were introduced. Also, many programming systems, called digital simulators, have been written (continuous-system simulation, industrial dynamics, probability simulation, servicing and queueing, discrete system simulation, etc.).

Deterministic Data Simulation

In this chapter we concentrate only on the techniques suitable for data simulation and for experiment design. Generation of continuous and discrete deterministic processes is shown first. Simulation of random data and systems is explained next. Special attention is given to the most widely used Monte Carlo techniques. Such techniques may be used to recognize and possibly minimize experimental difficulties or to prove interrelations between different parameters of the system.

7.1 Deterministic Data Simulation

Sine-Wave Generation Sine-wave generation has many applications in the laboratory. Sine waves are produced by any physical system that can be described through a differential equation of the harmonic oscillator:

$$\frac{d^2x}{dt^2} = -w^2 \cdot x \qquad (7.1)$$

Differential equation 7.1, with initial conditions $x = U$; $dx/dt = 0$, at the moment $t = 0$, has the solution

$$x(t) = U \cdot \cos wt \qquad (7.2)$$

Such an equation describes the motion of the mass hanging on the spring without friction; or electrical voltage, in loss less inductance-capacitance circuit; or elastic deformation of the solid, or internally controlled behavior and so on.

Although there are many different ways to build sine-wave generators, the most instructive way will be by making an electronic model of equation 7.1. Such a model is shown in Figure 7.1a.

Let us suppose that the function d^2x/dt^2 is available. Using this function as input to the integrator, we obtain $-dx/dt$ at the output of integrator number 1. Integration of $-dx/dt$ yields x at the output of integrator 2. Applying x as the input to the amplifier with the gain w^2, we obtain the output $-w^2x$. According to equation 7.1, $-w^2x$ equals d^2x/dt^2. Hence by closing the loop we form the harmonic oscillator described by Equation 7.1.

Initial conditions may be set on either integrator or on both. Integrator 1 generates dx/dt. No initial condition circuit is necessary for integrator 1, since the initial condition is $dx/dt = 0$. Integrator 2 generates x. Hence the initial condition circuit is necessary to supply the voltage $x(0) = U$. At $t = 0$, the contact $\tilde{S}$ should be open, and, in the same time, the contact S should be closed. The output x, according to equation 7.2, will be the cosine wave. Hence, the output $-dx/dt$ will be the sine wave.

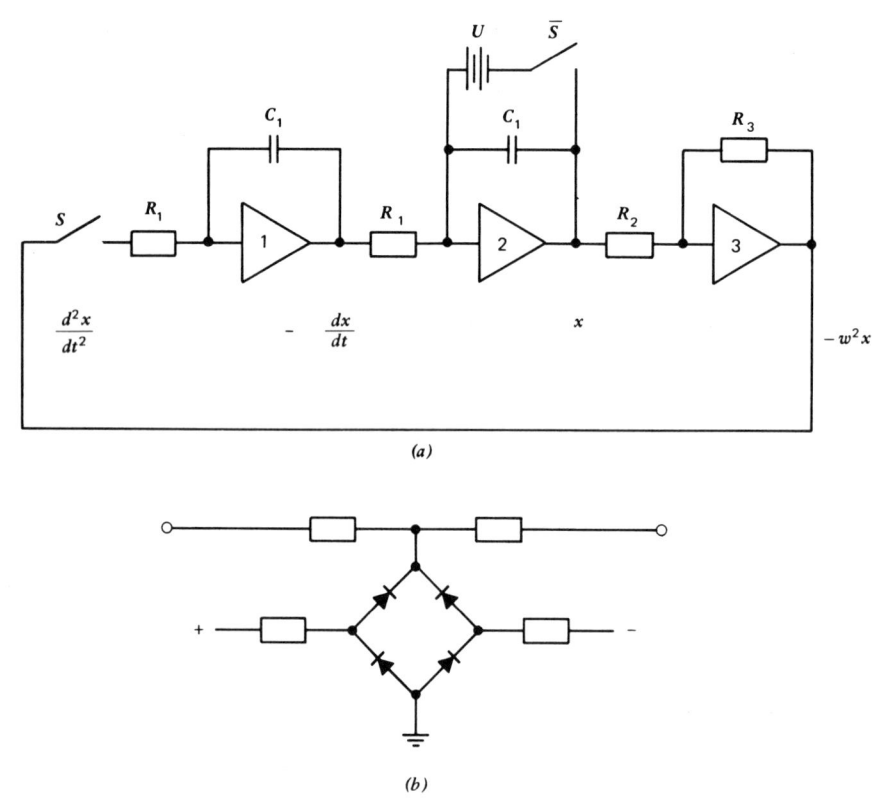

Fig. 7-1 Simulation of the harmonic oscilator, using operational amplifiers and integrators. (*a*) Linear part, (*b*) Nonlinear part.

In the above example gains of the integrators have been chosen to be 1. In a general case, integrator 1 will have a gain $1/R_1C_1$ and integrator 2 will have a gain $1/R_1C_1$. Hence the circuit will be described by the equation

$$\frac{d^2x}{dt^2} = \frac{1}{R_1C_1} \cdot \frac{1}{R_1C_1} \cdot \frac{R_3}{R_2} \cdot x \qquad (7.3)$$

producing the sine wave of the frequency

$$w_0^2 = \frac{1}{R_1C_1} \cdot \frac{1}{R_1C_1} \cdot \frac{R_3}{R_2}$$

For the proper operation of the circuit, the time constants of the integrators must be $1/R_1C_1 \gg 2\pi/w_0$; and the cut-off frequency w_c of the amplifiers 1, 2, and 3 must be $w_c \gg w_r$.

Deterministic Data Simulation

Further modifications of the circuit can improve sine-wave accuracy: slight regeneration can be achieved by adding the feedback resistor between output 3 and input 2; amplitude control can be achieved through the degenerative feedback from the output 1 to the input 1, including a nonlinear element in the loop.

The circuit is interesting in that it provides a way to generate a sine or cosine wave for use as a forcing function in analyzing the response of systems.

Computer Simulation of Harmonic Oscillator In digital computer simulation, the time scale t is divided into discrete intervals $T, 2T, \ldots, NT$. To make the equivalent of the integrator, we have to design the system with the following properties.

If at the instant NT the input IN (NT) is presented to the system, with the initial condition OUT$[(N - 1) \cdot T]$, then the new output of the system must be

$$\text{OUT}(NT) = \text{IN}(NT) + L \cdot \text{OUT}[(N - 1)T] \qquad (7.4)$$

Equation 7.4 presents actively the simplest definition of a digital integrator. The value OUT$[(N - 1)T]$ presents the accumulated result up to the instant $(N - 1)T$. The value IN (NT) presents the contribution of the signal arriving at the instant NT. The coefficient L presents the leakage from the integrator. Typically L is slightly less than 1, showing that the integrator would eventually discharge due to the leakage. Many physical and biological integrating processes could be simulated with integrators, as defined in equation 7.4. Ideal integrators, which would only accumulate the result and never leak, could be described with $L = 1$.

Figure 7.2a shows the symbol for the integrator. It is fully described by its initial condition OUT$[1 \cdot T]$, and by the leakage L. Figure 7.2b shows the digital equivalent for the harmonic oscillator, and it follows the basic structure of Figure 7.1.

Equation 7.4 could be programmed in the following way

$$\text{OUT} = \text{IN} + L * \text{OUT} \qquad (7.5)$$

This is typical programming equivalence statement: new value of OUT equals to the sum of new value of IN and of the old value of OUT multiplied by L.

The whole system in Figure 7.2 can be described by three such equations: two for integrators (units 1 and 2), and one for amplifier (unit 3, with the gain w^2):

Output of the first integrator (unit 1):

$$X1 = X2 + L1 \cdot X1 \qquad (7.6a)$$

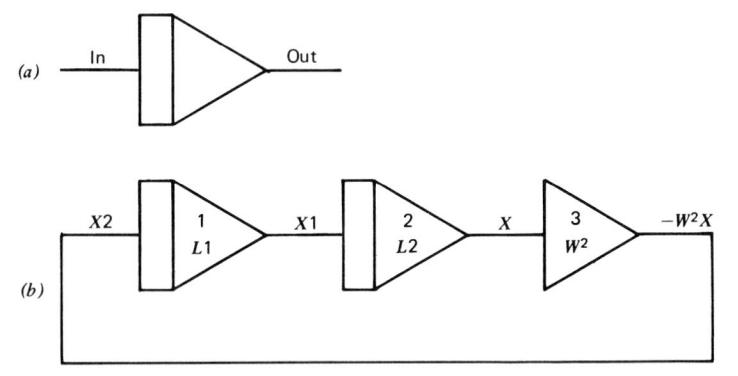

Fig. 7-2 Block diagram of the harmonic oscillator simulation. (a) Integrator, (b) Simulation chain.

Output of the second integrator (unit 2):

$$X = X1 + L2 \cdot X \qquad (7.6b)$$

Output of the amplifier (unit 3):

$$X2 = -(w^2) \cdot X \qquad (7.6c)$$

The program based on equations 7.6a, b, and c is shown in Table 7.1. This program simulates the harmonic oscillator, for 300 time intervals. The produced signal X as a function of time is displayed on the scope, and shown in Figure 7.3.

Figure 7.3a shows the result of the first run, with parameters W1 = 0.1, L1 = 1, L2 = 1, see the second part of the Table 7.1. Because the integrators have no leakage, ideal sine wave is generated.

Figure 7.3b shows the result of the second run, with parameters W = 0.1, L1 = 0.99, L2 = 0.99. Because of the leakage in the integrators, dumped sine wave is generated.

Note the following:

1. System simulation is the technique of solving a problem by following the changes over a period of time of a dynamic model of a system. The simulation does not attempt to solve the equations of a model analytically. Instead, it observes the way in which all variables of the model change with time.

2. Sometimes the programmer is not interested in the structure of the system, and he treats the system as a unit box. If he knows that such unit always generates, say, a sine wave, he will program the output using sine function, which is available as a part of language library.

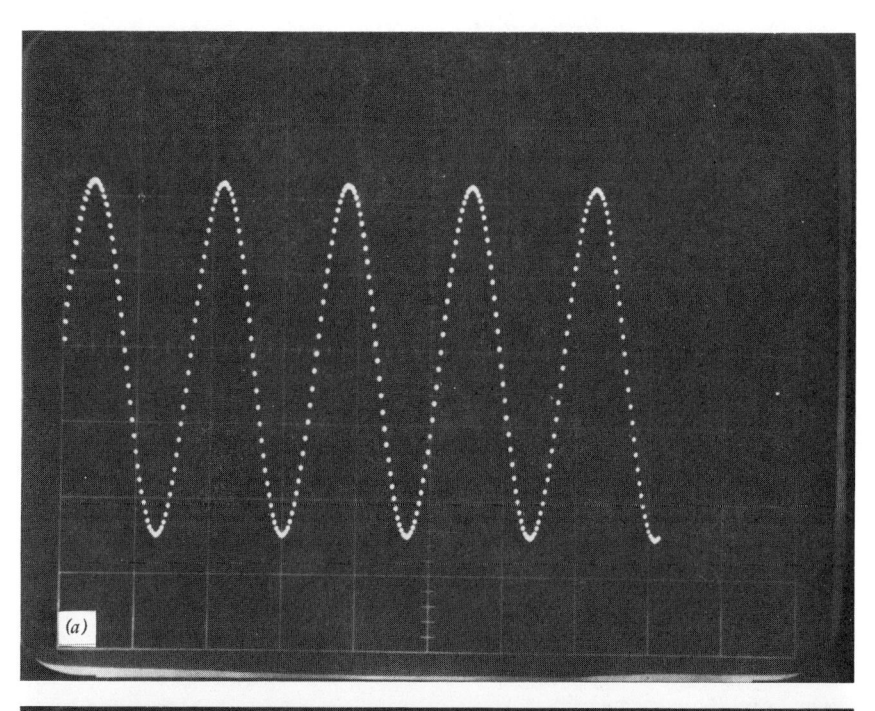

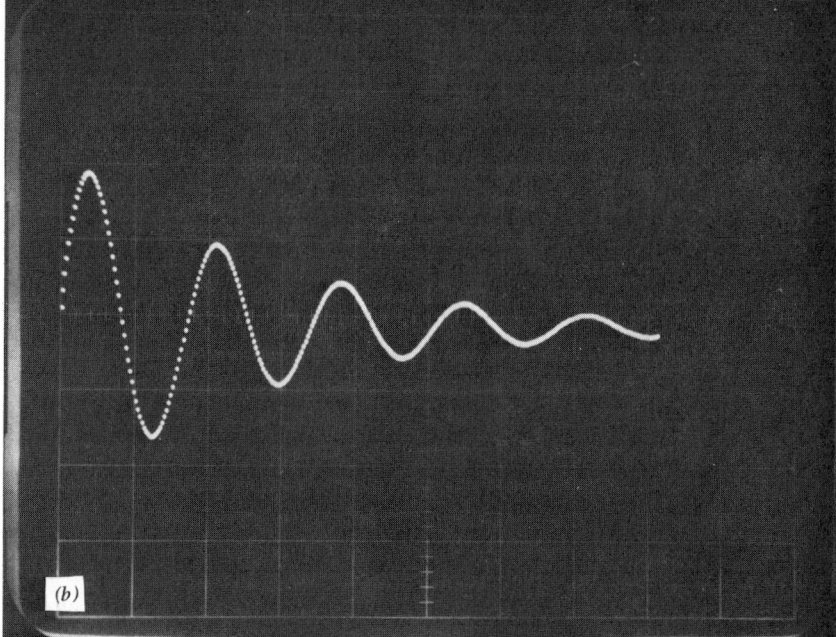

Fig. 7-3 Sine wave simulated on a digital computer. (*a*) Constant amplitude, no leakage, (*b*) Dumped amplitude due to the leakage.

TABLE 7.1

```
LIST
10 REM HARMONIC OSCILLATOR AS SIN WAVE GENERATOR
20 REM TIME UNIT T=1
30 REM INITIAL CONDITIONS
60 X2=.05
70 X1=0
80 X=0
90 REM CIRCULAR FREQUENCY W
95 PRINT "ENTER W,L1,L2"
100 INPUT W,L1,L2
110 W2=W↑2
120 REM INTERVAL COUNTER N
130 N=0
150 REM TIME INCREMENTING LOOP
160 REM FOR 300 TIME INTERVALS
200 FOR N=1 TO 300
210 X1=X2+L1*X1
220 X=X1+L2*X
230 X2=-W2*X
300 REM SCALING DATA FOR DISPLAY INTO 0 TO 1 RANGE
310 T=3.300000E-03*N
320 S=.5+.5*X
330 PLOT T,S
340 DELAY
350 NEXT N
400 STOP

RUN
ENTER W,L1,L2
?.1,1,1
READY.

RUN
ENTER W,L1,L2
?.1,.99,.99
READY.
```

7.2 Random Data and Probability Distributions

Data representing a random physical phenomenon cannot be described by an explicit mathematical relationship, because each observation of the phenomenon will be unique. In other words, any given observation will represent only one of many possible results which might have occurred.

Figure 7.4 represents a sequence of pulses with random amplitudes produced by an experiment. Measuring the amplitude of one particular pulse

Random Data and Probability Distributions

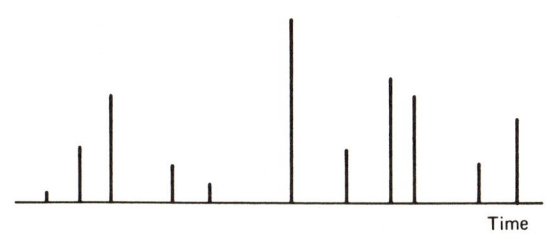

Fig. 7-4 Sequence of pulses with random amplitude.

will not produce significant information because amplitudes vary in an unpredictable way. What one can measure is the distribution of amplitudes. The distribution is obtained by counting how many times each possible amplitude occurs in a sample. The fraction of all observations that took a particular value is called its probability. If a stochastic variable can take on different values x_i, $(i = 1.2.3\cdots n)$ and the probability of the value x_i being taken is p_i, the set of numbers $p_i(i = 1,2\cdots n)$ is said to be a discrete probability function. Since the variable must always take one of the values, p_i, it follows that

$$\sum_{i=1}^{n} p_i = 1$$

Table 7.2, for example, represents data gathered on the amplitudes of pulses produced in experiment for a sample of 1000 pulses. The third column in Table 7.2 is the estimate of the probability of a pulse having amplitude x_i derived by dividing the number of pulses having amplitude x_i by the total number of observations.

TABLE 7.2

Amplitude	Number of Pulses	Probability P	Cumulative Probability F
1	2	0.002	0.002
2	18	0.018	0.020
3	30	0.030	0.050
4	103	0.103	0.153
5	157	0.157	0.310
6	210	0.210	0.520
7	180	0.180	0.700
8	151	0.151	0.851
9	119	0.119	0.970
10	30	0.030	1.000

In practice, there is more interest in deriving the cumulative distribution, F_r, which defines the probability that the observed value is less than or equal to x_r. The cumulative distribution function at the point x_r ($r = 1, 2, \ldots, n$) is given as

$$F_r = \sum_{i=1}^{r} p_i \tag{7.7}$$

The fourth column in Table 7.2 represents the cumulative distribution for the case shown. The probability distribution and the cumulative distribution are presented graphically in Figure 7.5.

In the experiment shown amplitudes have been grouped into only ten classes. If the number of classes is very large or infinite, then the amplitude x

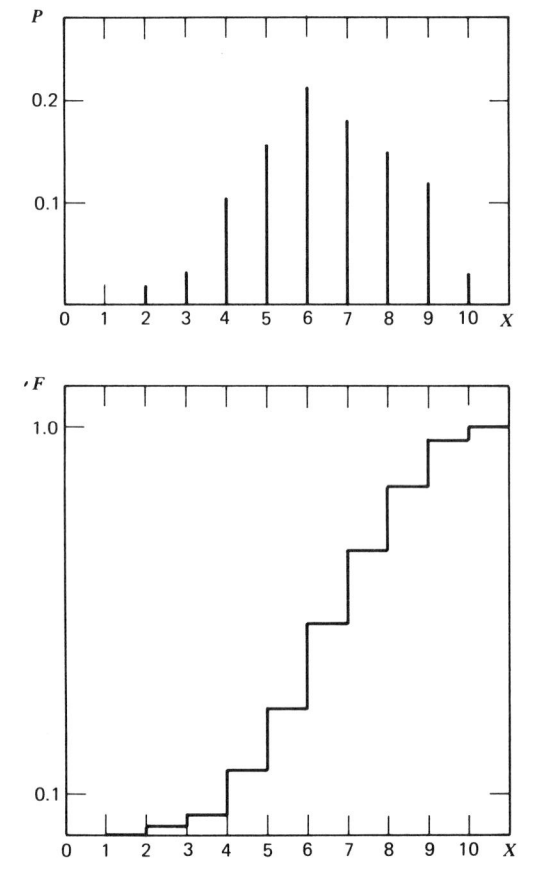

Fig. 7-5 The probability distribution P and the cumulative distribution F.

becomes a continuous variable that can be described by its probability density function $f(x)$. The probability that x falls in the range x_1, x_2 is given by

$$\int_{x1}^{x2} f(x)dx \qquad f(x) \geq 0 \tag{7.8}$$

The integral of the probability density function taken over all possible values, according to the intrinsic definition of a probability density function, must be

$$\int_{-\infty}^{\infty} f(x)dx = 1 \tag{7.9}$$

A related function is the cumulative distribution function that defines the probability that an observed value is less than or equal to x. If $F(x)$ is the cumulative probability distribution function, then

$$F(x) = \int_{-\infty}^{x} f(x)dx \tag{7.10}$$

From its definition, $F(x)$ is a positive number ranging from 0 to 1, and the probability of x falling in the range x_1, x_2, is $F(x_2) - F(x_1)$. Figure 7.6 illustrates a probability density function and the corresponding cumulative or integral distribution function.

The formation of the probability distribution function from experimental data has been shown. When the function is formed, it is of interest to see if it can be expressed by an analytical expression. There are large numbers of analytical expressions for probability density functions. Three basic analytical expressions for probability functions, which are of particular interest, are now given.

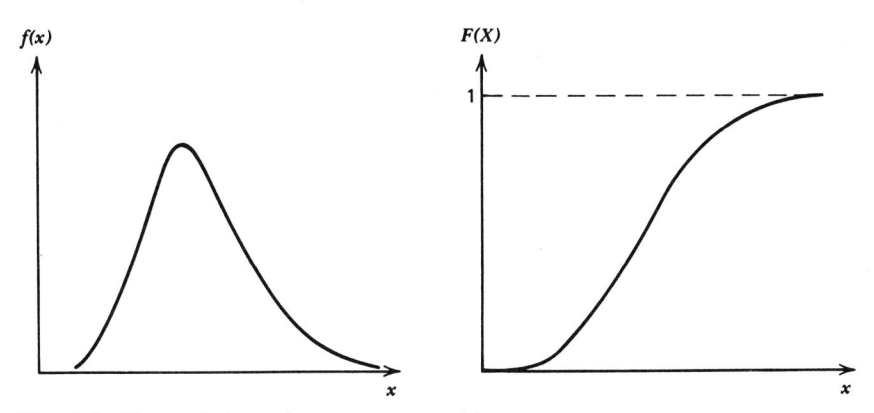

Fig. 7-6 The probability density function $f(x)$ and the cumulative or integral distribution function $(F(x))$.

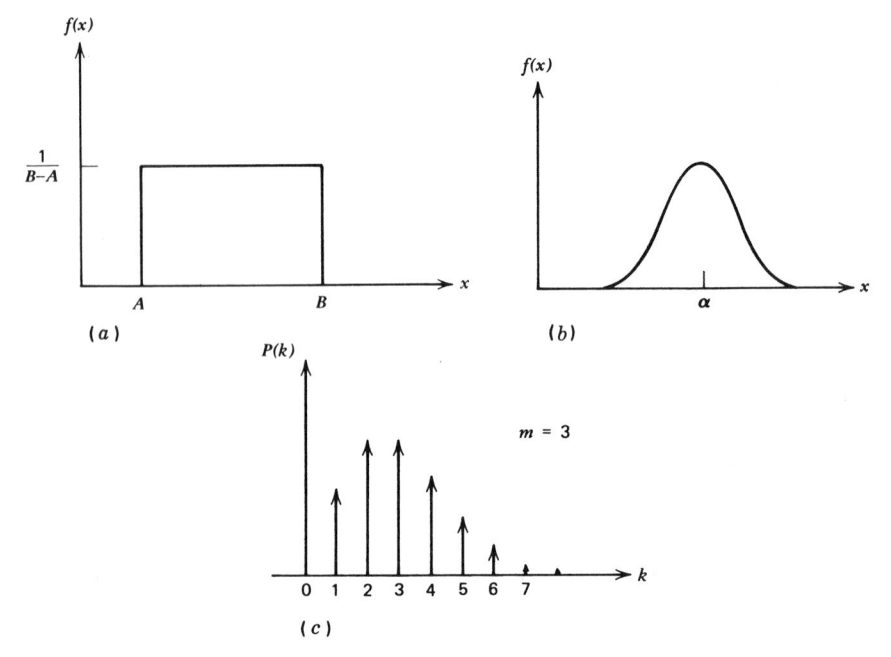

Fig. 7-7 Basic probability density functions. (*a*) Uniform. (*b*) Normal or gaussian. (*c*) Poisson distribution.

The simplest probability density function is a uniform distribution, as shown in Figure 7.7*a*. The uniform distribution describes the case when there is equal probability for the observed quantity to have any value between A and B. The uniform distribution can be expressed by

$$f(x) = \frac{1}{B - A} \text{ for } A < x < B \tag{7.11}$$

$$F(x) = 0 \quad \text{otherwise}$$

A special case is the uniform distribution with parameters $A = 0$, $B = 1$.

$$f(x) = 1 \text{ for } 0 < x < 1 \tag{7.12}$$

$$f(x) = 0 \quad \text{otherwise}$$

Such a distribution, as will be shown, presents the basis for the Monte Carlo simulation.

The most frequently used continuous probability distribution function is the normal or gaussian distribution as shown in Figure 7.7*b*. The gaussian distribution has a bell-like shape and is defined by two parameters, α (mean value) and σ (standard deviation), through the expression

$$f(x) = \frac{1}{\sigma\sqrt{2\pi}} \exp[-(x-\alpha)^2/2\sigma^2] \tag{7.13}$$

The most frequently used discrete probability distribution is a Poisson distribution, as shown in Figure 7.7c. The Poisson distribution is defined by one parameter, m (mean value), through the expression

$$P(k) = \frac{e^{-m}}{k!} m^k \tag{7.14}$$

A probability distribution describes the behavior of a stochastic variable. It shows the result of a number of observations, considering only the values assumed by the variable and not the sequence in which they occurred. In carrying out a simulation that involves stochastic variables, the reverse problem arises. It is necessary to generate a sequence of numbers in which the successive values are random but have the distribution that describes stochastic variables. In other words, the numbers should be generated in a random sequence, as in the real experiment shown in Figure 7.4.

7.3 Monte Carlo Techniques*

The Monte Carlo method is a technique for direct simulation of random phenomena and random data. Direct computer simulation of random phenomena permits estimation of a much wider variety of statistics than the analytical methods, and is not restricted to linear systems.

Modern experiments often investigate complex systems which produce a large quantity of data, often of a random character. A variable, representing the outcome of random activity, is said to be a stochastic variable. Although the exact sequence of values taken by a stochastic variable is not known, the range of values over which it can vary and the probability with which it will take the values, may be known or assumed to be known. Stochastic variables are, therefore, discussed in terms of functions which describe the probability of the variable taking various values.

As an example, one can consider the experiment producing a sequence of pulses, amplitudes of which fluctuate in a random fashion. Such random pulses are known to the physicists and chemists working with radiation detectors as well as to the biologists measuring neuroelectrical signals. Such

---

*Section 7.3 adapted from B. Souček, *Minicomputers in Data Processing and Simulation*, Wiley, New York, 1972, pp. 304–308.

random signals pass through the system and produce random outputs. Direct simulation of a system with random inputs is frequently the only possible method for conducting realistic laboratory-process studies and for estimate of the output.

The Monte Carlo Method and generation of random sequences can be introduced in the following way.

Various devices and techniques exist for producing such sequences. The simplest random process is the one described by a uniform discrete probability distribution where the choice is between n different numbers. An honest roulette wheel that has n sections will generate such a sequence. Because of this analogy, the name Monte Carlo, with its connection with roulette, has become a general term used to describe any computational method using random numbers.

A Monte Carlo simulation requires sequences of random numbers which are drawn from a distribution that, in general, is not uniform. Methods for directly generating random numbers with a particular distribution are not usually available. Methods do, however, exist for generating random numbers with a uniform distribution. Fortunately there is a simple way for transforming uniform distribution into the required distribution. As a result, most of the nonuniformly distributed sequences used in Monte Carlo simulation, are generated through the transformation of uniformly distributed sequences.

Transformation of Random Variable The system with input x and output y is described by its transfer function $y = g(x)$. For a given x, knowing a transfer function, one can determine the value of y. If the input x is a random variable, the output y will also be a random variable. If input x has a distribution function $f_1(x)$, then output y presents a new random variable whose distribution function $f_2(y)$, is a function of $f_1(x)$ and $g(x)$. The easiest way to determine the output distribution function, $f_2(y)$, is as follows: if there is a one-to-one correspondence between x and y, then the probability that the input variable is in the range $x, x + dx$, must be equal to the probability that the output variable is in the range $y, y + dy$ (Figure 7.8).

$$f_1(x)dx = f_2(y)dy$$

(7.15)

$$f_2(y) = f_1(x) \cdot \frac{1}{dy/dx} = f_1(x) \cdot \frac{1}{|g'(x)|}$$

The output distribution, $f_2(y)$, is a function of $f_1(x)$ and of the derivative of the transfer function, $g(x)$. The absolute value of the derivative is taken because the distribution function cannot have negative values. From the

Monte Carlo Techniques

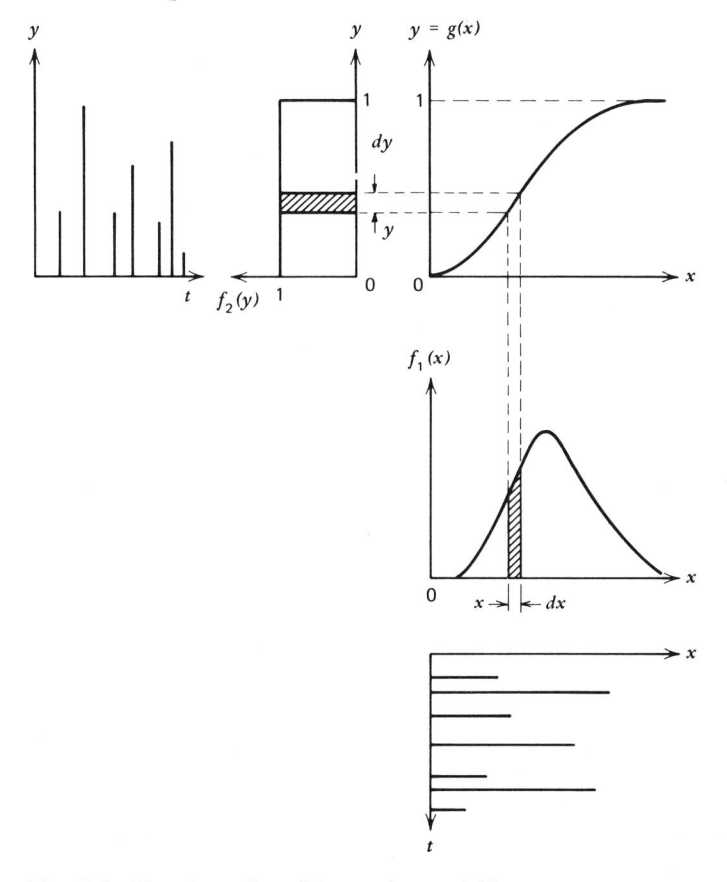

Fig. 7-8 Transformation of the random variable.

above expression, one can find the transfer function $g(x)$, which will transform a given distribution $f_1(x)$, into a new desired distribution $f_2(y)$.

$$g'(x) = \frac{f_1(x)}{f_2(y)}$$

$$g(x) = \int \frac{f_1(x)}{f_2(y)} dx$$

(7.16)

Of special interest is the generation of a given distribution through the transformation of a uniform distribution and vice versa. If

$$f_2(y) = K = \text{const}$$

then

$$g(x) = \frac{1}{K} \int f_1(x)dx = \frac{1}{K} F_1(x) + C \qquad (7.17)$$

$F_1(x)$ is the cumulative or integral distribution function of the variable x. The constant C can be obtained from boundary conditions for $g(x)$. If the uniform distribution is defined in the interval between 0 and 1, then $K = 1$, and from the boundary condition

$$g(x)_{x=\infty} = 1 = 1.1 + C$$
$$C = 0$$
$$g(x) = F_1(x) \qquad (7.18)$$

The transformation can go both ways. Usually a uniformly distributed sequence of random numbers is available. Applying such a sequence y, on a transfer function $g(x) = F_1(x)$, a new sequence x is generated, with the probability density function $f_1(x)$. It is important to remember that the transfer function $g(x)$, is the same as the cumulative or integral distribution $F_1(x)$.

7.4 Simulation of Experimental and Theoretical Data

Uniform Distribution Generate a series of pulses at time intervals $T = 1, 2, 3, \ldots$ The pulses should have random amplitudes, uniformly distributed in the range 0 to 1. The program is showned in Table 7.3.

Statements 60 to 90 serve to allocate the space for the array D. This array will be used to sort the generated pulses and to prove that the distribution is uniform.

Statement 100 initiates the loop to generate 5000 random pulses. Actual generation of one random number is done in statement 120.

Statements 140 to 160 display on the scope first 20 random pulses. For further pulses the time scale is out of the scope range and display is stopped. The display is shown in Figure 7.9a.

Statement 200 scales random amplitudes into integers in the range 0 to 25. Statement 210 uses those integers to address the array D. When a random pulse addresses the array, one is added to the content of a selected array channel. In this way the frequency of occurrence or amplitude distribution is formed. This part of the program (statements 100 to 230) is repeated 5000 times, and 5000 random amplitudes are generated, and after that sorted.

The rest of the program is used to display the distribution D(I). The display is shown in Figure 7.9b. Note that the array D is dimensioned for 50 items, and that statement 200 sorts the amplitudes in only 25 amplitude

Simulation of Experimental and Theoretical Data

TABLE 7.3

```
LIST
10 REM GENERATION OF RANDOM PULSES
20 REM AT TIMES T=1,2...5000
30 REM WITH RANDOM AMPLITUDES UNIFORMLY DISTRIBUTED
40 REM BETWEEN 0 AND 1
50 REM SPACE ALLOCATION FOR DISTRIBUTION
60 DIM D(50)
70 FOR I=1 TO 50
80 D(I)=0
90 NEXT I
95 REM PULSE GENERATING LOOP
100 FOR T=1 TO 5000
120 R=RND(1)
130 REM SCALING T AND R FOR DISPLAY
140 X=T/20
145 Y=.5*R
150 PLOT X,Y
160 DELAY
170 REM SORTING PULSES INTO 25 AMPLITUDE CHANNELS
180 REM TO FORM AMPLITUDE DISTRIBUTION
200 I=INT(25*R)
210 D(I)=D(I)+1
230 NEXT T
240 INPUT A
250 CLEAR
300 REM DISPLAY OF THE DISTRIBUTION FUNCTION
310 FOR I=1 to 50
320 X=1/50
330 Y=2.000000E-03*D(I)
340 PLOT X,Y
350 DELAY
360 NEXT I
400 STOP
```

channels. The display shows that random pulses are uniformly distributed over 25 amplitude channels.

Simulation of the Experiment Generate a series of pulses at time intervals $T = 1, 2, 3, \ldots$. The pulses should have random amplitudes, with the amplitude distribution as shown in Table 7.2. The program is shown in Table 7.4.

The statements 10 to 400 present virtually the same program as in Table 7.3.

Statements 410 to 455 are used to form the transformation curve, which is identical to integral or cumulative distribution. The cumulative distribution

TABLE 7.4

```
LIST
10 REM GENERATION OF RANDOM PULSES
20 REM AT TIMES T=1,2...5000
30 REM WITH RANDOM AMPLITUDES FOLLOWING
40 REM EXPERIMENTAL DISTRIBUTION (DATA STATEMENT)
50 REM SPACE ALLOCATION FOR DISTRIBUTION
60 DIM D(50)
70 FOR I=1 TO 50
80 D(I)=0
90 NEXT I
91 GO TO 410
95 REM PULSE GENERATING LOOP
100 FOR T=1 TO 5000
120 R=RND(1)
121 GO TO 460
130 REM SCALING T AND R FOR DISPLAY
140 X=T/20
145 Y=.1*R
150 PLOT X,Y
160 DELAY
170 REM SORTING PULSES INTO AMPLITUDE CHANNELS
180 REM TO FORM AMPLITUDE DISTRIBUTION
200 I=R
210 D(I)=D(I)+1
230 NEXT T
240 INPUT A
250 CLEAR
300 REM DISPLAY OF THE DISTRIBUTION FUNCTION
310 FOR I=1 TO 50
320 X=I/50
330 Y=2.000000E-03*D(I)
340 PLOT X,Y
350 DELAY
360 NEXT I
400 STOP
402 REM
406 REM
408 REM
410 REM FORMING THE INTEGRAL OR CUMULATIVE DISTRIBUTION
420 DIM C(10)
425 RESTORE
430 FOR I=1 TO 10
440 READ C(I)
450 NEXT I
455 GO TO 95
456 REM
457 REM
```

(Continued)

Simulation of Experimental and Theoretical Data

TABLE 7.4 (*continued*)

```
458 REM
460 REM TRANSFORMATION OF RANDOM VARIABLE
470 FOR I=1 TO 10
480 IF P<=C(I) GO TO 510
490 NEXT I
500 REM TRANSFORMED RANDOM NUMBER IS
510 R=I
520 GO TO 130
550 REM
560 REM
570 REM
600 REM EXPERIMENTAL DISTRIBUTION
610 DATA 2.000000E-03,.02,.05,.153,.31,.52,.7,.851,.97,1
```

from Table 7.2 is described with 10 points in statement 610. Statement 440 reads these points and places them into array C(I) (cumulative distribution). This part of the program presents initialization, and it is called from the main program at statement 91. The main program then continues at statement 95.

The major point of the program is transformation of a random variable. Statement 120 generates a random number between 0 and 1. Statement 121 switches the program to the line 460, where the actual transformation is performed in the following way: The random number R is compared with the cumulative distribution C(I). The values C(I) grow from 0.002 to 1, the right-hand column of Table 7.2. Comparison proceeds through the loop, as long as $R \leq C(I)$. When the value of R intersects the C(I) value, the index I is read out. The transformed value of R is equal to the index, $R = I$. In this way transformed random number is formed. The statement 530 returns the control back to the main program, at line 130.

Figure 7.10*a* shows first 20 transformed values displayed on the scope. Figure 7.10*b* shows sorted pulses. Note that the distribution shown in Figure 7.10*b* is the same as the distribution shown in the *P* column in Table 7.2. This step presents the checking procedure, to make sure that all steps have been performed correctly.

Random Time Intervals Generate a series of random time intervals. The probability distribution of the time intervals should follow the Poisson distribution.

The basic properties of Poisson arrival pattern are the following: The arrivals are random and independent. There are no simultaneous arrivals. The mean arrival rate L is constant.

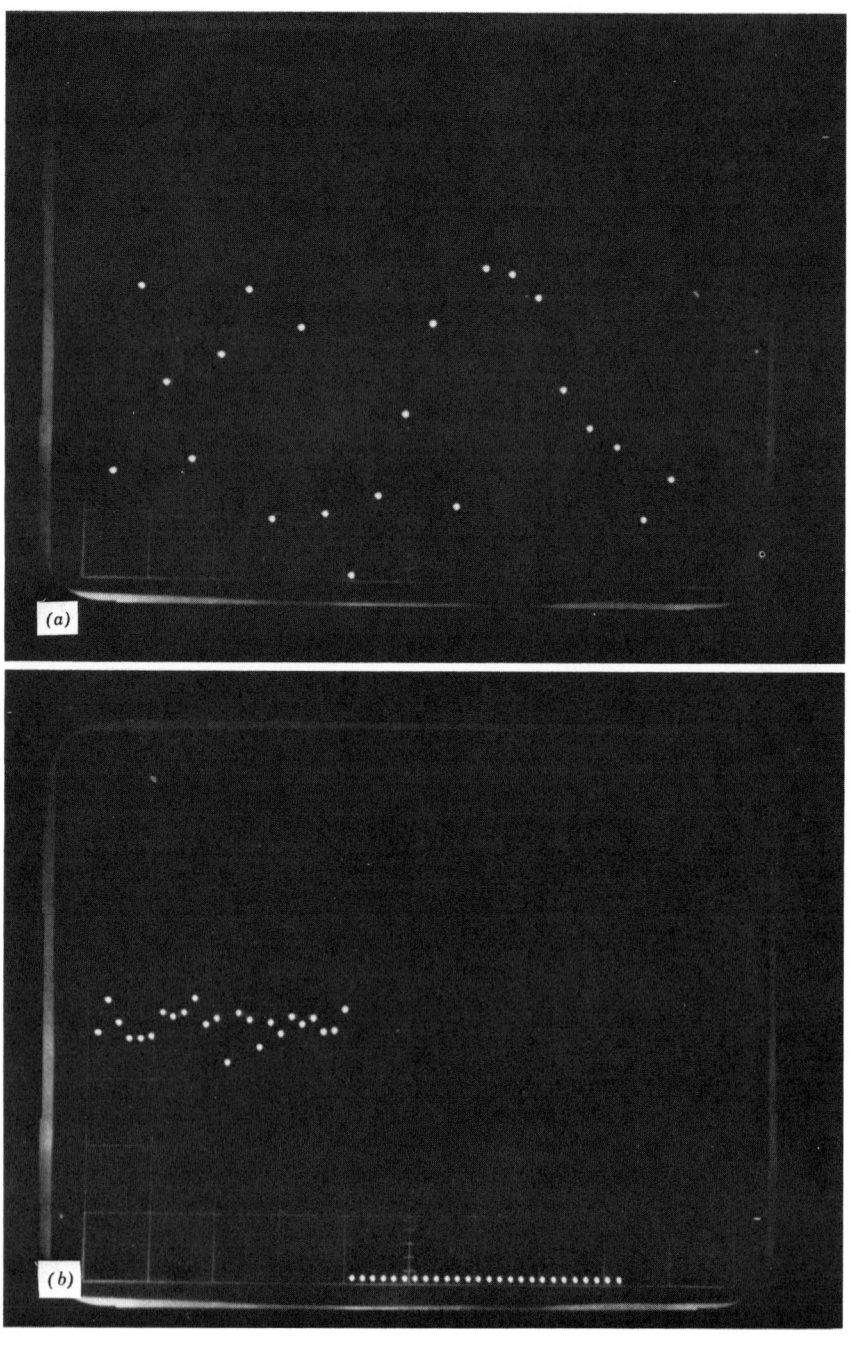

Fig. 7-9 Simulation of random pulses with amplitudes uniformly distributed between 0 and 25. (*a*) First 20 pulses. (*b*) Probability distribution measured into 50 amplitude channels.

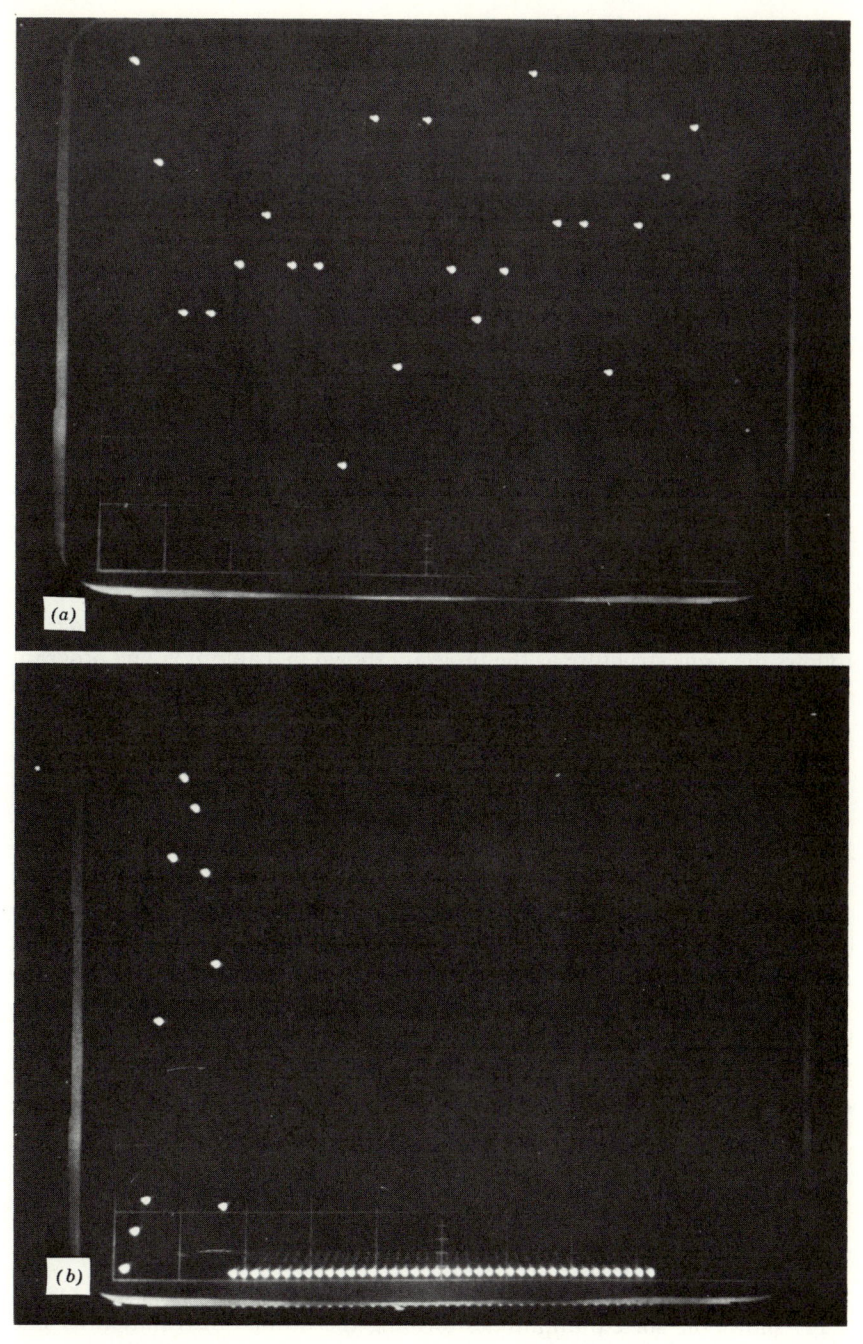

Fig. 7-10 Simulation of the experiment. (a) First 20 pulses with random amplitudes. (b) Probability distribution.

The probability distribution of intervals is given by

$$f(t) = L \cdot \exp(-Lt) \tag{7.19}$$

By integrating equation 7.19 we obtain integral distribution

$$F(t) = Y = 1 - \exp(-Lt) \tag{7.20}$$

Equation 7.20 should be used to transform uniform distribution into Poisson distribution. This time the transformation equation is a simple one and yields an analytical answer. Solving equation 7.20 for t gives

$$t = -\frac{1}{L} \log(1 - Y) \tag{7.21}$$

Equation 7.21 reads as follows: Generate a uniformly distributed number Y. Substitute the number into equation 7.21. The result is a random interval t, with a Poisson distribution.

Table 7.5 shows a simple program, based on equation 7.21. This program follows the same structure as the program shown in Table 7.3. The program reads in the mean value L of the Poisson distribution and generates 1000 random intervals. Transformation operation based on equation 7.21 is programmed in statement 125.

Figure 7.11a shows the display of first 20 intervals. Figure 7.11b shows the interval distribution function. As expected, the distribution follows Poisson law equation 7.19.

Normal Distribution Normal or Gaussian distribution is very frequently needed in simulation. Fortunately, it can be easily generated, on the basis of the central limit theorem: Form the sum X, of random variables $X1, X2, \ldots XN$. It could be proved that the sum X will have Gaussian distribution, regardless of the distributions of the variables $X1, X2 \cdots XN$.

We shall show the generation of Gaussian distribution through a triple summation of the uniform distribution. The program is the same as in Table 7.3, except for the statement 120:

$$120 \text{ R} = \frac{1}{3}(\text{RND}(1) + \text{RND}(1) + \text{RND}(1))$$

Figure 7.12a shows first 20 amplitudes generated by the program. Figure 7.12b shows the sorted amplitudes forming the normal distribution. If one uses more then three random numbers to form the sum, a still better Gaussian distribution will result.

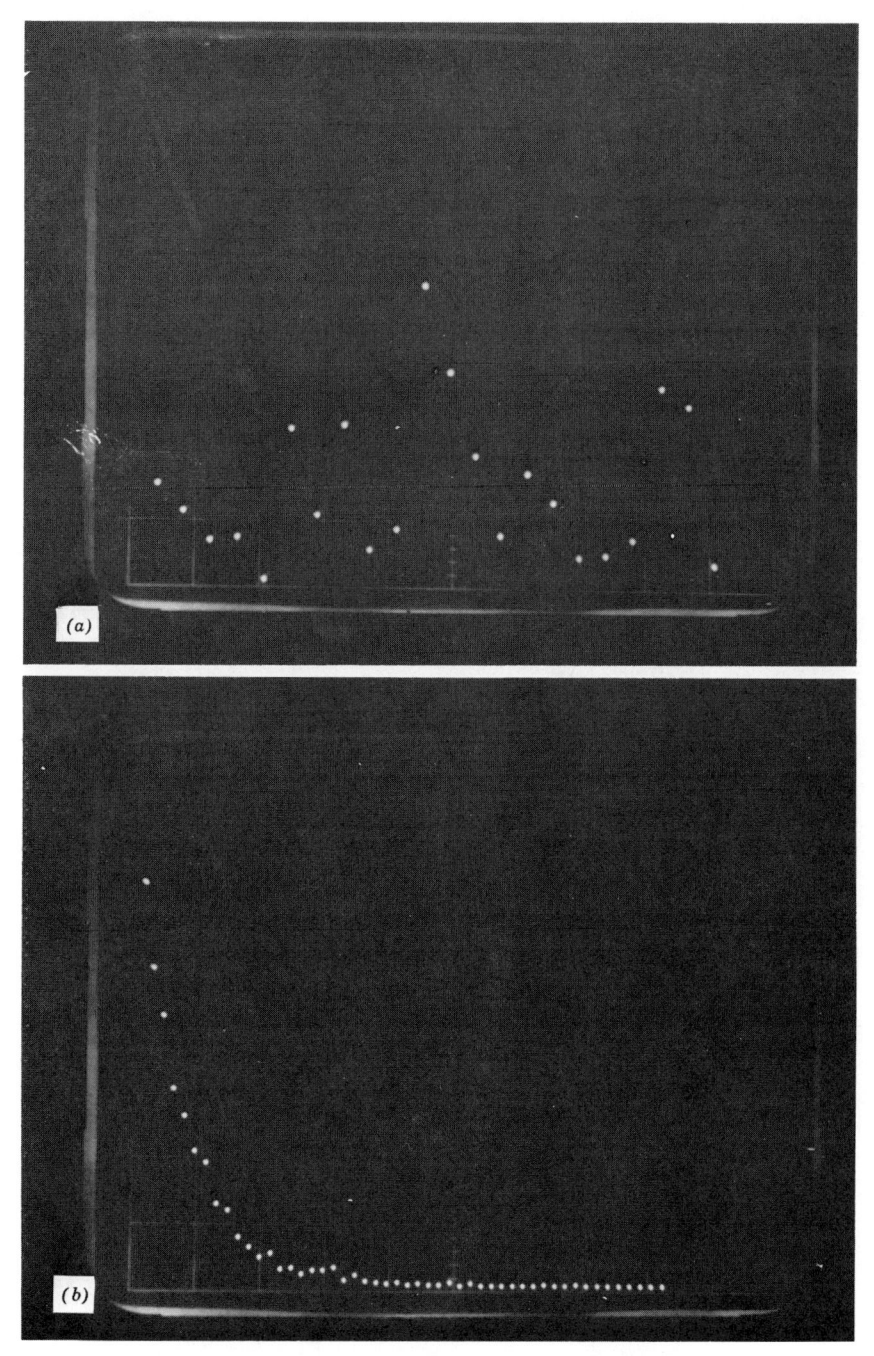

Fig. 7-11 Simulation of random time intervals, following Poisson arrival pattern. (*a*) First 20 random intervals. (*b*) probability distribution of the intervals.

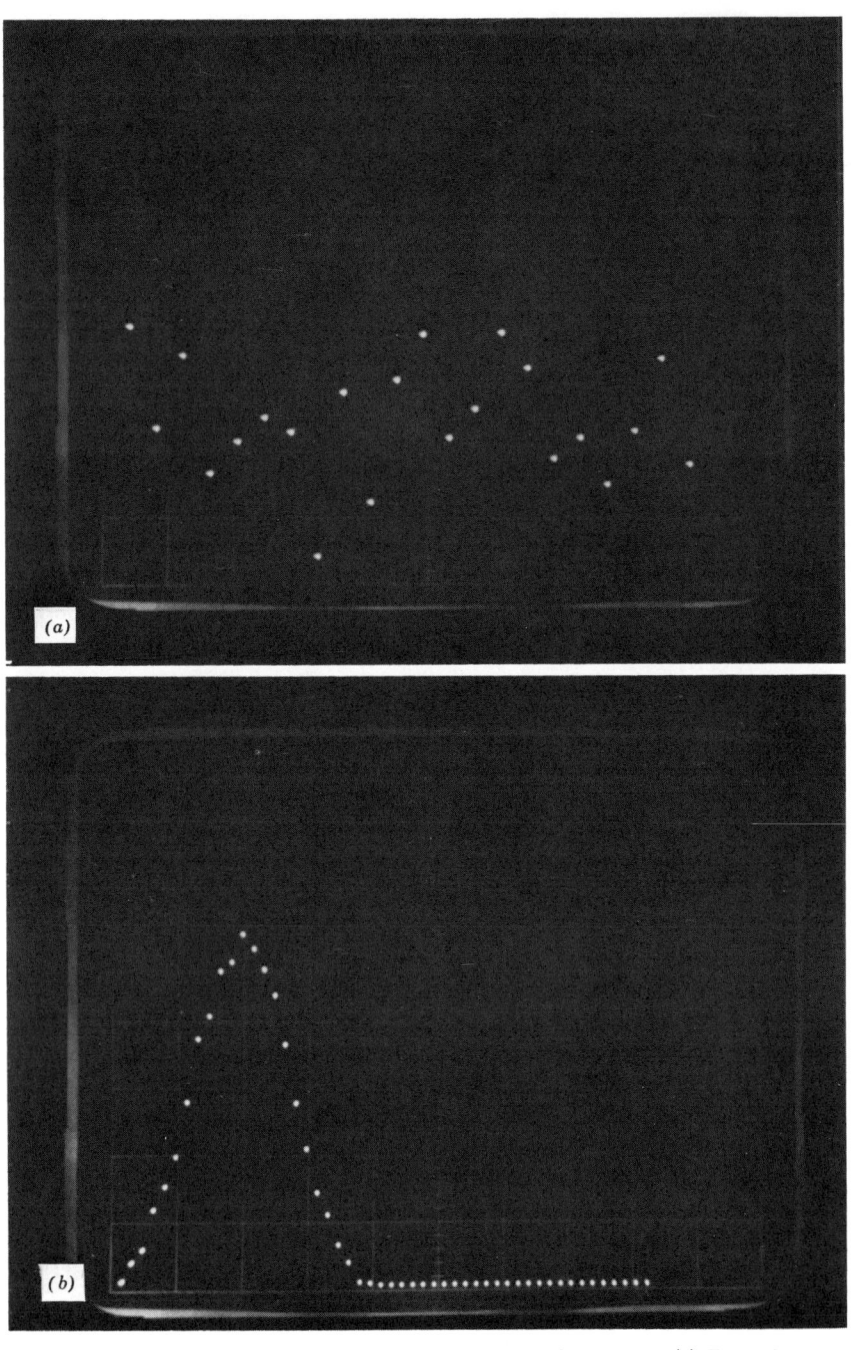

Fig. 7-12 Simulation of random pulses with gaussian distribution (*a*) First 20 pulses. (*b*) Probability distribution.

TABLE 7.5

```
LIST
10 REM GENERATION OF RANDOM INTERVALS,
20 REM WITH POISSON DISTRIBUTION
30 REM MEAN VALUE L
40 INPUT L
50 REM SPACE ALLOCATION FOR DISTRIBUTION
60 DIM D(50)
70 FOR I=1 TO 50
80 D(I)=0
90 NEXT I
95 REM INTERVAL GENERATING LOOP
100 FOR N=1 TO 1000
120 R=RND(1)
125 T=(LOG(1-R))/L
130 REM SCALING T AND N FOR DISPLAY
140 X=N/20
145 Y=.2*T
150 PLOT X,Y
160 DELAY
170 REM SORTING INTERVALS INTO CHANNELS
180 REM FOR INTERVAL DISTRIBUTION
200 I=INT(5*T)
210 D(I)=D(I)+1
230 NEXT N
240 INPUT A
250 CLEAR
300 REM DISPLAY OF INTERVAL DISTRIBUTION FUNCTION
310 FOR I=1 TO 50
320 X=I/50
330 Y=2.000000E-03*D(I)
340 PLOT X,Y
350 DELAY
360 NEXT I
400 STOP
```

References

1. Souček, B.: *Minicomputers in Data Processing and Simulation*, Wiley, New York, 1972.
2. Gordon, G.: *System Simulation*, Prentice-Hall, Englewood Cliffs, N. J., 1967.
3. Deutsch S.: *Models of the Nervous Systems*, Wiley, New York, 1967.

Chapter 8

AMPLITUDE AND LATENCY HISTOGRAMS

Introduction and Survey

A substantial increase in the efficiency in the data collection can be achieved if data can be sorted on the basis of some parameter. In this case, hundreds or thousands of data, instead of being individually recorded, can be accumulated into one counting location. Each value of the parameter should have its own counting location. The histogram accumulated this way presents, in fact, the probability distribution function of the measured parameter. This kind of data collection is used in a large number of experiments, and it can be readily done with a laboratory computer. Data collecting and sorting systems are usually called analyzers.

This chapter describes the basic data analysis techniques. The probability distribution measurement principles are reviewed first. The most important analysis techniques are described next. Different kinds of data analyzers are in wide use in physics, chemistry, biomedicine, and engineering (pulse-height analyzers, time-of-flight analyzers, Mossbauer-effect analyzers, neuron-activity analysis, industrial product testing, communication-signal analysis, etc.).[1-71] Detailed description of different analyzing techniques is presented in Souček.[72]

8.1 Probability Density Function*

In analyzing nature one finds that many results indicate that the subjects of measurements belong to the class of random physical phenomena. In a

---
*Sections 8.1–8.3 adapted from B. Souček, *Minicomputers in Data Processing and Simulation*, Wiley, New York, 1972, pp. 387–397.

Probability Density Function

descriptive manner, randomness means that the data are nonperiodic; exhibit no explicit time trend, bias, or regularity, cannot be precisely defined for all the time by any simple analytic function. Many random processes belong to the class of stationary processes. Stationary means that certain statistical properties of the data do not change with time, and will be the same in the future as they are at present.

There are three main types of general statistical analysis which should be carried out for the data verified as random and stationary:

- Amplitude probability density functions.
- Correlation functions.
- Power spectral density functions.

Those three classes of functions describe a random process in a similar way as the amplitude, waveform, and Fourier frequency spectrum describe a deterministic process.

In this chapter we shall deal with methods for measuring amplitude probability density functions.

Five important examples of processes that could occur in practice, singly or in various combinations, will now be considered.

1. Sine wave.
2. Narrow-band noise.
3. Wide-band noise.
4. Discrete random pulses.
5. Discrete random time intervals.

For each of these waveforms the following discussion will develop its appropriate probability density functions. Five waveforms are pictured in Figure 8.1.

For the sine wave, Figure 8.1a representation of the waveform is described by

$$x(t) = A \sin (2\pi f_0 t + \theta) \tag{8.1}$$

Equation 8.1 is a well-known sine wave function, where A = maximum amplitude, f_0 = frequency and θ = initial phase angle. With these three parameters the sine function is completely described.

For narrow band noise it is more difficult to give analytical representation for instantaneous amplitude $x(t)$. Narrow band noise, Figure 8.1b can be considered in some cases as a combination of two waves: sine wave and noise. In mathematical notation we have

$$x(t) = A(t) \cdot \sin [2\pi f_0 t + \theta(t)] \tag{8.2}$$

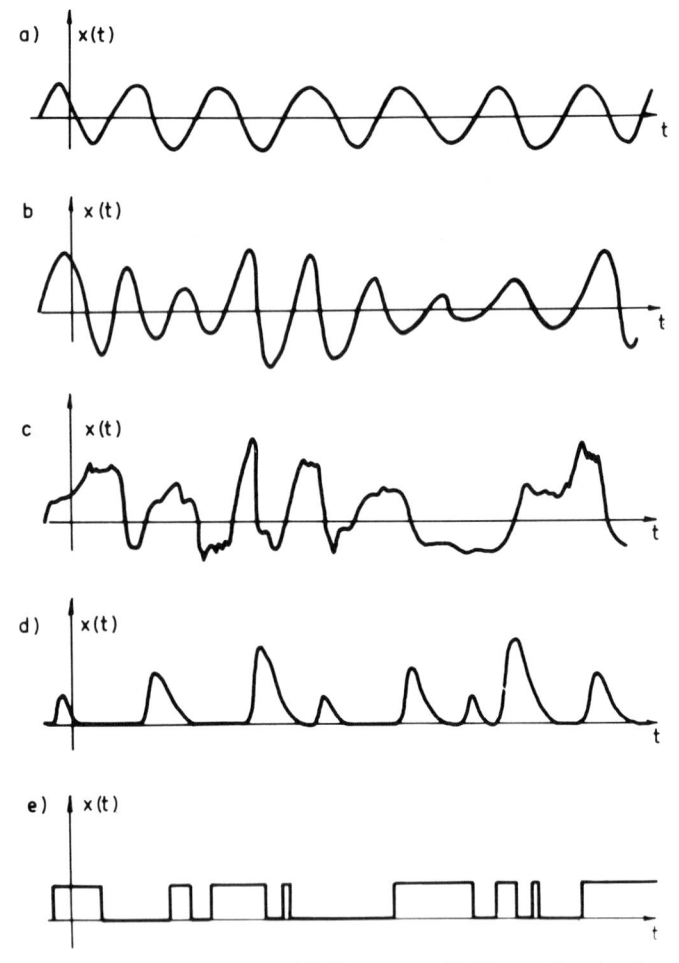

Fig. 8-1 Five special time histories. (a) Sine wave. (b) Narrow-band noise. (c) Wideband noise. (d) Discrete random pulses. (e) Discrete random time intervals.

where $A(t)$ and $\theta(t)$ indicate that the amplitude factor and phase factor vary (relatively slowly) in some random fashion with time. However, the center frequency f_0 is still assumed to be the same as it was for the original sine wave. The frequency spread associated with $x(t)$ is assumed to be small compared with f_0.

Another possible representation for the narrow band noise is the additive type

$$x(t) = s(t) + n(t) \tag{8.2a}$$

Probability Density Function

Here $s(t)$ is the fixed sine wave, as described by Equation 8.1. This wave is mixed with a random noise $n(t)$, of a relatively narrow frequency spectrum and with the central frequency close to f_0.

In practice we can obtain narrow band noise by passing a random noise through a narrow band filter. If the noise is mixed with the useful signal, this method can be used to improve the signal to noise ratio, provided that the two are not inside the same frequency band.

For wide-band noise no analytical representation is possible. All frequencies are theoretically possible in the record. The proportion of time spent in any frequency band is variable. The knowledge of instantaneous amplitude values associated with any narrow frequency band generally gives no information about the amplitudes to be associated with any adjacent narrow frequency band. The typical wide band noise is presented in Figure 8.1c. Discrete random pulses, Figure 8.1d, and discrete random time intervals, Figure 8.1e, are often produced in nuclear and biomedical experiments. Generally there is no way to describe such waveforms by simple analytical expression.

The probability density functions to be described will be applied to instantaneous amplitude values for the first three processes in Figure 8.1, for the peak values of pulses in the fourth process, and for the duration of pulses in the last process. Similar probability density functions can be developed for other parameters of interest.

Pictures of appropriate probability density functions for a sine wave, narrow-band noise, and wide-band noise, where the mean value is assumed to be zero, are shown in Figure 8.2a, b, c.

The probability density function in Figure 8.2a is for the sine wave. This well known distribution states the fact, that the sine wave instantaneous amplitudes are distributed between $x = \pm A$, and that the most probable amplitudes are $+A$ and $-A$ (the amplitudes on which the sine wave spends most of the time).

The probability density function shown in Figure 8.2b is for narrow-band noise when a sine wave component is present. This curve is double peaked, with a minimum at $x = 0$ and with tails rapidly approaching zero. The probability density function shown in Figure 8.2c for wide-band noise is of the well-known normal (gaussian) type, with tails that are asymptotic to the x-axis as x approaches $\pm \infty$.

Figure 8.2d shows the probability density function of the peak values for discrete random pulses obtained from radio-active source Co[60]. Figure 8.2e shows the capture gamma ray events versus time of flight. Such "spectra" gives an insight into the nuclear process under consideration.

From the knowledge of a probability density function one can compute the probability that the amplitude values will lie in any specified range, and

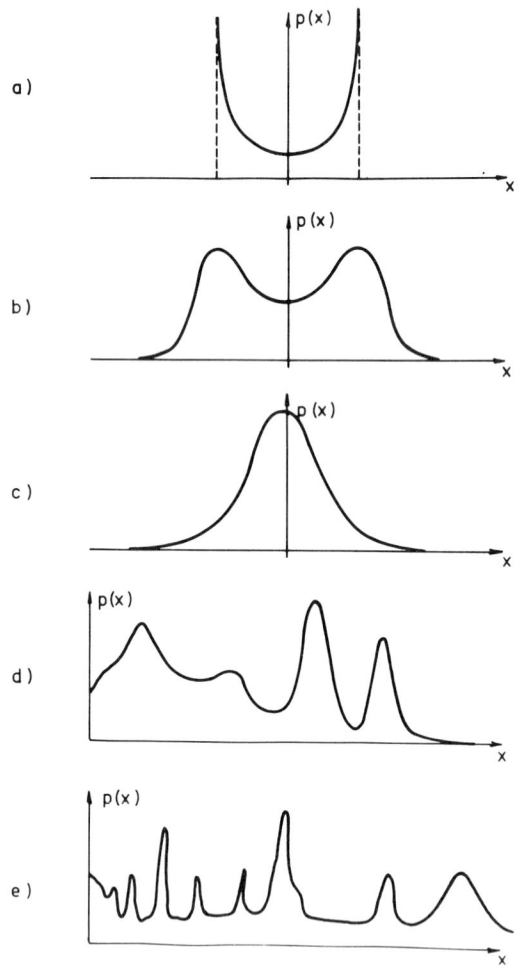

Fig. 8-2 Probability density functions. (*a*) Sine wave. (*b*) Narrow-band noise. (*c*) Wide-band noise. (*d*) Discrete random events. (*e*) Discrete random time intervals.

this may be a very significant parameter for particular applications. For example, one may be interested in estimating the probability of excessive high amplitudes as an indication of an abnormal condition in an electroencephalogram data analysis and in predicting structural failures under random vibration, discovering clipping, or nonlinearities of a system. Of special importance is the application in nuclear pulse spectrometry. Most of measuring systems are developed primarily for nuclear fields but can usually be used for other applications as well.

Direct Recording Analyzers 145

Many discussions of work in this area are available in the literature. In particular, one can consult excellent books by Chase[1], Korn[2], Bendat and Piersol[3]. A review paper by Bendat[4] covers the application in biomedical electronics. Similar papers by Souček and Spinrad[5,6] are concerned with the application in nuclear spectrometry.

8.2 Direct Recording Analyzers

A computer analyzer can have different modes of operations, the most important being single-parameter, sample-voltage, multiplex-mode, and multi-parameter analysis.

Single-Parameter Analyzer The input to the analyzer is a train of pulses, such as generated by a nuclear detector. The analog-to-digital convertor changes each pulse amplitude into a binary word that is sent to the computer. In the computer, the binary word represents a memory address. Each time a given core memory-address word is received from the convertor, the memory location is incremented by 1. The transformation from pulse amplitude to memory channel is demonstrated in Figure 8.3a. The pulse amplitude is projected horizontally onto the conversion gain slope, then vertically into the appropriate memory channel. The time of arrival of pulses is dictated by the physical source of pulses. Each pulse initiates analog-to-digital conversion and data accumulation.

Sample Voltage Analysis This kind of analysis should be used for continuous waveforms. The analog-to-digital convertor is controlled by the external sample clock which determines the sampling rate. When the sample is taken, the amplitude at the sample time is digitized and becomes a memory address. The memory word contents corresponding to the address are increased by 1 count and becomes accumulated data. The sampling analysis of a sine wave is shown in Figure 8.3b. The accumulated value in each memory location is proportional to the probability density at that amplitude.

Multiplex-Mode Analysis In this mode, few analog-to-digital convertors can be connected to the same computer and can operate under multiplex scanner control. In this way, few independent measurements can be carried out at the same time. Each convertor has its device address and responds only to that address for which it was programmed. The transfer of data between convertors and memory can be either in scanning programmed mode, or on the basis of interrupt request.

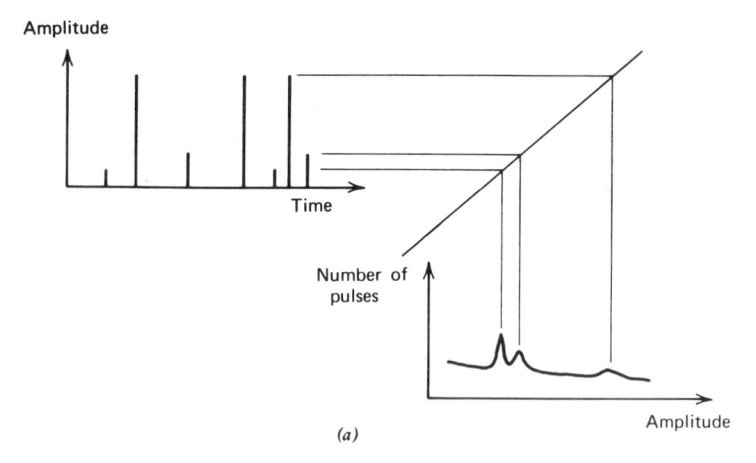

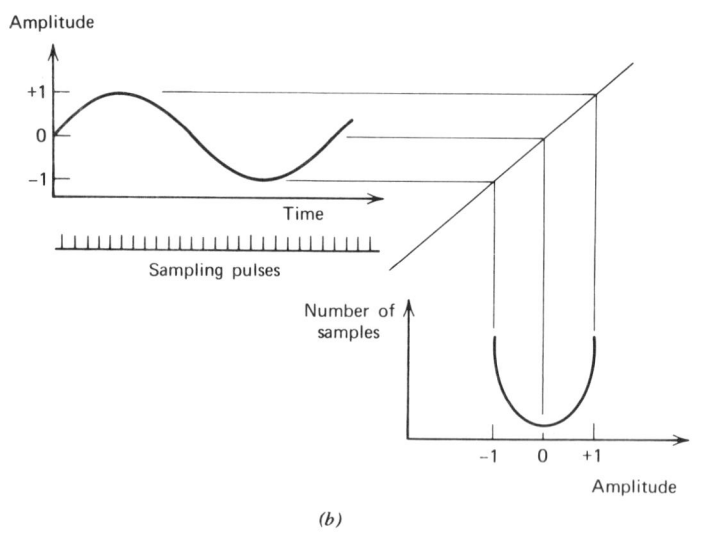

Fig. 8-3 (a) The transformation from pulse amplitude to the memory channel. (b) The sampling analysis of a sine wave.

Multiparameter Mode In this mode, two or more inputs are processed to form two or more dimensional probability density functions. For example, two parameter pulse-height analysis is a coincidence analysis of two pulses from separate detectors occurring in the same time interval. The output of one convertor becomes the X parameter, the output of the other becomes the Y parameter, and the number of times each XY combination occurs becomes the Z parameter of a three-dimensional field.

8.3 Basic Minicomputer-Analyzer System

The predominant technique for the quantization of pulse height (voltage or charge) derives from a method originated by Wilkinson[15] in 1950. It consists of first converting the height into a time interval and then counting clock pulses to measure that interval. The count is then a direct proportion of the original pulse height to the precision of clock intervals, within the accuracy of the pulse-to-time convertor. After conversion the obtained information is used to address the memory.

The first memories were acoustic delay line, Hutchinson and Scarrot[16]. At a later time electrostatic memories and magnetic drums appeared.[18,16] Ferrite core memories first came into use in 1955 and 1956. Recently, real-time computers have been directly connected to the analog-to-digital convertors and work as multichannel analyzers. One of the major advantages of the computer-oriented stored-program system is that both the sequencing of the functions and their detailed makeup can be readily altered by programming to meet the individual requirements of any given situation. The first computer-analyzer systems[20-23] were published in 1962-1963.

Figure 8.4 shows a typical pulse-height analyzer using the Wilkinson-type convertor directly connected to the computer equipped with a CRT display.

The pulse-to-time conversion is accomplished by the technique of charging a capacitor to a voltage proportional to the maximum excursion of the signal pulse, and then discharging it linearly in time using a constant current. The counting of clock pulses is started at the beginning of the run down and stopped when the capacitor voltage reaches zero. Those pulses are counted in an ADC register, which is used to address the computer memory.

The computer is used to display the data on CRT. It should be mentioned that x coordinates come here from the memory address A_x, but the y coordinate is the contents of the A_x address. The flow diagram of the display program consists of two loops: an inner loop for point display and an outer loop for picture display. After the abscissa has reached the maximum value x_{max} the x value is reset, that is, $x = 0$, and the picture display starts again.

During the display of every point, interrupt is made possible. The display program is interrupted with the appearance of new data, and the computer jumps to another, interrupt program. The computer saves the accumulator contents and commences the datum transfer, that is, the taking of a new item. The ADC transfers the register contents (it has a new item accepted at the end of ADC conversion) to the computer accumulator, and afterwards these contents are used as memory address. The contents of that location are incremented, which means that a new descriptor is stored together with some others. Afterward the computer restores the AC contents, clears the ADC register, and tests to see if a sufficient number of data has

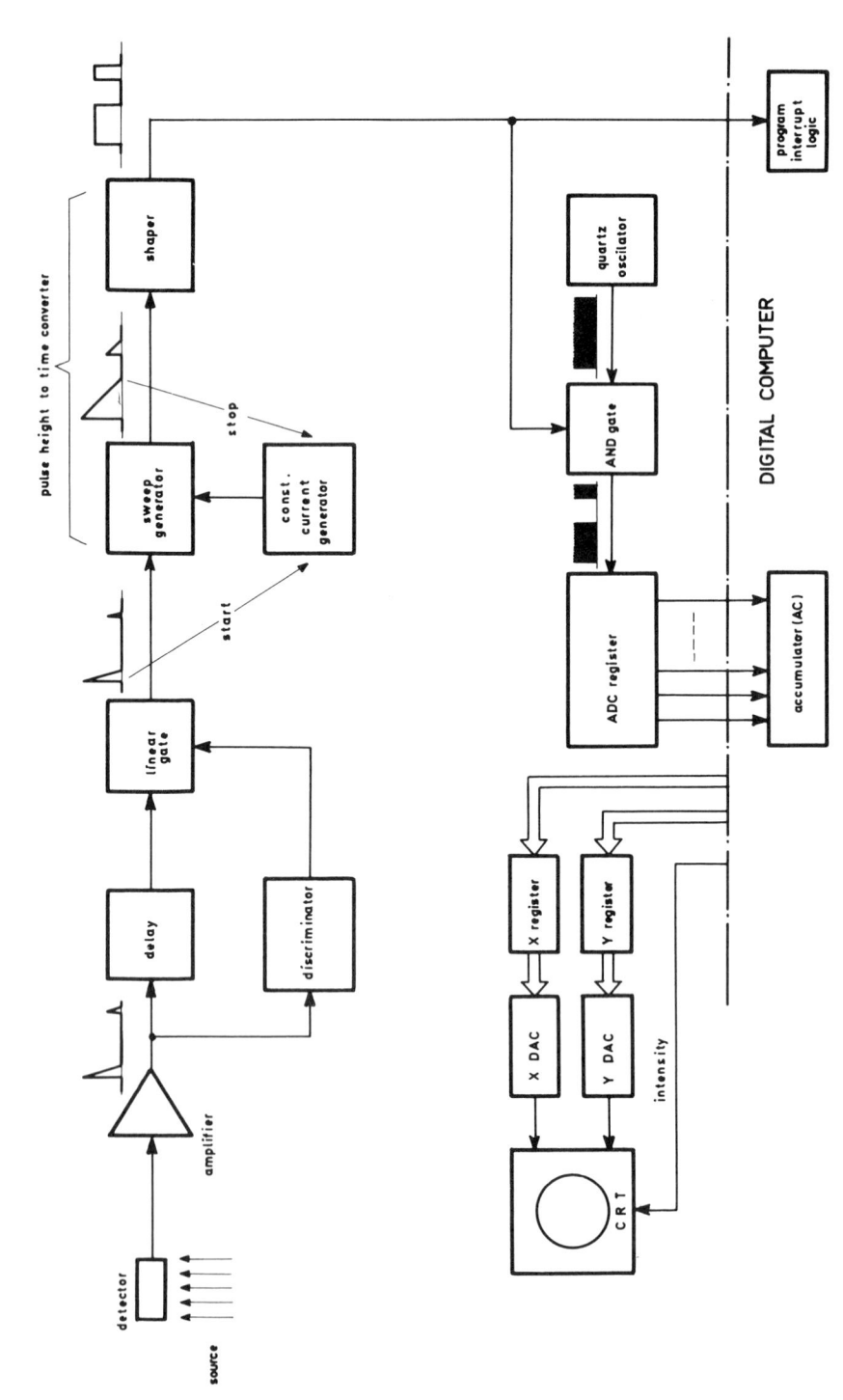

Fig. 8-4 Typical pulse-height analyzer system using amplitude-to-time-to digital convertor directly connected to the computer. The probability distribution is displayed on the CRT scope.

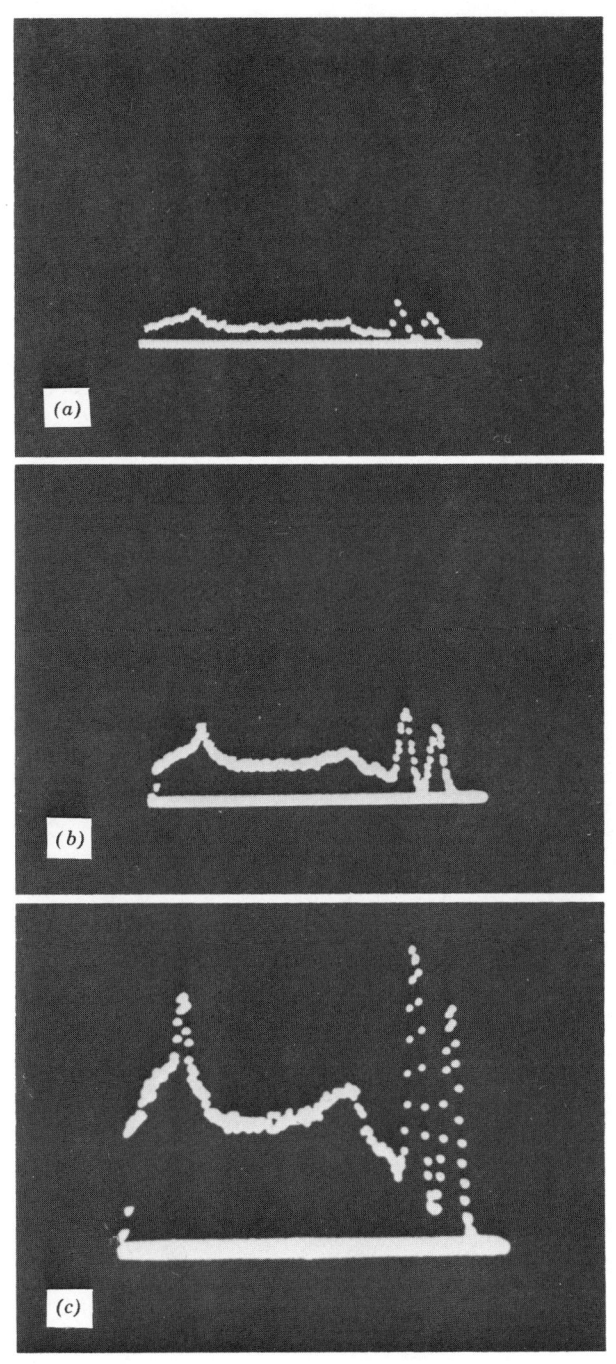

Fig. 8-5 Course of measurement of the probability distribution, as displayed on CRT screen after 1, 3, and 10 time intervals.

been taken. If so, the computer continues displaying the collected data. If not, a new interrupt is made possible, and the display continues waiting for new data.

Figures 8.5a, b, c shows the progressing in measurement of 1024-channel Co^{60} spectra, using a PDP-8 laboratory computer.[34] It is obvious that the spectrum is increased with time by an increasing number of input data.

TABLE 8.1

```
LIST
10 REM REAL-TIME HISTOGRAM MEASUREMENT
20 REM SAMPLE DATA FROM CHANNEL TWO
30 REM AT A RATE OF 10 MILISECONDS
35 FEM SIGNAL FREQUENCY IS 10 Hz
40 REM HISTOGRAM ARRAY H, DATA ARRAY D
50 DIM H(50),D(100)
60 FOR K=1 to 50
70 H(K)=0
80 NEXT K
100 REM INITIALIZE CLOCK
110 SET RATE 3,10
120 REM SAMPLE 100 DATA POINTS FOR DISPLAY
130 FOR I=1 TO 100
140 WAITC
150 D(I)=ADC(2)
160 NEXT I
170 FOR I=1 TO 100
180 X=.01*I
185 Y=.5+.5*D(I)
190 PLOT X,Y
195 DELAY
198 NEXT I
200 INPUT A
205 CLEAR
210 REM SAMPLE DATA FOR HISTOGRAM
220 FOR I=1 TO 1000
230 WAITC
240 S=ADC(2)
250 S1=.5+.5*S
260 K=INT(50*S1)
270 H(K)=H(K)+1
280 NEXT I
300 REM DISPLAY HISTOGRAM
310 FOR K=1 TO 50
320 X=K/50
330 Y=5,000000E-03*H(K)
340 PLOT X,Y
350 DELAY
360 NEXT K
400 STOP
```

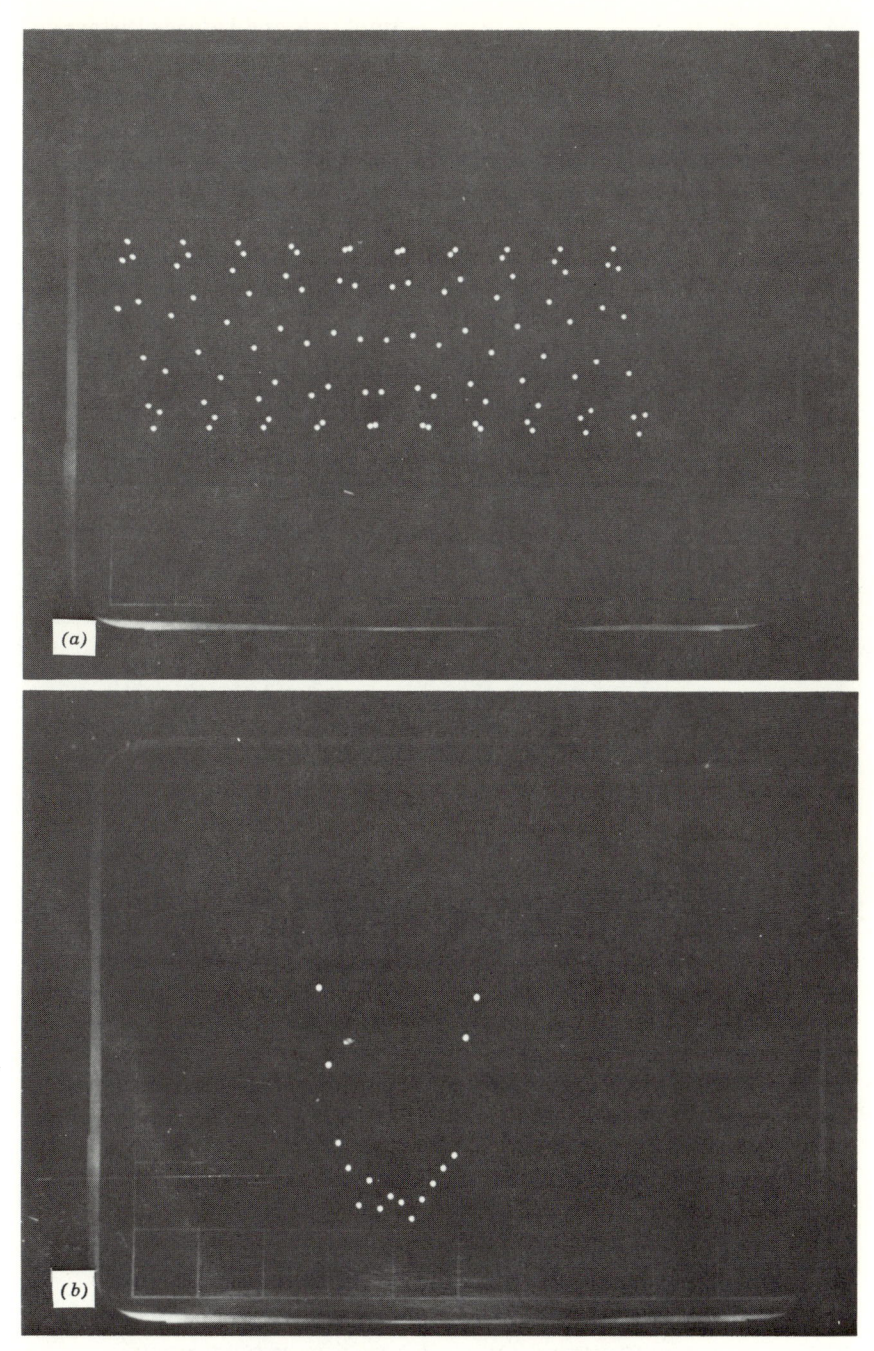

Fig. 8-6 Real-time experiment programmed in BASIC language. (*a*) Sampling and display of the sine voltage. (*b*) Measured amplitude probability distribution.

8.4 Sample Example

The sine-wave generator is connected to the second channel of the multiplexer. Write the program to sample this channel, to convert the signal into digital form, and to sort the data. In this way the amplitude histogram or probability density function will be formed. Display the data and the histogram on the scope. The program is presented in Table 8.1.

Two arrays are prepared: array H for final histogram, and array D for 100 data to be used for data display.

- Statement 100 sets the clock at a 10-msec rate.
- Statement 120 to 160 sample the signal at channel 2/100 times. Wait until statement synchronizes the ADC sampling with clock interrupts.
- Statements 170 to 200 display the data. Produced display is shown in Figure 8.6a.
- Statements 220 to 280 sample the signal at channel 2 1000 times. The data are sorted into histogram array H.
- Statements 300 to 400 display the histogram. Produced display is shown in Figure 8.6b.

References

1. Chase, R. L.: *Nuclear Pulse Spectrometry*, McGraw-Hill, New York, 1961.
2. Korn, G. A.: *Random-Process Simulation and Measurements*, McGraw-Hill, New York, 1966.
3. Bendat, J. S., and Piersol, A. G.: *Measurement and Analysis of Random Data*, Wiley, New York, 1966.
4. Bendat, J. S.: Interpretation and Application of Statistical Anaysis for Random Physical Phenomena, *IRE Transac. Bio-Med. Electron.* (January 1962) 31–43.
5. Souček, B., and Spinrad, R. J.: Megachannel Analyzers, *IEEE Trans. on Nucl. Sci.*, NS-13, No. 1, pp. 183–192, 1966.
6. Spinrad, R. J.: Data Systems for Multiparameter Analysis, in *Annual Review of Nulear Science*, Vol. 14, 1964.
7. Chase, R. L.: *Proc. Conf. Utilization Multiparameter Analyzers Nucl. Phys.*, Grossinger, N. Y., 1962, CU(PNPL)-227, pp. 79–82.
8. Bonitz, M.: *Nucl. Instr. Methods*, 22 (1963) 238–252.
9. Cottini, C., Gatti, E., and Svelto, V.: *Nucl. Instr. Methods* (letter), 24 (1963) 241–242.
10. Hrisoho, A.: *Proc. Internat. Con.fon Nucl. Electr.*, Versailes, 1968.
11. Moody, N.: *Electron Eng.*, 24 (1952) 289
12. Cottini, C., Gatti, E., and Gianelli, G.: *Nuovo Cimento*, 4 (1956) 156. Cottini, C., and Gatti, E.: *Nuovo Cimenta*, 4 (1956) 1550.
13. Lefevre, H., and Russel, J.: *Rev. Sci. Inst.*, 30 (1959) 159–166.
14. Chase, R. L., and Higinbotham, W. A.: *Rev. Sci. Instr.*, 28 (1957) 448–451.
15. Wilkinson, D.: *Proc. Cambridge Phil. Soc.*, 46 (1950) Pt. 3, p. 508.

References

16. Hutchinson, G., and Scarrot, G.: *Phil. Mag.*, **42** (1951) 792.
17. Higinbotham, W. A.: *Proc. Intern. Conf. Peaceful Uses At. Energy*, Geneva **4** (1955) 53–61.
18. Byington, P. W., and Johnstone, C. W.: A 100-Channel Pulse Height Analyzer Using Magnetic Core Storage, *IRE Conv. Record*, **10** (March 1955) 204–210.
19. Schumann, R. W., and McMahon, J. P.: *Rev. Sci. Instr.*, **27**(9), (1956) 675–685.
20. Bromley, D. A., Goodman, C. D., and O'Kelley, G. D.: *Proc. Conf. Utilization Multiparameter Analyzers Nucl. Phys.*, Grossinger, N. Y., 1962, CU(PNPL)–227, pp. 35–48.
21. Deinert, R. H., and Koerts, L. A.: *Nuclear Electro*, II, International Atomic Energy Agency, Vienna, 1962, pp. 197–204.
22. Kenney, R. W.: *Nucl. Instr. Methods*, **20** (1963) 342–344.
23. Kirsten, F. A., and Mack, D. A: *Nuclear Electro.*, II pp. 127–141. International. Atomic Energy Agency, Vienna, 1962.
24. Nakamura, M., and Simonof, G. S.: *Proc. Conf. Instr. Tech. Nucl. Pulse Analysis*, Monterey, Calif., 1963.
25. Collinge, B., and Marciana, F.: *Nucl. Instr. Methods*, **16** (1962) 145–152.
26. Groom, D. E., and Marshall, J. H.: *Rev. Sci. Instr.* **33**(11), (1962) 1249–1255.
27. Matalina, L. A., Chubarev, S. I., and Tishechkina, A. S.: *Nuclear Electr.*, II, pp. 121–125, International Atomic Energy Agency, Vienna, 1962.
28. Spinrad, R. J.: *Nucleonics*, **21**(12) (1963) 46–49.
29. Spinrad, R. J.: *Trans. IEEE PT, Group Nucl. Sci.*, **11**(3), (1964).
30. Brun, J. C., Verroust, G, and Victor, C.: *J. Phys. Radium Phys. Appl.*, **23** (1962) 129A–133A.
31. Norbeck, E.: *Proc. Conf. Utilization Multiparameter Analyzers Nucl. Phys.*, Grossinger, N. Y., 1962 CU(PNPL)–227, pp. 56–58.
32. Kane, J. V., and Spinrad R. J.: *Proc. Conf. Utilization Multiparameter Analyzers Nucl. Phys.*, Grossinger, N. Y., 1962 CU(PNPL)–227, pp. 149–154; and *Nucl. Instr. Methods*, **25** (1963) 141–148.
33. Krüger, G., and Dimmler, G.: *Proc. Intern. Symp. Nucl. El.*, Paris, 1963.
34. Souček, B., Bonačić, Čuljat, K.: *Proc. IV Yugoslav Intern. Data Processing Symposium*, Ljubljana, 1968.
35. Hooton, I. N., and Best, G. C.: *Nucl. Instr. Meth.*, **56** (November 1967) 284.
36. Best, G. C., and Hooton, I. N.: *AERE Report R 5425*, Harwell, England, 1967.
37. Chase, R. L.: *IRE Natl. Conv. Record*, Pt. 9, pp. 196–201, 1959.
38. Westman, A., Petrusson, E.: and Tove, P. A.: *Proc. Int. Conf. on Nucl. Electr.*, Versailles, 1968.
39. Stüber, W.: *Proc. Int. Conf. on Nucl. Electr.*, Versailles, 1968.
40. Widley, P. T.: *Computer J.*, **3** (1960) 84.
41. Souček, B.: *Rev. Sci. Instr.*, **36** (June 1965) 750–753.
42. Souček, B.: *Nucl. Instr. Meth.*, **36** (1965) 181–191.
43. Hooton, I. N.: *IEEE Trans. on Ncul. Sci.*, NS, **13**, No. 3 (1966) 553.
44. Best, G. C., Hickman, S. A., Hooton, I. N., and Prior, G. M.: *IEEE Trans. on Nucl. Sci.*, NS, **13**, No. 3 (1966) 559.
45. Best, G. C.: *IEEE Trans. on Nucl. Sci.*, NS, **13**, No. 3 (1966) 566.

46. Hooton, I. N.: *Proc. EAUC Conf.*, Karlsruhe, 1964.
47. Rosenblum, M.: Brookhaven National Laboratory, private communication, 1966.
48. Souček, B.: *IEEE Trans. Nucl. Sci.*, NS, **13**, No. 3 (1966) 571.
49. Souček B., Bonačić, V., Čuljat, K., and Radnić, I.: *Internat. Conf. on Nucl. Electr.*, Versailles, 1968.
50. Souček, B., Bonačić, V., and Čuljat, K.: *Nucl. Instr. Method.*, **66**, No. 2 (1968) 202.
51. Bonačić, V., Souček B., and Čuljat, K.: *Nucl. Instr. Meth.*, **66**, No. 2 (1968) 213.
52. Durand, P., and Giraud, P.: *Proc. Intern. Symp. Nucl. Electr.*, Paris, 1963, p. 643.
53. Thenard, J., and Victor, G.: *Nucl. Instr. Meth.* **26** (1964) 45.
54. Poole, M. A.: Brookhaven National Laboratory, *Informal Report IH-363*, April 1965.
55. Colling, F., and Stüber, W.: *Nucl. Instr. Meth.*, **64** (1968) 52.
56. Spilling, P., Gruppelaar, H., and Van den Berg, P. C.: *Proc. Intern. Conf. Nucl. Electr.*, Versailles, 1968.
57. *Proc. EANDC Conf.*, Karlsruhe, 1964.
58. *Proc. Intern. Conf. Nucl. Electr.*, Versailles, 1968.
59. Gere, E. A., and Miller, G. L.: *IEEE Trans. Nucl. Sci.*, NS, **13** (June 1966) 508.
60. Bhat, M. R., Borrill, B. R., Chrien, R. E., Rankowitz, S.: Souček, B., and Wasson, O.A.: *Nucl. Instr. Methods*, **53** (1967) 108–122.
61. Adam, J. P., Brun, J. C., Faucher, L., and Victor, C.: *Proc. Int. Symp. on Nucl. Electr.*, Versailles, 1968, p. 110.
62. Adam, J. P., Brun, J. C., and Cuzon, J. C.: *Proc. Int. Symp. on Nucl. Electr.*, Versailles, 1968, p. 143.
63. Adam, J. P.: *Nucl. Instr. Meth.*, **73** (1969) 89–92.
64. Tilger, C. A.: Digital Equipment Company, private communication, 1969.
65. Mollenauer, J. F.: Bell Telephone Laboratory, internal report, 1967.
66. Mollenauer, J. F.: *Proc. Conf. Computer Systems Exptl. Nucl. Phys.*, Skytop, Pena., 1969.
67. Kletsky, E. J.: *DECUS Conference Proc.*, 1970, p. 287.
68. Mishelevich, D. J.: *IEEE Trans. Bio-Med. Eng.*, BME-17, 2 (1970) 147.
69. Carlson, A., and Souček, B.: *IEEE Trans. Nucl. Sci.*, Feb. 1971.
70. Souček, B.: *IEEE Trans. Nucl. Sci.*, NS17 (August 1970) 20.
71. Souček, B.: *IEEE Trans. Nucl. Sci.*, NS16 (Oct. 1969) 36.
72. Souček, B., *Minicomputers in Data Processing and Simulation*, Wiley, New York, 1972.

Chapter 9

CORRELATION MEASUREMENT

Introduction and Survey

The correlation function is simple yet efficient tool for analyzing the random data through statistical averaging. The correlation function describes the time dependencies of a random data record; it could be measured using special instruments, as well as using, a laboratory computer.

In this chapter the basic definitions and properties of the correlation function are described. Two different methods for the correlation function calculation are explained. It is shown that the classical method, although good for analysis of random amplitudes, produces meaningless results if used to analyze random intervals. Both computer simulated data and real experimental data have been used to check the correlation algorithms. Modified method for interval analysis is explained. Computer programs for both the methods are shown and discussed.

9.1 Correlation Function

Definitions The correlation function describes the general dependence of the values of the data at one time on the value at another time. Correlation is usually applied for random data analysis, and it can be used to detect periodic signals buried in a noise. Also, correlation provides a measure of similarity between two waveforms. Two correlation functions are generally used: autocorrelation and cross correlation. Autocorrelation measures the similarity of a signal to a time delayed version of itself, whereas cross-correlation measures the degree of similarity of one waveform (source, input, stimulus) to a second waveform (output, response).

Mathematical definition of the autocorrelation function is given in equation 9.1.

$$R_{xx}(C) = \lim_{T \to \infty} \frac{1}{T} \int_0^T x(t)x(t+C)dt \qquad (9.1)$$

Figure 9.1 shows a small part of a random waveform. Equation 9.1 takes the value $x(t)$ at time t and the value $x(t+C)$ at the time $(t+C)$ and makes the product of the two. This operation is repeated for every value of t, $0 < t < T$. The integral presents the summation of all products, and the operation $1/T$ presents the averaging over the observation time T. The resulting average product approaches an exact autocorrelation function, as T approaches infinity.

Mathematical definition of the crosscorrelation function is given in equation 9.2.

$$R_{xy}(C) = \lim_{T \to \infty} \frac{1}{T} \int_0^T x(t)y(t+C)dt \qquad (9.2)$$

The only difference between equations 9.1 and 9.2 is that in the crosscorrelation calculation the signal $x(t)$ is multiplied by a time-delayed version of a second signal $y(t)$, rather than a time-delayed version of itself.

Properties and Applications of Correlation Function Figure 9.2 shows a typical plot of autocorrelation $R_{xx}(C)$ versus the time displacement C for four waveforms, sine waves, sine waves plus random noise, narrow-band random noise, and wide-band random noise. The autocorrelation function has the following properties:

1. Autocorrelation is an even function with a maximum at $C = 0$. Because the even functions are symmetrical around the value $C = 0$, it is enough to compute the correlation function only for positive values of C.

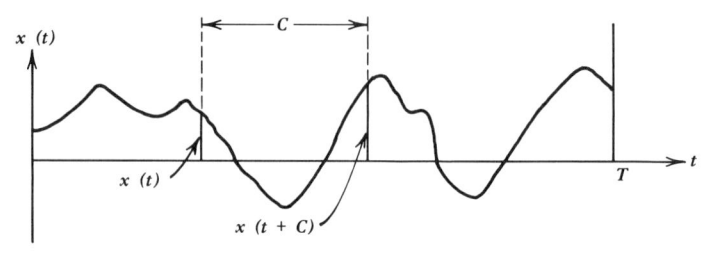

Fig. 9-1 Random waveform. To form autocorrelation function for the lag C, the product $x(t) \cdot x(t+c)$ is formed for every value of t.

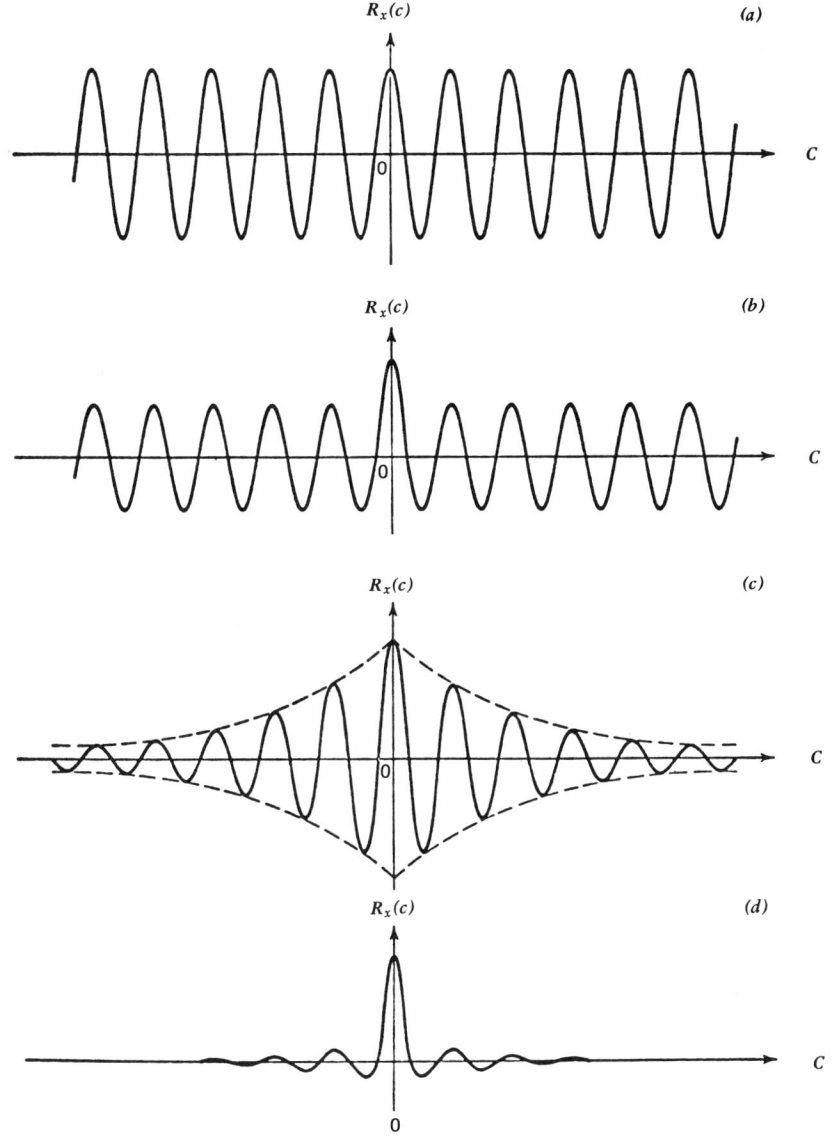

Fig. 9-2 Autocorrelation functions for some typical signals. Signals are (*a*) Sine wave. (*b*) Sine wave plus random noise; (*c*) Narrow-band random noise; (*d*) Wide-band random noise.

2. The maximal value $R_{xx}(0)$ of the autocorrelation function is equal to the mean square value of the time function. For display purposes it is convenient to normalize maximum value to 1, by displaying the function

$$\frac{R_{xx}(C)}{R_{xx}(0)}$$

3. The value of correlation function at $C > \infty$ is equal to the square of the mean value of the time function.

4. If the time function contains periodic components, the autocorrelation function will contain components having the same period, Figure 9.2a.

5. If the time function contains only random components, the autocorrelation function will exponentially approach zero as C increases, Figure 9.2d.

6. If the time function is composed of two or more components, the correlation function will be the sum of the correlation functions of each individual components. Figure 9.2b presents the correlation function for sine wave plus random noise; it is obtained through the summation of the functions displayed in Figures 9.2a and d.

7. The crosscorrelation function has also properties 3 to 6. However, the crosscorrelation function is not necessarily an even function. The maximum value of $R_{xx}(C)$ will occur for that value of the time shift C for which the two signals, x and y, are most alike.

Typical applications for correlation include:

1. Determination of the transmission path and propagation delay of electrical, mechanical, acoustical, or seismic waves.

2. Detection of very weak signal buried deep in noise.

3. Indication of epilepsy through comparison of electroencephalograms from the two halves of the brain.

4. Measurement of the impulse-response function of complex systems in the presence of noise.

5. Parkinson's disease study and tremor frequency analysis.

6. Communication and speech research.

9.2 Amplitude Correlations

Equation 9.1 has been modified and programmed for digital computer, to calculate autocorrelation functions for different signals. The program is presented in Table 9.1. To apply digital computing techniques, equation 9.1 has been modified in the following way: The signal is sampled at regular time intervals, $I = 1, 2 \cdots T$, producing digital data $x(1), x(2) \cdots x(T)$.

Interval Correlation

The integration is performed through a summation for I going from 0 to T. The parameter T is the number of sampled values used for calculation. Large values of T are needed for accurate calculation.

Statements 20 to 100 are used to allocate the space for sampled data X, and for correlation function R and also to enter the parameters of the program: total number of data available M; number of desired correlation intervals C1; and the number of samples to be used for calculation T. The value is $T = M - C1$, to make sure that every sample will have a chance to be multiplied with another sample C1 positions farther down the array.

This program can be used for real data generated by an external source. To show the features of the correlation function, we shall simulate few typical waveforms, using an auxiliary computer program. Statements 120 to 220 generate a sine wave plus noise, with the amplitudes S1 and N1, respectively. Statements 230 to 310 will produce the display of the generated waveforms. Figures 9.3a to 9.6a shows four waveforms, for four different signal to noise ratios, as displayed on the scope.

Actual correlation function calculation is programmed with statements 410 to 470. For each value of the time lag C the autocorrelation is obtained through the multiplication $x(I)\, x(I + C)$, and the products are added together, for values of $0 \leq I \leq T$. Statements 510 to 600 will produce the display of the calculated correlation function. Figures 9.3b to 9.6b shows four autocorrelation functions for the four waveforms presented in Figures 9.3a to 9.6a.

Figure 9.3 presents a sine wave and its autocorrelation function. As expected, correlation function is periodic, with the same period as the signal.

Figure 9.4 presents random noise and its autocorrelation. This time the correlation function has exponential shape, approaching zero for large values of the time lag C.

Figures 9.5 and 9.6 present a mixture of sine wave and noise. Because the signal is composed of two components, the autocorrelation function is the sum of the autocorrelation functions of each individual component.

Figure 9.6 shows how the correlation can be used to find periodic signal buried in noise. The periodic signal is small, and it cannot be observed directly Figure 9.6a. However, the correlation function clearly shows the periodic component, Figure 9.6b.

9.3 Interval Correlation

Simulated Data Frequently in biological and biomedical signals, the basic information is carried in the form of a time interval (interspike latency, or response interval). One typical train of spikes is presented in Figure 9.7.

TABLE 9.1

```
LIST
20 REM CORRELATION,AMPLITUDES
30 DIM D(100)
40 USE D
50 DIM X(300)
60 DIM R(50)
70 PRINT "#OF DATA"
80 INPUT M
90 PRINT "#OF CORR"
100 INPUT C1
110 T=M-C1
111 REM
112 REM
115 REM
120 REM INPUT OR SIMULATION OF DATA
130 REM SIMULATE SINE PLUS NOISE
135 REM SINE AMPL.=S1,NOISE AMPL.=N1
140 PRINT "ENTER S1,N1"
150 INPUT S1,N1
155 F=0
160 FOR I=1 TO M
170 E=RND(1)+RND(2)+RND(3)-1.5
180 F=.5*F+E
190 S=SIN(.5*I)
200 X(I)=S1*S+N1*F
220 NEXT I
230 PRINT "DISPLAY DATA"
240 FOR I=1 TO C1
250 Z=I/C1
260 W=.5+X(I)/3
270 PLOT Z,W
280 DELAY
290 NEXT I
300 INPUT V
310 CLEAR
311 REM
312 REM
315 REM
400 REM CORRELATION LOOP
410 REM CORRELATION LAG IS C
415 FOR C=0 TO C1
420 R(C)=0
430 FOR I=1 TO T
440 R(C)=R(C)+X(I)*X(I+C)
450 NEXT I
460 R(C)=R(C)/T
470 NEXT C
```

(Continued)

Interval Correlation

TABLE 9.1 (*continued*)

```
500 PRINT "DISPLAY CORRELATION"
510 FOR C=0 TO C1
520 Z=C/C1
530 W=R(C)/R(0)
540 W=.5+W/5
550 PLOT Z,W
560 DELAY
570 NEXT C
600 STOP
```

In this case, one has to be careful in calculating the correlation function. Direct substitution of data values, $x1$, $x2$ $\cdots$ into equation 9.1 might lead to the totally wrong conclusions. Another approach is needed to measure and calculate correlation function for the signal in which the information carrier is the time interval. The correct procedure could be explained in the following way, starting from equation 9.1 and Figure 9.7.

The signal in Figure 9.7 presents the train of spikes with random interspikes intervals. Usually, the amplitudes of the spikes are of no particular interest to the experimenter, whose only concern is to find out the properties of the interspike intervals. Hence we shall normalize all amplitudes to the value 1.

For a given time lag C, the product from equation 9.1 can have only one of two values: $x(t) \cdot x(t + C) = 1$, if the second spike is found inside the window G at a distance C from the first spike; or $x(t) \cdot x(t + C) = 0$, if there is no second spike inside the window G on the distance C from the first spike.

To calculate the correlation function for a given lag C, the above procedure should be applied on all spikes in the train. By adding together so formed partial products, the correlation function is obtained. The program for latency correlation, based on this procedure is shown in Table 9.2.

Statements 30 to 114 allocate the space and initialize the parameters of the program. Statements 120 to 230 simulate a random train of intervals. Each interval is composed of two components: constant period of the duration S1, and random period with the mean duration N1. Statement 210 generates random numbers following Poisson interval distribution, as explained in the chapter dealing with simulation. Statement 220 adds the intervals which means that the interval x is always measured relative to the beginning of the interval train as shown in the bottom part of Figure 9.7.

Statements 240 to 320 will produce the scope display. Ordinate of the display is the interval x, measured relative to zero, whereas the apscisa is a serial number of the interval. Figures 9.8, 9.9, and 9.10a show three interval displays for three different ratios of constant period and random period.

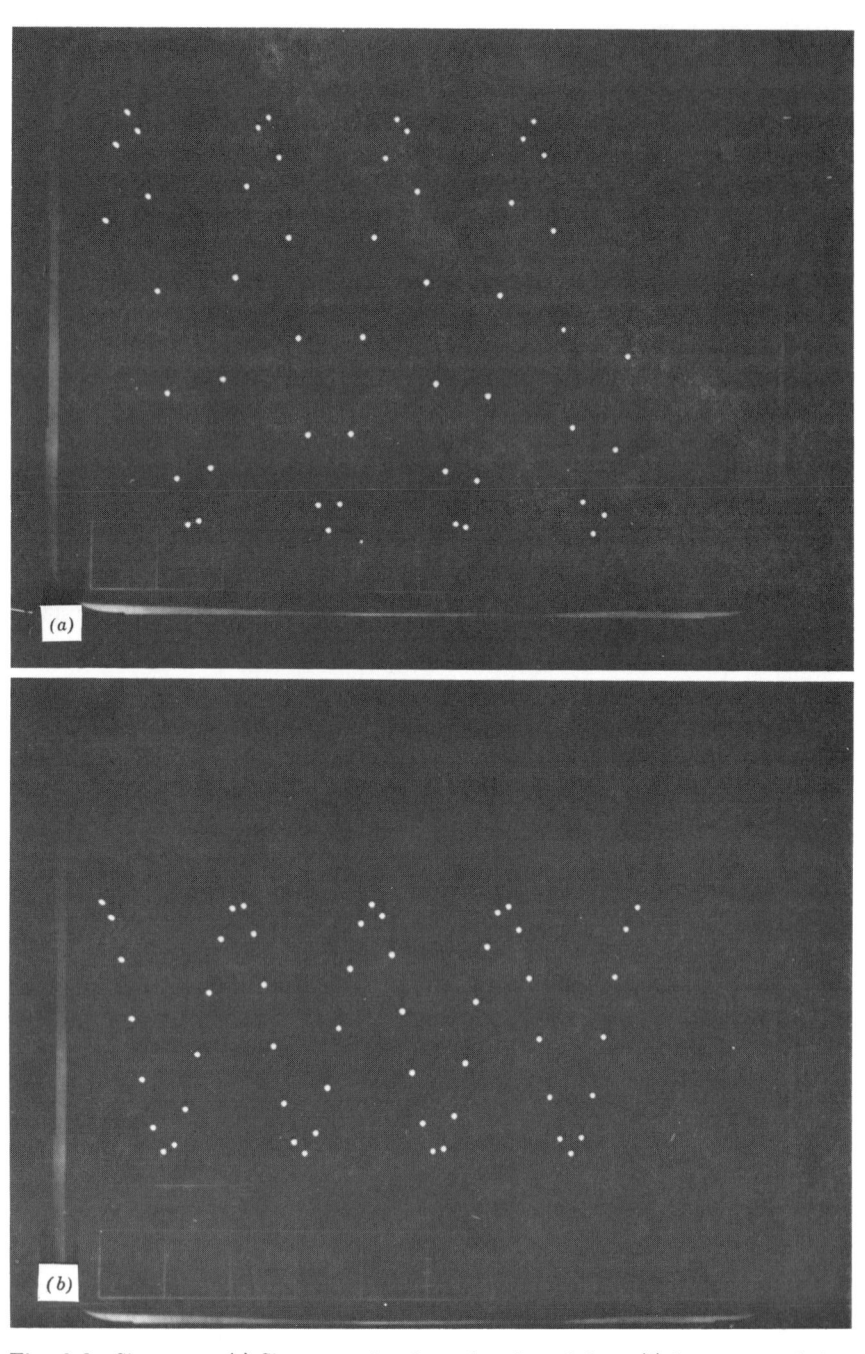

Fig. 9-3 Sine wave. (a) Sine wave signal as a function of time; (b) Its autocorrelation function as a function of the time lag C.

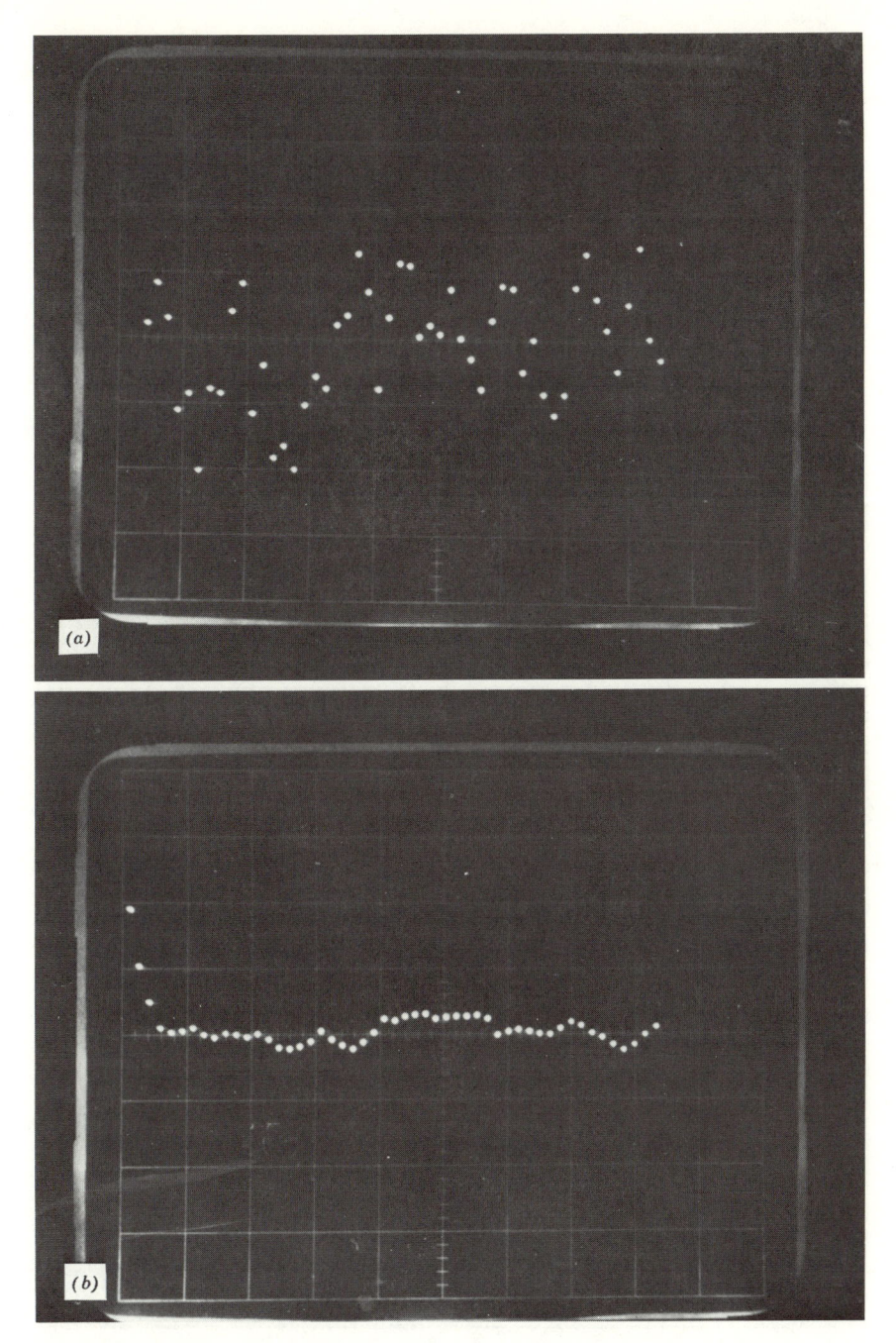

Fig. 9-4 Random noise. (*a*) Signal as a function of time. (*b*) Its autocorrelation function as a function of the time lag C.

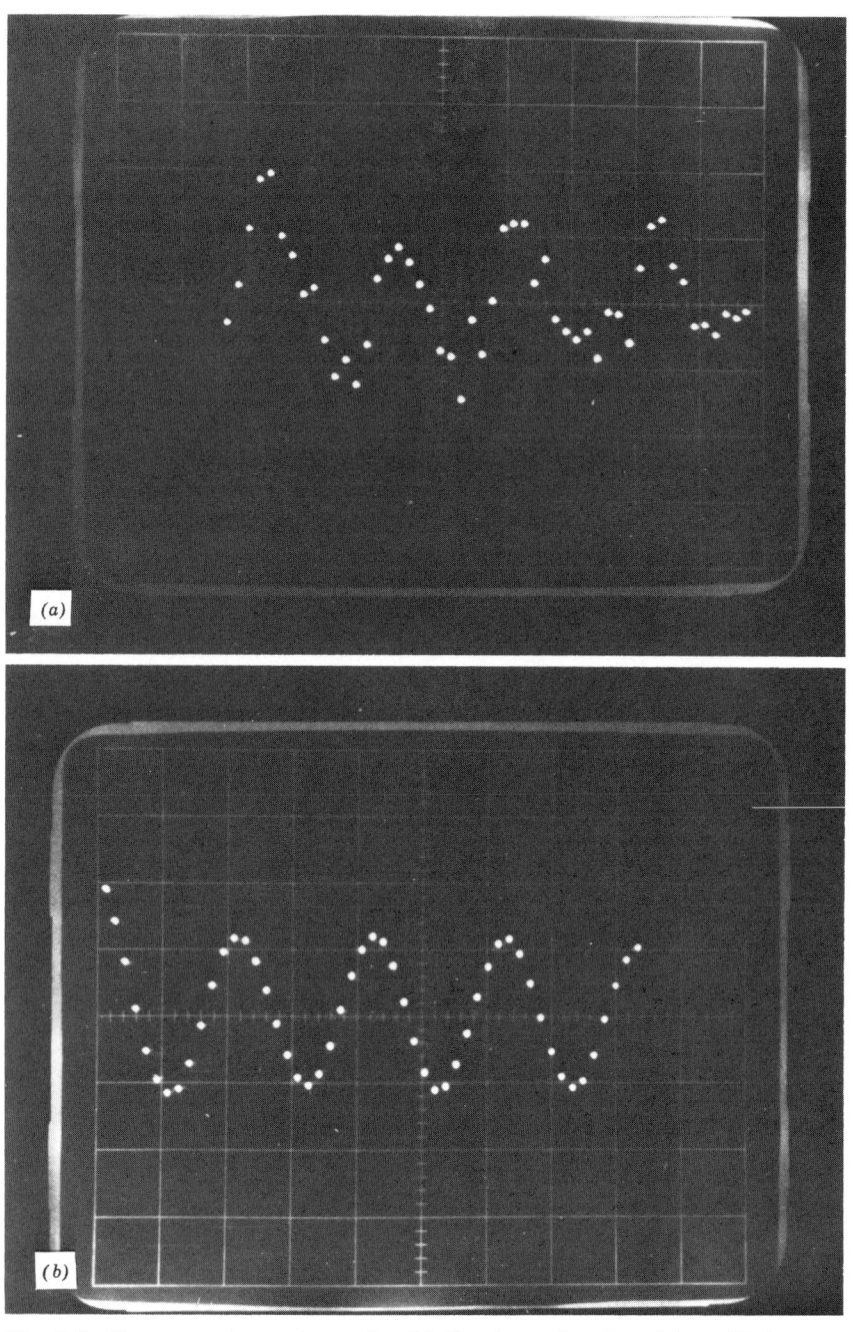

Fig. 9-5 Sine wave plus random noise. (*a*) Signal as a function of time. (*b*) Its autocorrelation function as a function of the time lag *C*.

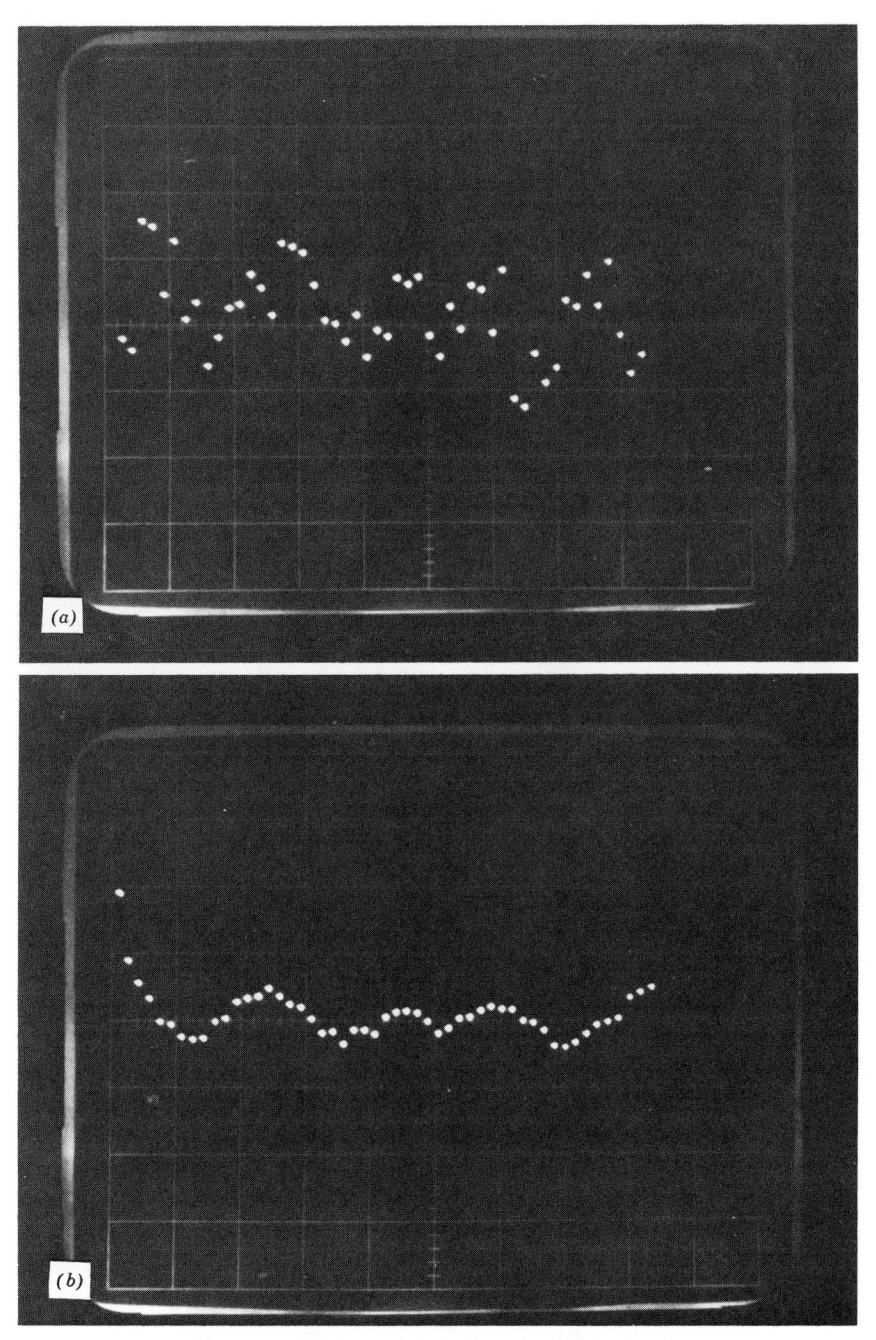

Fig. 9-6 Sine wave plus random noise. (*a*) Signal as a function of time. (*b*) Its autocorrelation function as a function of the time lag *C*.

TABLE 9.2

```
LIST
20 REM CORRELATION,INTERVALS
30 DIM D(100)
40 USE D
50 DIM X(300)
60 DIM R(50)
70 PRINT "#OF DATA"
80 INPUT M
90 PRINT "OF CORR"
100 INPUT C1
110 T=M-C1
112 PRINT "TIME WINDOW C=1"
114 C=1
116 REM
118 REM
119 REM
120 REM INPUT OR SIMULATE DATA
130 REM SIMULATION
140 REM INTERVAL=CONSTANT PERIOD+RANDOM PERIOD
150 REM CONSTANT AMPLITUDE=S1,RANDOM AMPL.=N1
160 PRINT "ENTER S1,N1"
170 INPUT S1,N1
180 X(1)=0
190 FOR I=2 TO M
200 E=PND(1)
210 F=-LOG (1-E)
220 X(I)=X(I-1)+S1+N1*F
230 NEXT I
240 PRINT "DISPLAY INTERVALS"
250 FOR I=1 TO C1
260 Z=I/C1
270 W=1.000000E-03*X(I)
280 PLOT Z,W
290 DELAY
300 NEXT I
310 INPUT V
320 CLEAR
350 REM
351 REM
352 REM
400 REM CORRELATION LOOP
410 REM CORRELATION LAG IS C
418 FOR C=0 to C1
420 R(C)=0
422 P=C
424 Q=C+.95*G
430 FOR I=1 TO T
```

(Continued)

Interval Correlation

TABLE 9.2 (*continued*)

```
435 K1=3+INT(.1*C)
440 FOR K=0 TO K1
450 IF X(I+K)>=(X(I)+P)GO TO 470
460 NEXT K
465 GO TO 500
470 IF X(I+K)>(X(I)+Q)GO TO 500
480 R(C)=R(C)+1
500 NEXT I
510 R(C)=R(C)/T
520 NEXT C
600 PRINT "DISPLAY CORRELATION"
610 FOR C=0 TO C1
620 Z=C/C1
630 W=.5+.2*R(C)
640 PLOT Z,W
650 DELAY
660 NEXT C
700 STOP
```

The actual correlation calculation starts at the statement 418. For each time lag C, the following operations are performed. Upper and lower boundaries (P and Q) of the time window are calculated. For each spike I, the ten spikes that follow are examined. If any of them arrives during the time window, one is added to the accumulated value of the correlation function. Otherwise, accumulated value of the correlation function stays unchanged.

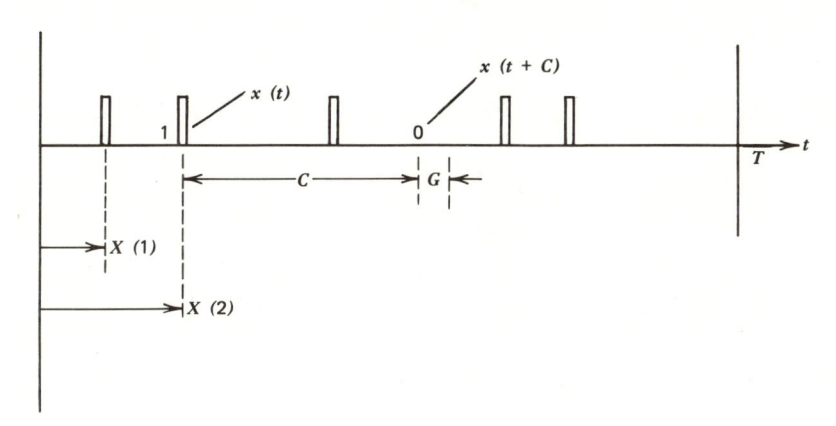

Fig. 9-7 Random interspike intervals. To form the autocorrelation function for the lag C the product $x(t) \cdot x(t+c)$ is formed for each spike. The product could take only the value zero or one.

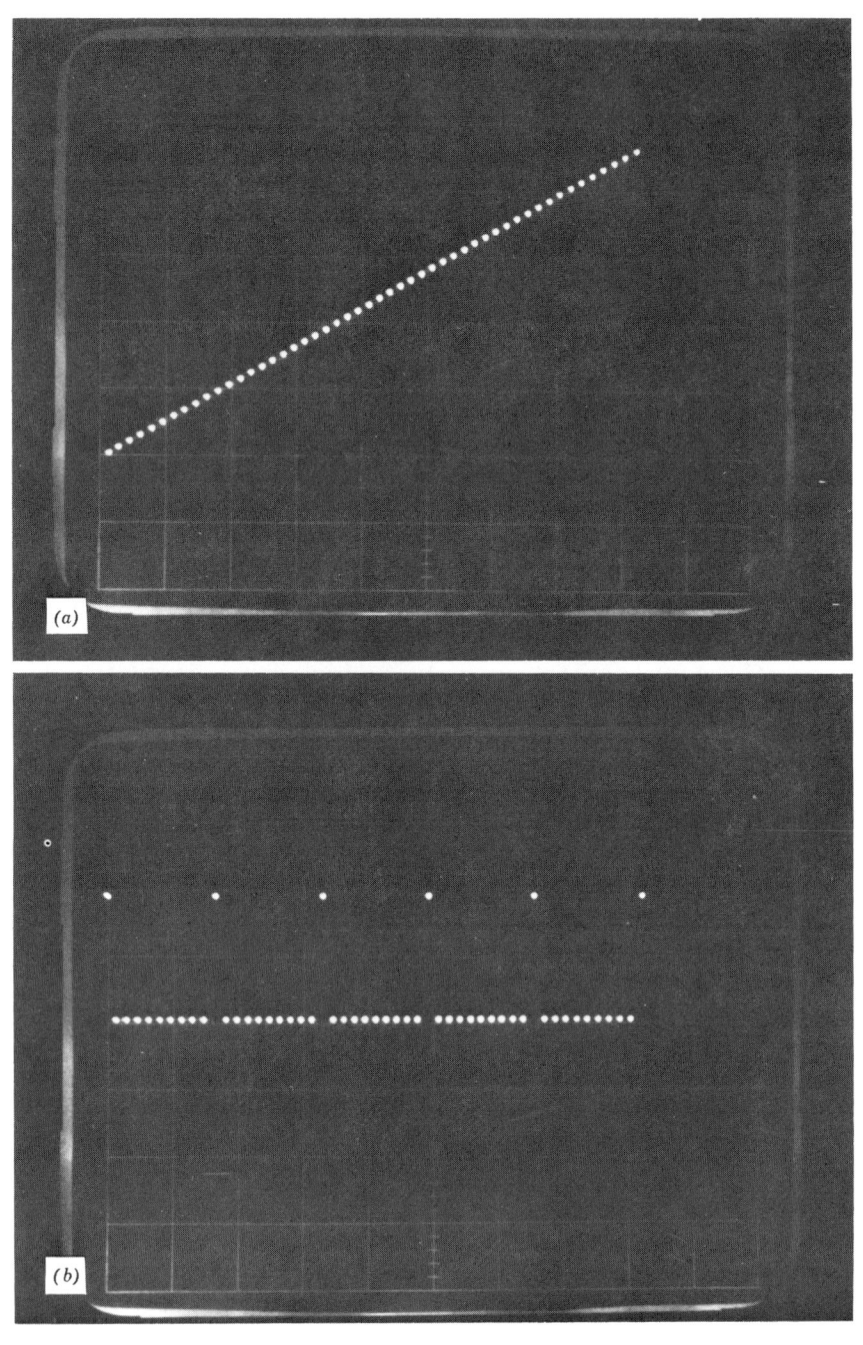

Fig. 9-8 Intervals composed only of constant period, of 10 time units. (*a*) Spike arrival instants *x*, measured relative to time $t = 0$, and displayed as a function of the serial number of the spike. (*b*) Autocorrelation function. Note the period of 10 time units.

168

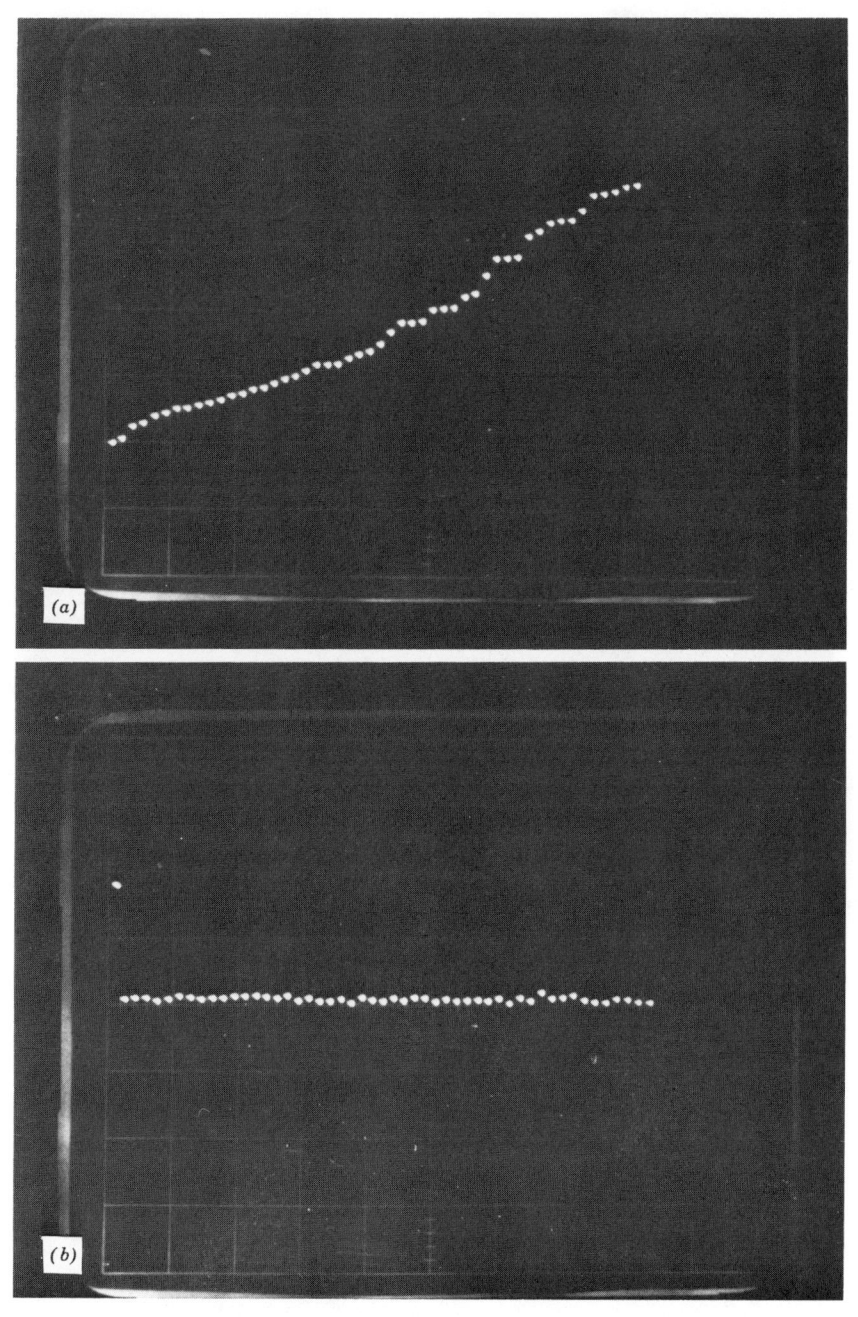

Fig. 9-9 Intervals composed only of the random component. (*a*) Spike arrival instants *x*, measured relative to time $t = 0$ and displayed as a function of the serial number of the spike. (*b*) Autocorrelation function.

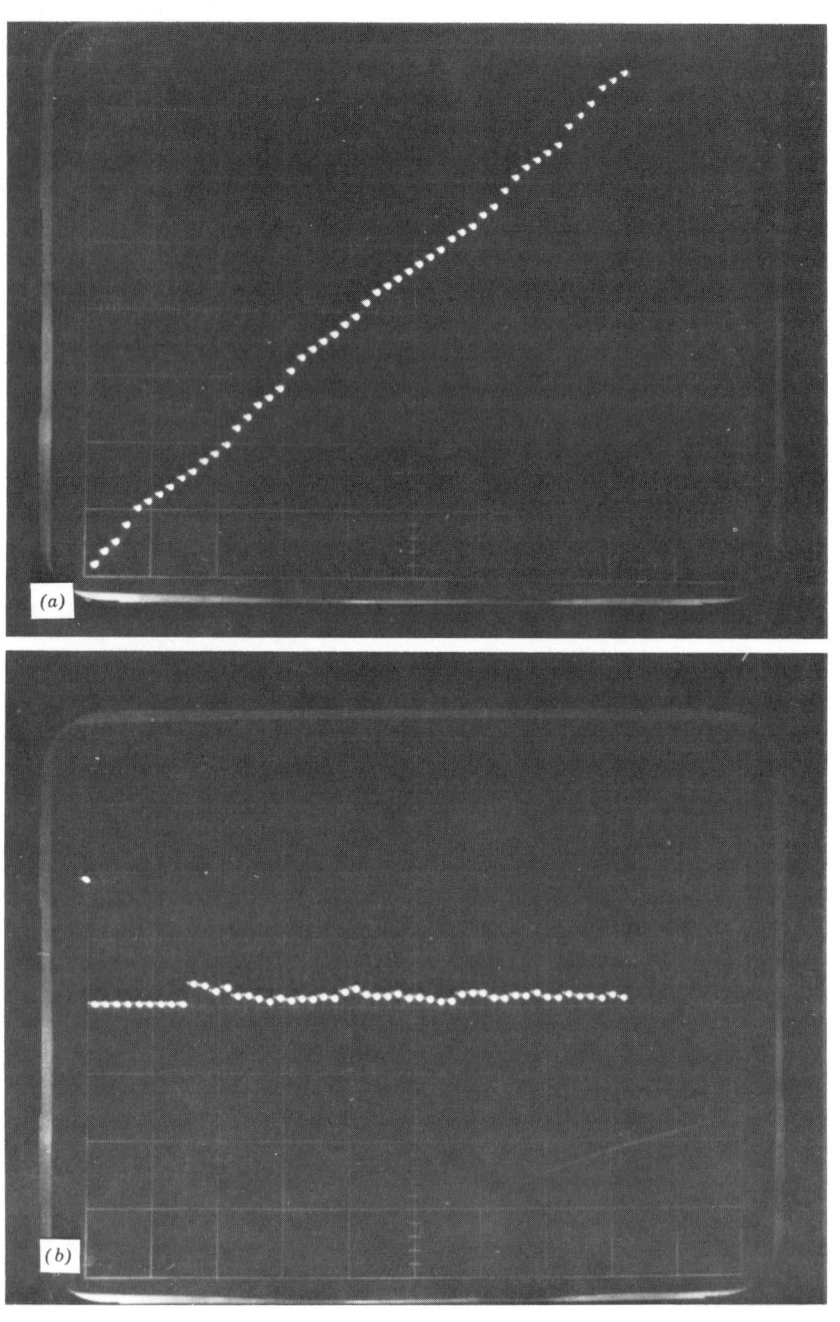

Fig. 9-10 Mixture of constant period and random period. (*a*) Spike arrival instants. (*b*) Autocorrelation function.

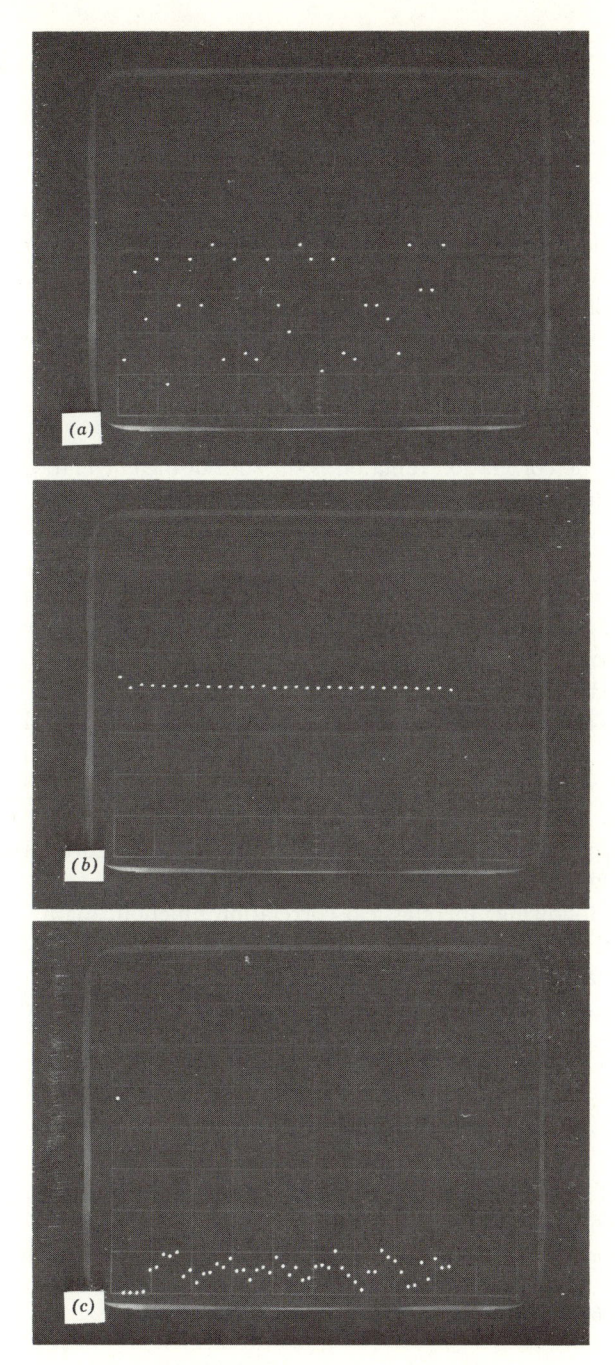

Fig. 9-11 Insect calls analysis. (a) Interchirp intervals from katydid calls displayed as a function of the serial number of the chirp. (b) autocorrelation analysis based on Eq. 9-1. and Fig. 9-1. (c) Autocorrelation analysis based on Fig. 9-7.

171

Statements 600 to 660 produce the display of the correlation function on the scope. Figures 9.8, 9.9, and 9.10b show three autocorrelation functions for the three waveforms presented in Figures 9.8, 9.9, and 9.10a.

Figure 9.8 presents the intervals composed only of constant period. As expected, the correlation function is periodic with the same period as the signal. In this example, the constant period S1 = 10 time units. Note that the period of the correlation function is also 10.

Figure 9.9 presents the interval composed only of the random component. This time the correlation function is delta function at time lag $C = 0$. For all other values of C the correlation function is approximately zero.

Figure 9.10 presents a mixture of a constant period and a random period. As a result, the correlation function is the sum of the correlation function of each individual component. Note the delta function $C = 0$ as well as periodic component with the period of 10.

Real Data: Insect Calls The katydid chirps have been measured on magnetic tape. Interchirp intervals have been measured and used as a raw data for correlation programs.

First the data have been read into the program based on equation 9.1, and shown in Table 9.1. The results are shown in Figure 9.11. Figure 9.11a displays the measured intervals as a function of the intervals serial number. Figure 9.11b shows the correlation function.

Next the data have been read into the program based on the latency correlation procedure, Table 9.2. Figure 9.11c shows the correlation function produced by this program.

Note the following: the classical correlation function, Figure 9.11b shows meaningless result. The modified correlation function, Figure 9.11c clearly shows that a katydid interchirp intervals must be composed of two components: random and periodic. Also, one could directly read the period of the periodic component.

References

1. Bendat, J. S., and Piersol, A. G.: *Measurement and Analysis of Random Data*, Wiley, New York, 1966.
2. *Lab 8 Software System User's Manual*, Digital Equipment Coproration, 1972.
3. Cox, D. R. and Lewis, P. A. W.: *The Statistical Analysis of Series of Events*, Wiley, New York, 1966.

Chapter 10

FOURIER ANALYSIS AND POWER SPECTRA

Introduction and Survey

The most obvious presentation of the signal is in the time domain: the amplitude of the signal is displayed as a function of time. Another important presentation of the signal is in the frequency domain. For this purpose a Fourier series is used. The signal waveform may be expanded into a series of sine waves. By finding the amplitudes of these sine waves, the power spectra is formed. The power spectral density function describes the general frequency composition of the data. The frequency composition of the data, in turn, bears important relationships to the basic characteristics of the physical or biomedical system involved.

In this chapter the Fourier series expansion is explained. The computer programs are shown to calculate Fourier coefficients for a periodic signal, as well as to calculate the power spectra for a nonperiodic signal. Also, some typical, frequently occurring waveforms are analyzed, and their power spectra are displayed and discussed. The chapter ends with the discussion of special machines used to measure the power spectra and their variation with time. Such "sonagrams" are widely used in acoustic communication study. Sonagrams are used for speech study as well as for animal communication study.

10.1 Periodic Data and Fourier Series

The most elementary periodic data is sine wave

$$x(t) = X \sin(2\pi f t + \theta) \qquad (10.1)$$

where X = amplitude
f = frequency in cycles per second
θ = initial phase angle in radians
$x(t)$ = instantaneous value

Three numbers are enough to completely describe the sine wave: X, f, and θ. Also, the parameter called period T is often used

$$T = \frac{1}{f} \tag{10.2}$$

The period presents the time required for one full fluctuation or cycle of sinusoidal data.

More complex periodic data cannot be described in such a simple way. Complex periodic data repeats itself at regular intervals, such that

$$x(t) = x(t + T) = x(t + 2T) \cdots = x(t + nT) \tag{10.3}$$

In equation 10.3, T is again the period, that is the time required for one full fluctuation. The value $f = 1/T$ is called basic frequency.

Complex periodic data may be treated as being composed of a number of sinusoidal waves with different amplitudes and with frequencies f, $2f$, $3f \cdots nf$. By choosing the proper mixture of such waves, one could sintetize any periodic waveform with the fundamental frequency f. This procedure is called Fourier series expansion. Thus the periodic waveform $x(t)$, may be expanded into a series according to the following formula

$$x(t) = \frac{A_0}{2} + \sum_{N=1}^{\infty} (A_n \cos Nwt + B_n \sin Nwt) \tag{10.4}$$

where $w = 2\pi/T$ is fundamental circular frequency and A_n and B_n are amplitudes of sine and cosine waves at frequency Nw.

One can prove that the right values for coefficients A_n and B_n are

$$A_n = \frac{2}{T} \int_0^T x(t) \cos Nwt \, dt \qquad N = 0, 1, 2 \cdots \tag{10.5}$$

$$B_n = \frac{2}{T} \int_0^T x(t) \sin Nwt \, dt \qquad N = 1, 2, 3 \cdots \tag{10.6}$$

By combining sine and cosine waves of the same frequency in one single cosine, but in a phase shifted way, equation 10.4 changes into

$$x(t) = C_0 + \sum_{N=1}^{\infty} C_n \cos (Nwt - \theta) \tag{10.7}$$

Frequency Spectrum Program

where

$$C_0 = \frac{A_0}{2} \tag{10.8}$$

$$C_n = \sqrt{A_n^2 + B_n^2} \tag{10.9}$$

$$\theta = \tan^{-1}\frac{B_n}{A_n} \tag{10.10}$$

Equation 10.7 may be discussed in the following way: Complex periodic data $x(t)$ is composed of a static component C_0 and an infinite number of cosine waves called harmonics. The frequency of the Nth cosine wave is Nw, the amplitude is C_n, and the phase is θ_n. Frequently the phase angles θ_n are ignored.

Substituting equations 10.5 and 10.6 into equation 10.9 and using Euler formula leads to

$$C_n = \int_0^T x(t) \cdot e^{-j2\pi f t} dt \tag{10.11}$$

Equation 10.11 is in a form that allow a natural extension of Fourier analysis to nonrepetitive waveforms. One can view a nonrepetive waveform as a repetitive wave in which the period T approaches infinity. For finite T the ratio N/T is the frequency of the Nth harmonic, and the frequency separation between harmonics equals the frequency of the fundamental, $1/T$. When T approaches infinity, the separation between harmonics approaches zero and N/T approaches the smooth frequency variable f.

10.2 Frequency Spectrum Program

Based on equations 10.5, 10.6 and 10.9, a frequency spectrum program has been developed; it is shown in Table 10.1. This program could be used for both periodic and nonperiodic waveform analysis. For periodic waveform analysis, the integration period T should be equal to the basic period of the waveform, Figure 10.1. For nonperiodic waveform the integration period T theoretically should approach infinity. If the waveform is defined for $0 < t < D$ and is zero for $t > D$, then for practical application, it is usually enough to take $T = (5 \div 10)*D$, Figure 10.2.

The program in Table 10.1 divides the integration period T into 100 time steps (Statement 40), hence $W = 2\pi/T = 2\pi/100$ (Statement 110).

The first part of the program (statements 150 to 227) is used to simulate and display a different waveforms. In this example statement 170 is used to simulate a waveform.

TABLE 10.1

```
LIST
20 REM FOURIER ANALYSIS
25 DIM D(100)
26 USE D
30 DIM X(100)
40 T=100
100 P=3.141503
110 W=2*P/T
130 REM
140 REM
145 REM SIMULATE SIGNAL
150 PRINT "ENTER AMPLITUDE,FREQUENCY,DAMPING"
151 INPUT L,C,H
152 G=C*W
155 FOR J=1 TO T
170 X(J)=COS(G*(J-1))*EXP(-H*J)
200 D=J/T
210 R=.2*X(J)+.5
215 PLOT E,F
220 DELAY
221 X(J)=L*X(J)
225 NEXT J
226 INPUT E
227 CLEAR
228 REM
229 REM
230 DIM A(11),B(11),C(111)
240 REM NUMBER OF HARMONICS Z=10
250 Z=10
300 FOR N=0 TO Z
310 A(N)=0
329 B(N)=0
330 FOR J=1 TO T
340 R=N*W*(J-1)
350 A(N)=A(N)+X(J)*COS(R)
360 B(N)=B(N)+X(J)*SIN(R)
370 NEXT J
400 A(N)=A(N)/T
410 B(N)=B(N)/T
430 C(N)=SQR(A(N)↑2+B(N)↑2)
440 NEXT N
500 REM DISPLAY THE SPECTRUM
512 FOR N=0 TO Z
520 E=N/Z
530 F=.1+C(N)
540 PLOT E,F
550 DELAY
560 NEXT N
570 STOP
```

Some Typical Spectra

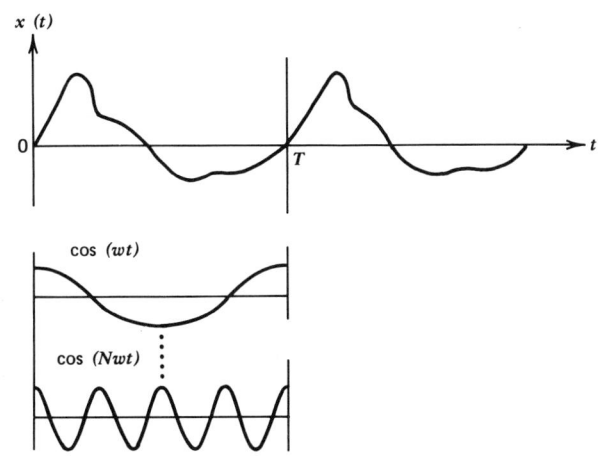

Fig. 10-1 Periodic waveform, with the period T. The Fourier integration period T should be equal to the basic period of the waveform.

The actual Fourier analysis program starts at statement 230. This particular program is fixed to calculate $Z = 10$ harmonic coefficients.

Statements 350 and 400 follow equation 10.5.
Statements 360 and 410 follow equation 10.6.
Statement 430 follow equation 10.9. The rest of the program is used to display the frequency spectrum.

10.3 Some Typical Spectra

Cosine Wave Different waveforms and their frequency spectra have been calculated using the program shown in Table 10.1. The waveforms could be controlled through proper selection of the parameters in the statement 151: L = amplitude of the signal; G = frequency; H = dumping coefficient for exponential part of the waveform. Note that this three parameters are used to calculate the signal, statements 170 and 221.

The cosine wave has a frequency G that is four times higher than the basic analyzing frequency W. The parameter are $L = 1, G = 4, H = 0$. The signal is shown in Figure 10.3a, and its frequency spectrum is shown in Figure 10.3b. Note that the unit step of the abcisa is W, whereas the basic frequency of analyzed signal is $4 \cdot W$. Hence the unit step of the abcisa is one fourth the basic frequency of the signal. Obviously, only the basic frequency of the periodic cosine wave is found in the spectrum and it is displayed after four unit steps. This is the simplest discrete spectrum, showing only basic frequency G of the signal.

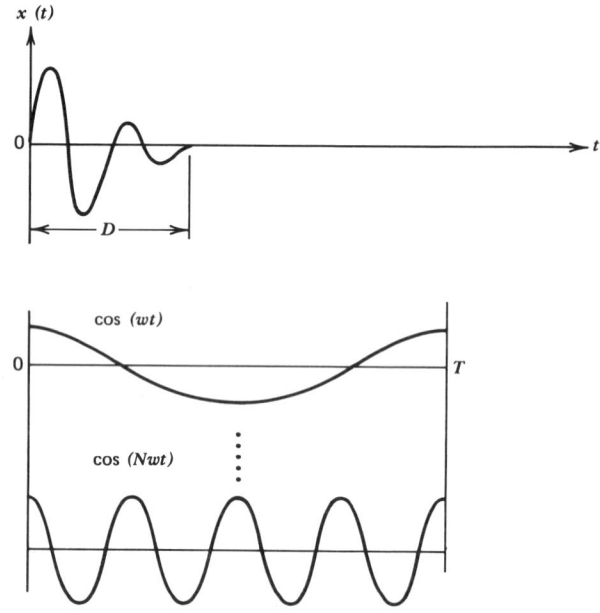

Fig. 10-2 Nonperiodic waveform. Theoretically the Fourier integration period T should be infinity. Usually it is enough to take $T = (5 \div 10) \cdot D$.

Dumped Cosine Wave The parameters are $L = 30, G = 4, H = 0.25$. The signal is shown in Figure 10.4a, and its frequency spectrum is shown in Figure 10.4b. Again, the unit step of the abcisa is one fourth the basic frequency of analyzed signal. Note that this time the spectrum is "continuous." The harmonics presented have frequencies $G/4, 2G/4, \ldots$. The spectrum shows the maximum at the frequency $4G/4 = G$. The ratio between the signal frequency and the unit frequency is G/W. In this example $G/W = 4$, and each basic frequency is displayed with four frequency unit steps. For larger values of G/W, more accurate frequency spectrum would be obtained, and one basic frequency range would be displayed with more points.

Exponential Wave The parameters are $L = 30, G = 0.1, H = 0.1$. The signal is shown in Figure 10.5a and its frequency spectrum is shown in Figure 10.5b. This is another example of "continuous" frequency spectra.

Pulse Waveform The pulse waveform is programmed in the following way

```
171 X (J) = 0
172 IF (J>25) GO TO 200
175 X (J) = 1
```

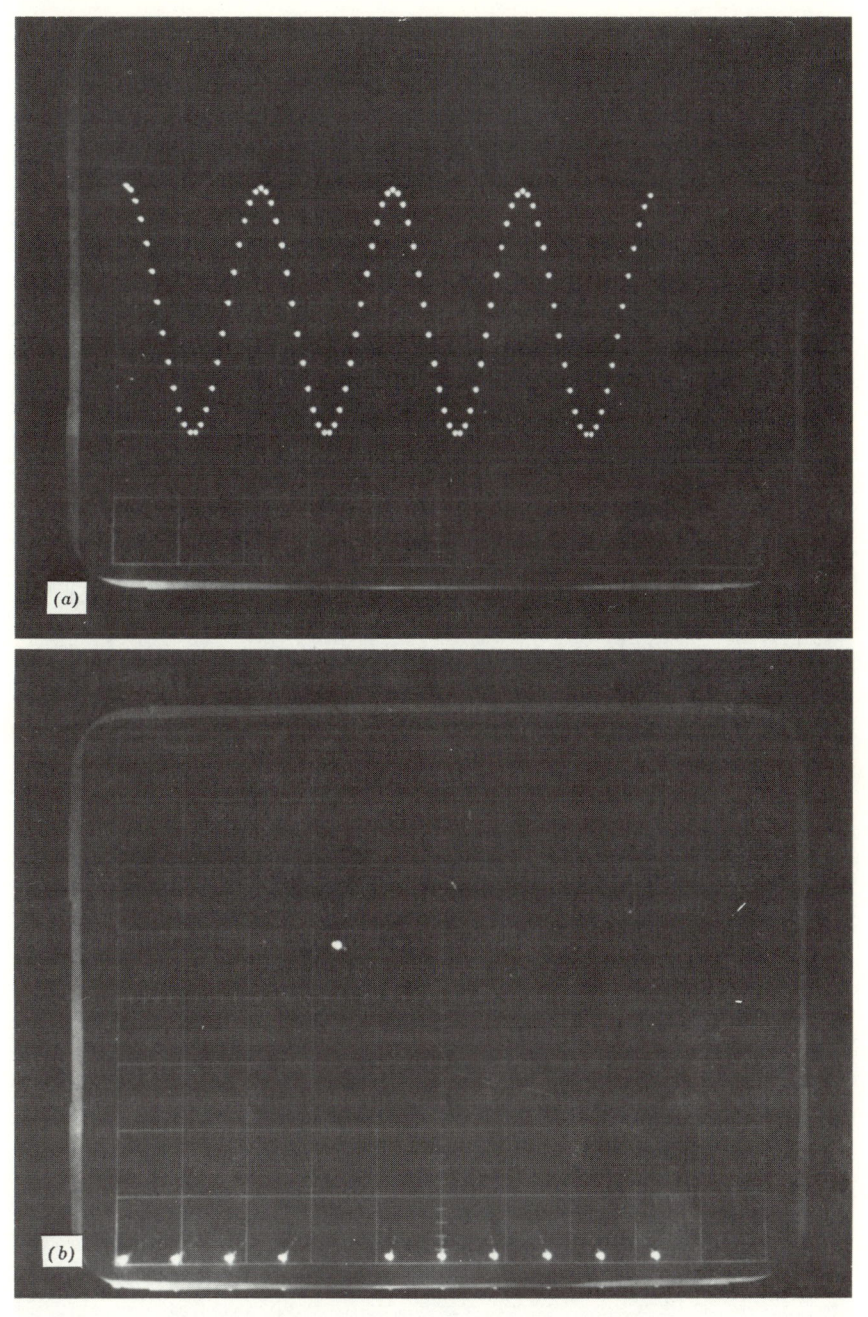

Fig. 10-3 Cosine wave analysis (*a*) Cosine wave at the the frequency $G = 4 \cdot W$, where W is basic analyzing frequency, (*b*) Fourier spectra showing the frequency $G = 4W$.

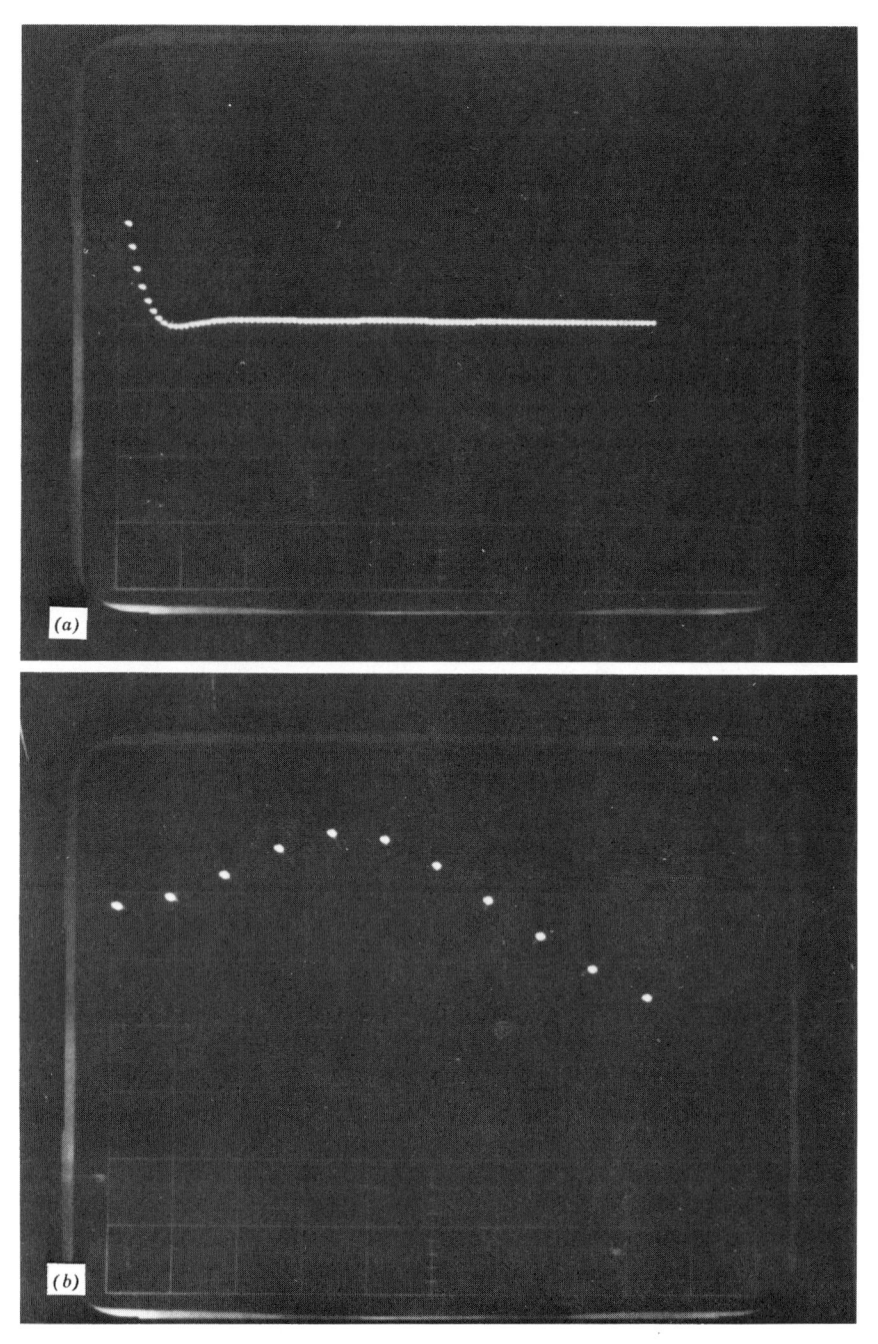

Fig. 10-4 (*a*) Damped cosine wave. (*b*) "Continuous" frequency spectra.

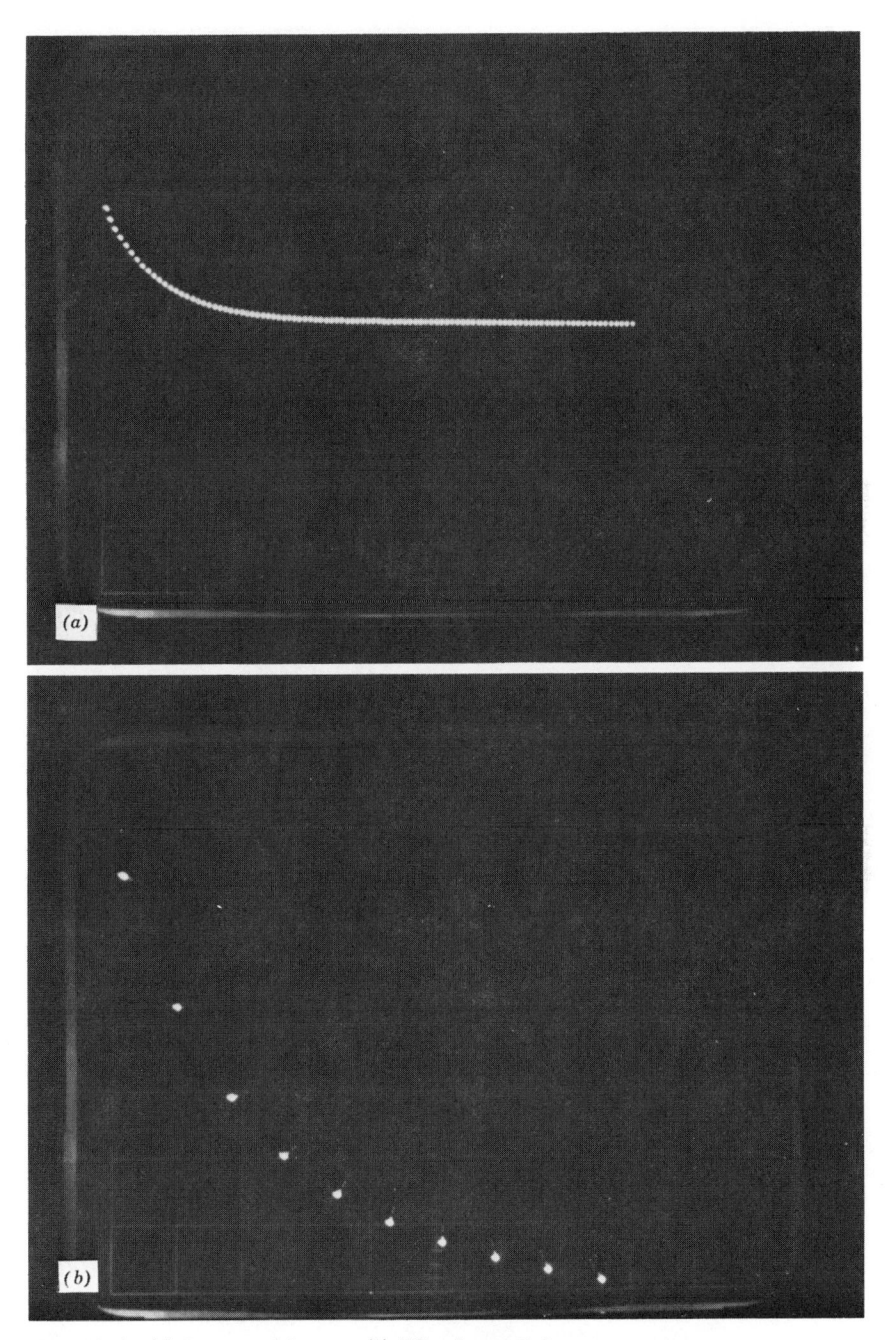

Fig. 10-5 (a) Exponential wave. (b) "Continuous" frequency spectra.

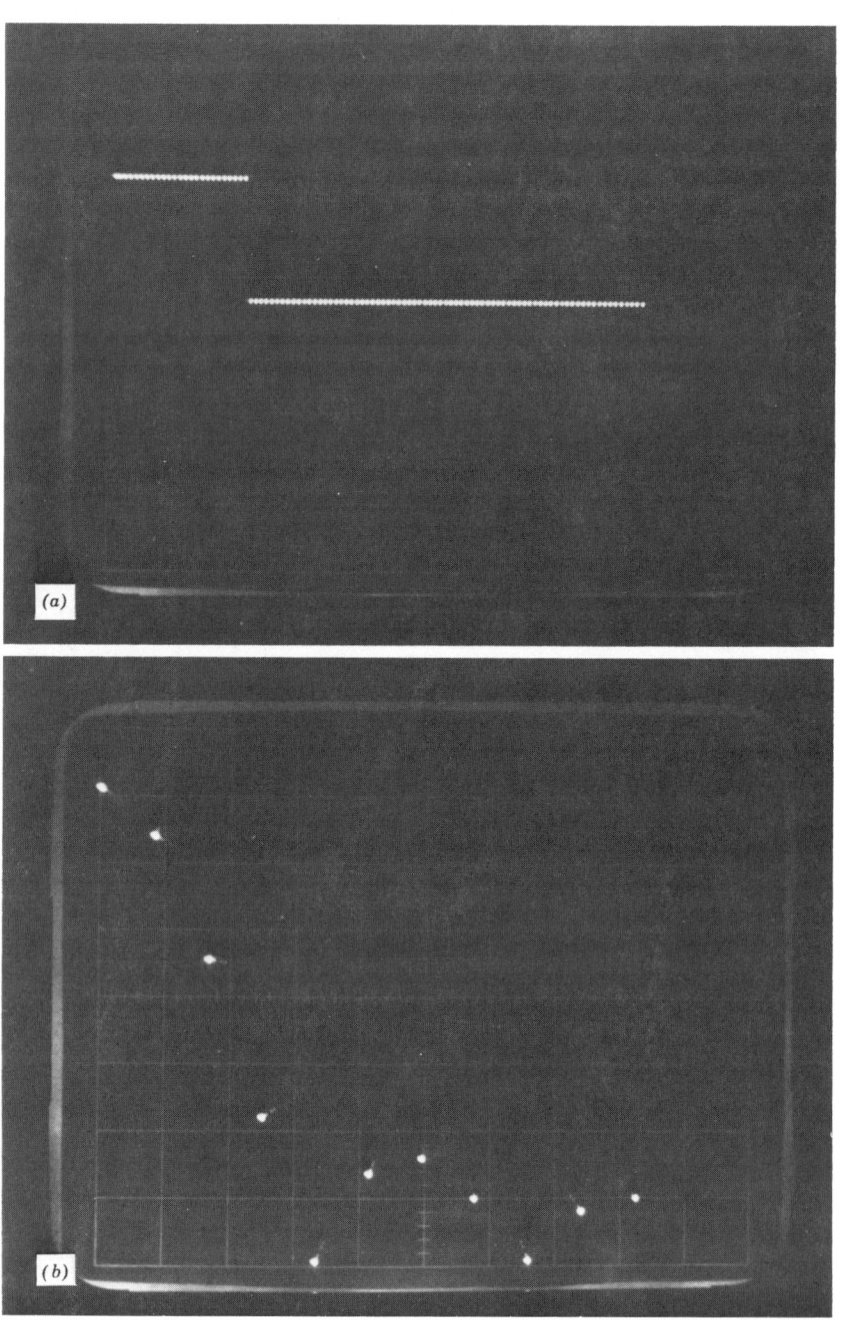

Fig. 10-6 (a) False waveform. (b) "Continuous" frequency spectra.

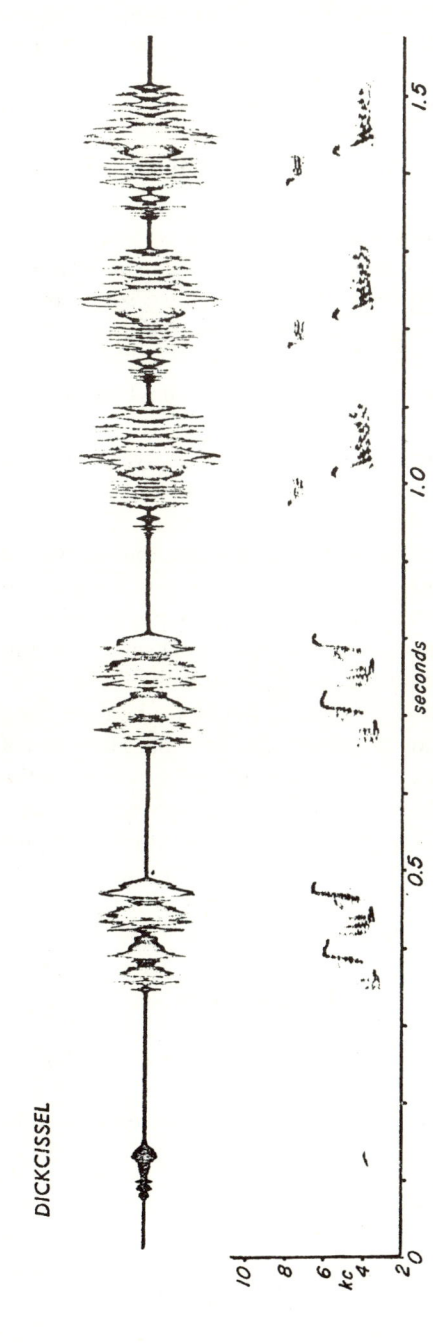

Fig. 10-7 Bird song. (*a*) Acoustical record showing the amplitude of the song as a function of time. (*b*) Sonagram of the same record as a function of time. (From C. H. Greenewalt: *Bird Song Acoustics and Physiology*, Smithsonian Institution Press, Washington, D.C., 1968.)

The waveform is shown in Figure 10.6a, and its frequency spectrum is shown in Figure 10.6b. This is another example of continuous frequency spectrum.

10.4 The Sonagraph Machine

The power spectra analysis is used in many areas of research. This kind of analysis presents also the basic tool in acoustic communication study. Typical examples are speech analysis and bird song analysis. Special machines have been developed for spectral analysis; one of the best known is sonagraph machine.

The instrument comprises a recording and playback unit providing storage on a single-channel magnetic drum of a sound sample 2.4 msec in duration. The basic part of the instrument is a wave analyzer covering the frequency range 0–8000 Hz. The analyzer has built in filter with bandwidth of 300 Hz. The center frequency of analysis changes linearly with time, 15 Hz per revolution of the recording drum. In this way the analyzer scans the waveform sample continuously, measuring the power in a given frequency window. The spectrum is then recorded on teledeltos paper, with spectral intensity roughly indicated by density of the gray-black marking. Figure 10.7 shows an example of a sonagram produced by this instrument.

Figure 10.7a shows a part of the bird song in the time domain: song waveform as a function of time. Figure 10.7b shows the sonagram of the same song. This is actually a three-dimensional display: the abcisa is time, the ordinate is frequency, and the darkness intensity is proportional to the spectral power. Hence, the sonagram is composed of a large number of power spectra, one for each 2.4-msec sample duration. The sonagram shows the variations in power spectra from one time sample to the next.

References

1. Bendat, J. S., and Piersol, A. G.: *Measurement and Analysis of Random Data*, Wiley, New York, 1966.
2. Korn, G.: *Random-Process Simulation and Measurements*, McGraw-Hill, New York, 1966.

PART II

COMPUTER MODELS OF NEURAL ACTIVITIES AND OF ANIMAL COMMUNICATIONS

B. Souček and A. D. Carlson

General Features of Animal Communication and Neural Processes and Their Computer Modeling, 187   Solo, Alternating, and Aggressive Communication, 203   Communication Based on Timing and Pulse Pattern Recognition, 224   Computer Simulation of Firefly Flash Sequences, 238   Neural, Communication, and Behavioral Sequential Patterns, 258   Communication Based on Frequency Pattern Recognition, 286   Models of Quantized Information Transmission on Neural Terminals, 300   Appendix, 317

Note

Figures, tables, and examples presented in Chapters 12 to 15 and in Chapter 17 are taken from our papers published in the *Journal of Theoretical Biology*. Courtesy of Academic Press, London.

Chapter 11

GENERAL FEATURES OF ANIMAL COMMUNICATION AND NEURAL PROCESSES AND THEIR COMPUTER MODELING

Introduction and Survey

The process of animal communication involves the exchange of information between animals of the same or different species in a code using one or more signal modalities. Communication can serve a number of widely varied functions. The investigator, observing the behavior of the communicating animals and recording the signals exchanged, attempts to determine what message has been transmitted and how it is translated from the signal code. The animals receive the signal with their sensory receptors and process the information at the receptor and at higher levels in the nervous system. Neural activity in the form of action potentials represents an internal communcation process, which utilizes a pulse-time code, and which can be analyzed in similar fashion to external communication systems. Communication systems have been shaped through evolution by the physical demands of the environment, the modifications of the anatomical and neurological substrates of the signal-reception process, and the behavioral patterns of the species.

This chapter describes the relationship between the investigator and the animals in the communication process. It discusses the functions that communication serves in controlling the behavior of animals. The modalities comprising visual, auditory, chemical, and tactile signals, as well as the patterns by which these signals form the communication codes, are described. The operation of the nervous system in receiving and processing signals is explained. Examples of the evolution of communication patterns are provided. Finally, the possible advantages and limitations that the computer brings to the analysis of neural and behavioral activities are discussed.

11.1 Description of Animal Communication Process

Communication is the process by which information is exchanged between individuals in an attempt to modify their subsequent behavior. Human language, perhaps, represents the ultimate in communication complexity and sophistication. Animal communication must fulfill many of the same functions that human language performs. This review restricts itself primarily to the analysis of communication between animals.

The partners in the communication process are usually animals of the same species, but the general alarm call of a bird to a circling hawk can alert birds of many different species within hearing and serve as a communication signal. The functions of animal communication are varied but are often tied to the reproductive process. Such functions include defense of breeding territory by males, sexual recognition, and stimulation of sexual partners.

The process of animal communication can be described by Figure 11.1, which outlines the relationships between the participating animals and the observer.

As shown in Figure 11.1, information is transmitted from the initiator (signaller) to the receiving animal in the form of a coded message. The signaller transmits the message in any one of a number of signal modalities such as sound (vocalization), visual displays, odors, or tactile stimuli. The signals, combined with noise that tends to degrade the message, are received by the sensory receptors of the receiver animal. The receptors are tuned to respond to a narrow range of stimuli, and they filter the signal and transduce or change it from its original modality to the electrical activity of the nervous system. If the recipient is in the proper behavioral state, the message is perceived and the animal's behavior is changed. In turn, the recipient's changed behavior may be perceived by the signalling animal which causes it to modify its behavior.

The investigator oversees the communication process mainly from two points: interception of the signal associated with noise and observation of the recipient's changed behavior. His job is to determine which components of the signal form the message; that is, which signal components have information value to the recipient. These components might include the frequency and structure of the signals or the relationship of the signal to the initiator's behavior. The investigator has a number of experimental options open to him. He can record both the signal and the behavioral response of the

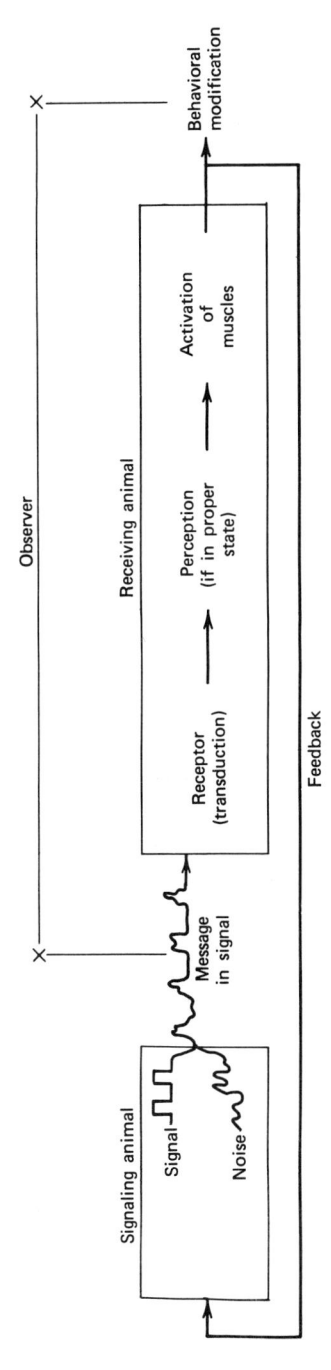

Fig. 11-1 Diagram of the communication process.

receiving animal, then by analysis of the signal elements to subsequent behavioral responses attempt to understand the message (break the code). He may record the signal, modify its parameters, give the modified signal to the receiver, and observe its effect.

The computer can aid this process because of its ability to manipulate large amounts of data in a short time. It can classify and analyze the recorded signals for patterns which may compose a message. It can correlate each possible signal parameter with a behavioral response. It can modify signal parameters and codes which can be used to test for possible informational content. This modification can occur during the course of the actual experiment itself. It can be used to design a functional model which mimics, with some level of fidelity, the communication process. From this model new experiments can be developed to further probe the communication system of the animal under study.

11.2 Functions of Animal Communication

The functions of animal communication are varied but are often tied to the reproductive process. Among animals that defend breeding territories, the resident males warn intruding males of the same species by signals. The redwinged blackbird displays its red epaulets, and male crickets produce a territorial call that repels other males. Communication is especially important between partners of the opposite sex. The entire courtship process is controlled by a series of signals between the male and female stickleback fish. Often vocalizations and visual displays are used to signal the readiness to mate of one of the partners. Head flicks by male ducks, mating calls by male frogs, and submissive postures by female monkeys, all serve a signal function. Odiferous chemicals called pheromones are released by female moths and can attract male moths of the appropriate species over wide distances.

Young animals produce signals that induce appropriate behavior in adults. The young herring gull pecks at a red spot on the bill of the adult, which induces the latter to regurgitate food for the young. Abandoned baby mice produce ultrasonic cries, which induce the adult to retrieve them. The grooming and fondling of young monkeys by adults functions to socialize the animals, which is an obvious form of communication.

Social insects communicate in a host of ways to ensure the success of the colony. Bees communicate the distance and direction of pollen sources by a complex dance on the hive surface. Foraging ants leave odor trails to indicate the location of food sources to other members of the colony. The role of odor signals in the organization of insect colonies appears to be vast, and we

Fig. 11-2 Two male ringnecked pheasant communicating at a territorial boundary (photograph by Douglas G. Smith).

and have just begun the process of analysis of this complex communication system.

Figure 11.2 shows the process of communication between birds. In nature this process may be obvious or subtle, but to the trained observer it is seen to occur constantly.

11.3 Modalities and Patterns of Communication

Stimuli of a certain kind elicit a sensory experience that can be defined with some precision and recognized whenever encountered. Each of these readily distinguishable classes is termed a sensory modality. In order to understand how the various modalities are utilized by animals, some general considerations of how the communication process shapes the signal patterns and the signals themselves may be helpful.

To establish the behavioral framework for analyzing the signal exchange between animals under observation two questions of importance are What is to be communicated? and What are the natural circumstances in which the signal is used? If the information is to code for species specificity or individual recognition the signals will probably have a high order of complexity and precision. If the message is merely a warning of imminent danger, the signal can be uncomplicated and variable. What parameters of a signal are most likely to be of communication value? Sensory receptors are adapted to respond most strongly to signal parameters that change rapidly with time; thus these parameters merit our principal attention.

Signal carriers have particular physical properties and are modified in known ways by physical aspects of the environment. Certain components of the signal are attenuated differently as they proceed from the source, which affects their efficiency as information-carrying units. The nature of the biological environment strongly affects how signals are used. In predator-prey interactions the predator is attempting to maximize the information coming from prey, while the prey is trying to minimize it. In sexual encounters both partners are presumably attempting to maximize the information exchange. The proximity of the communicating animals may strongly affect the signal parameters. Signals that are used over large distances tend to be stereotyped and discrete to overcome the degradation of the signal. At close range, species identification may have already occurred and the signals used can be graded with subtle variations. Animals that live in large, close-knit communities, such as social insects and colonial breeding birds, live in a signal environment of great density in which the noise covers exactly the same frequency band as that of the species signal emission. In these situations, described by Cherry[1] as the cocktail party effect, communicating animals must be tuned to each other in such a way that the extraneous background noise does not completely degrade the meaning of their signals.

Animals may use various parameters of a signal or qualitatively different signals for different parts of the message. By this means some components may be stereotyped, while others are graded and variable. In many cases signals exist in hierarchies in which one modality controls one aspect of the animal's behavior and another controls a competing aspect. For instance, the hen turkey will attack her devocalized chicks as predators in spite of their obviously youthful appearance.

The principal modalities of communication used by animals are visual (light), auditory (sound), olfactory (chemical), and tactile (mechanical stimuli). Auditory and visual communication have characteristics that lend themselves to computer analysis because of the ease of signal manipulation. Olfactory and tactile communication have been less adaptable to computer study.

Modalities and Patterns of Communication

Auditory Modality, Sound Signal Auditory communication involves the transmission of sound waves through a medium, usually air or water. The parameters of sound used in coding are

1. Modulation of frequency and amplitude.
2. Pattern of spectral energy distribution.
3. Coding of temporal pattern of emission.

Although vertebrate ears can discriminate sound frequencies (pitch), insect ears cannot perform frequency analysis, whereas they may be tuned to a particular frequency. Insects must use amplitude modulation and temporal patterning for acoustic communication.

The designs of the acoustic signal must take into account the physical properties of sound waves as well as the effect of the medium on their transmission. Higher sound frequencies attenuate more rapidly with distance, and signals that contain a spectrum of frequencies will change their spectral distribution with distance. Therefore animals that communicate by sound patterns of different spectral energy distribution must do so in close proximity. Sound attenuates most rapidly in dry air, which helps explain why desert animals develop large external ear structures. Water is the most ideal sound transmission medium, and whales are believed to transmit sounds up to 1000 miles under special ocean conditions. Localization of sound by binaural comparison of intensity or time differences is most effective with high and wide-ranging frequencies, but the high frequencies are most readily reduced by reflection, absorption, and diffraction in dense vegetation. To reduce this degradation of their signals birds sing from high perches or while on the wing.

Visual Modality, Light Signal Visual signals are used extensively by both vertebrates and invertebrates. The directional character of light provides necessary information on the location of the participating animals. Predator–prey interactions demand crypticity on the part of the prey in conflict with the need for conspicuousness in visual information exchange.

The variables of light stimulus that can be used in signaling are

1. Intensity or brightness.
2. Wave length or color.
3. Spatial pattern of stimulus delivery.
4. Temporal pattern of stimulus delivery.

Other than bioluminescent animals such as fireflies, which can generate their own light, visual signals utilize reflected light. The vertebrate eye, in particular, is designed to respond strongly to contrasting spatial patterns and rapidly changing intensity. Visual displays often involve movement of portions of the body designed to enhance visual responses.

Olfactory Modality, Chemical Stimuli Wilson and Bossert[2] have described the physiochemical features that would be affected in a chemical communication system. They have shown that the important signal characteristics of specificity and transmission over distance are strongly affected by molecular size. Below a minimum molecular size the signaling chemical loses specificity, whereas a large molecule loses volatility. Mixtures of chemicals used as signals can lose their information content because the particular compounds diffuse at different rates. The signal, therefore, changes composition with distance. Highly volatile substances quickly dissipate and thereby lose locatability. This is an obvious consideration for an ant leaving an odor trail. The chemical message must be sufficiently volatile to be perceived by other ants, but not so volatile as to dissipate before they can take advantage of it.

11.4 Neurological Signals

Information exchange within the nervous system occurs by electrical activity within and among cells called neurons. Neurons are composed of processes for receiving stimuli known as dendrites and usually a single transmitting element called an axon. Electrical activity is generated by changes in the flow of sodium and potassium ions that are at different concentrations across the neuron's membrane. Activity is spread by local, graded currents from the dendrites, across the cell body to the beginning of the axon. Here the currents may be enough to generate an action potential. This is a regenerating, all-or-none voltage change that is conducted at a constant velocity toward the axon terminals. Typical characteristics of an action potential in a moderately sized vertebrate axon would be 100 mV, of 1 to 2 msec duration and 5 m/sec velocity.

The axons of interneurons terminate on the dendrites of another neuron in a specialized junction called a synapse. The axon of motor neurons may terminate on muscle fibers in a similarly specialized junction called an endplate. At the nerve terminals the arriving action potential causes an increased rate of release of a substance called transmitter, which is contained in packets of uniform amount. The transmitter induces a potential in the postsynaptic membrane of the dendrite or muscle fiber and thereby transfers the electrical activity to the next unit. In the chain of elements that compose the nervous system the receptor or sensory neurons receive a signal, it is passed to interneurons which further process it, and the final result may be activation of a muscle by a motor neuron.

The continually regenerating action potential is useful for reliably transmitting electrical activity over distances in the central nervous system without appreciable degradation. The bulk of information processing is handled by graded, local potentials in the closely adjacent cells of the brain and lower centers. The nervous system uses a pulse-time code in which the action potential itself has little or no meaning; instead, the intervals between potentials carry the information. Computers are easily adapted to process the output of the nervous system in the form of the all-or-nothing action potential.

Figure 11.3 is a diagrammatic picture of a motor neuron (left side) and the oscilloscope display of three action potentials.

Figure 11.4 is a diagram of a giant Mauthner cell from the brain of a gold fish. This cell controls the flip of the animal's tail. This picture strongly emphasizes that the nerve cell can integrate large amounts of information from huge numbers of input sources.

11.5 Receptors and Neural Integration

Sensory receptors are transducers that convert the energy of the signal to which they are tuned into nerve activity, which is the mode of information exchange within the nervous system.

Receptors act as filters, selecting a particular parameter of the signal and rejecting other parameters. They can be highly sensitive to energies within their operating range. For example, the vertebrate eye responds to wavelengths of the electromagnetic spectrum from 400 to 760 nm and is so sensitive that as few as 54 light quanta incident on the cornea can be perceived. In the mammalian ear displacements of the tympanic membrane of 10^{-8} cm are effective. Not only are receptors sensitive within their response ranges, but they can respond over a wide range of stimulus intensities. In the receptor the stimulus normally induces a graded, generator potential that is proportional to the stimulus intensity. The generator potential in turn induces action potential discharge of the receptor cell neuron, the frequency of which is proportional to the generator potential. In general, most receptors adapt quickly to a constant stimulus so that their action potential frequency declines with time. Signals that change rapidly in time and space are enhanced. Visual receptors give enhanced responses to strongly contrasting patterns.

Electrophysiological studies of receptors that utilize carefully controlled stimulus conditions can aid in the analysis of their signal processing capabilities. Direct examination of the receptor can set the outer limits of its response sensitivity. It cannot determine which signal parameters are of importance in the message, because further processing occurs in the nervous system. In mammals much of the information received by the receptors is passed on with little integration to the highest brain centers. In lower verte-

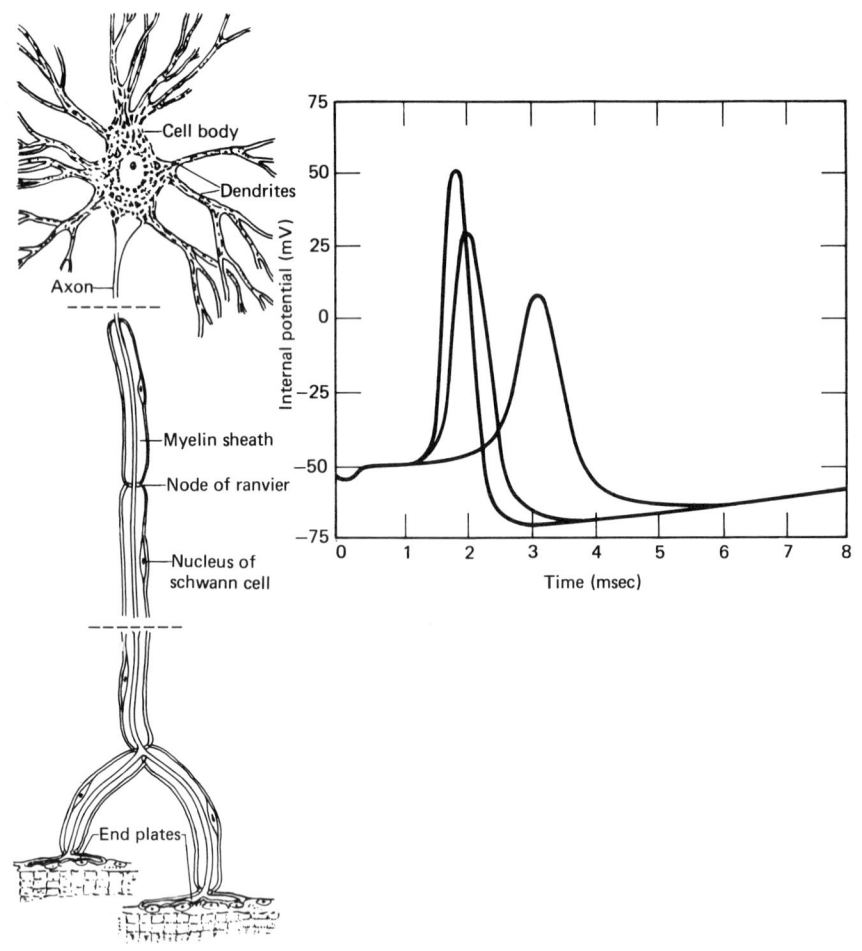

Fig. 11-3 Left. Motor neuron. The cell body fans out into a number of twigs, the *dendrites*, which make synaptic contact with other nerve fibers. Nerve action potentials arise in the initial, unsheathed region of the axon and travel to the end plate which is embedded in the muscle fibers. Transmitter is released at the end plate to activate the muscle fiber membrane. (From *How Cells Communicate*, by B. Katz, copyright © Sept. 1961, by Scientific American, Inc. All rights reserved.) Right. Three action potentials recorded on the oscilloscope screen. Potential (A) in axon with normal ionic medium. Potentials from axons in which a quarter (B) and half (C) of the potassium was replaced with sodium. (From *The Nerve Axon* by P. Baker, copyright © March, 1966, by Scientific American, Inc. All rights reserved.)

Receptors and Neural Integration

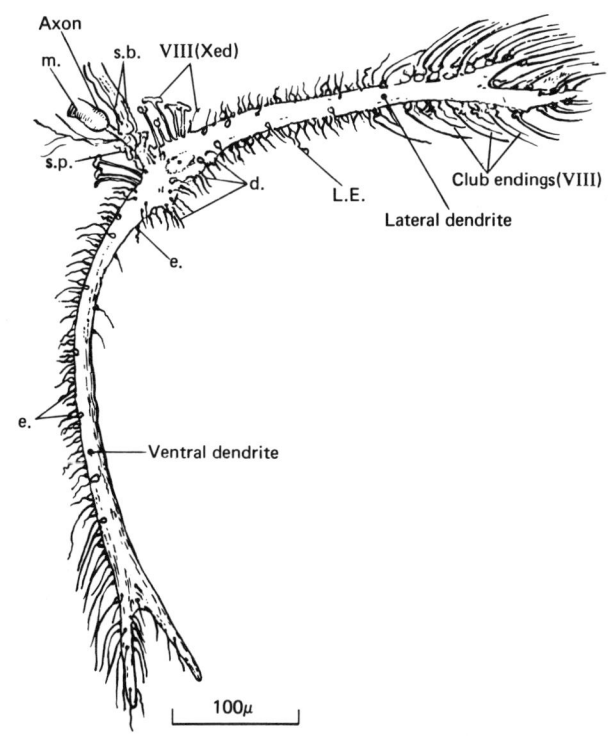

Fig. 11-4 Mauthner cell from brain of gold fish. An example of mosaic segregation of different types of synaptic endings from different sources on specific parts of the dendritic zone of a neuron. M-myelin sheath of Mauthner cell axon, s.b. - bundle giving origin to spiral fibers, h-axon hillock, d-small dendrites, e-small end bulbs. (From Bodian, D., in *Cold Spring Harbor Symp. Quant. Biol.*, 17, 1952).

brates and invertebrates the signals are highly processed before reaching the brain. This fact in part explains why insects, fish, and frogs can be induced to respond to bizarre models and inappropriate signals.

How the brain analyzes the information from its receptors and how it generates particular behavior in response to the signal is poorly understood. Observations on crickets and chickens suggest that neurological control centers exist that program each call of behavioral significance. Stimulation of the brain in particular locations can induce the animal to make a call of a certain type which is complete in all its components. Furthermore, the same stimulation induces the animal to adopt the appropriate postures and movements associated with that call. In higher mammals, however, the control centers are apparently more diffusely organized because localized stimuli do not induce entire behavioral patterns, which probably reflects the more variable, less stereotyped nature of mammalian behavior.

Although control centers of some organization do exist, behavior is not completely present at birth but changes with development. Particular signaling patterns can be varied by experience, and in many cases autofeedback is required in order for the animal to correctly shape its signal. Certain birds deafened at birth are no longer capable of developing the complete structure of the species-specific song and emit only the inbred portion which is almost unintelligible.

11.6 Evolution of Communication Patterns

The communication of animals is closely tied to their survival. It plays a crucial role in reproduction, predator–prey relationships, and social organization. The communication process should therefore be strongly influenced and directed by selection. How selection acts on the patterns of communication has been difficult to assess, because the signaling and receiving animals act in very different ways while communicating. The nervous system of the signaler generates a motor act that produces a signal. The receiver accepts the signal and processes it with its nervous system. One would anticipate that an animal that transmits a modified signal would be selected against because that signal would be an inefficient information carrier. The receiver that develops a change in its nervous system which modifies the analysis of a species specific signal should also be selected against because it is less efficient at deciphering the message. The observation of Hoy and Paul[3] on communication between hybrid crickets suggest that the genes affecting the neural elements that control signal production by the male and signal recognition by the female are in some way very closely associated. They found that the hybrid female preferred the hybrid male's song. It is possible that hybrids with modified signals are not as strongly selected against in nature as previously supposed and that selection can act on the overall communication process rather than only on its individual elements.

11.7 Computer Modeling of Neural and Behavioral Activities

The goal of analysis of neural and behavioral activities is to understand the underlying processes that guide and control these activities. Although behavioral activities can be described in completely nonneural terms, the nervous system is the underlying control, and explanations that ignore this fact have little regard for reality. An understanding of neurophysiology at the neuron level and at the system level is crucial to a realistic explanation of behavior. In the words of Harmon and Lewis,[4] "One may profitably explore the information-processing aspects of neural activity at several levels. One may seek understanding at the subcellular level, dealing with molecular organization, ionic dynamics, and membrane mechanics. At another level, one may consider questions pertaining to cellular input-output functions, seeking understanding of information transfer in single-unit action; this is the domain of signal integration and transmission. At a still higher level it is useful to explore the function of cell assemblies and networks. Further, one might wish to investigate the holistic properties of signal generation, interaction, and propagation by viewing gross electrical activity. Finally, it is important to attempt to understand the entire organism from a behavioral point of view."

The use of models has been enormously helpful in understanding the operation of the nervous system and behavioral interactions. By their use the investigator can develop functional analogs that have been simplified to essentials, which can be treated to mathematical analysis, which can be manipulated experimentally with greater ease and from which new relationships can be predicted. Inherent in any model, however, are underlying simplifying assumptions that may be incorrect or only superficially valid. As long as one realizes the limitations inherent in these analogs of the real world, models are essential tools in the analysis of neural function and behavior. The digital computer, with its great speed and storage capabilities, is a device of almost unlimited potential for modeling the massive amounts of information produced by nervous systems. Neural modeling with the use of computers is just beginning, but its promise for the future is assured.

One of the first attempts to develop a model neuron is the logical or McCulloch-Pitts neuron.[5] It is a device that produces an output if it receives a certain number of inputs. Threshold (θ) is the minimum number of inputs to produce an output, and in the simplest model it is a constant positive integer for that neuron. The neuron can only change its state at one of a discrete series of equally spaced times. This assumption allows one to predict its output at a known time. A network of these logical neurons acts in synchrony because they all have the same starting time and response period. A logical neuron is a binary device because it has two possible states (0 for inactive and 1 for active). It is possible to analyze the operation of a simple network, as shown in Figure 11.5, using binary notation and the use of the computer in this process is very evident.

The logical neuron has been criticized as being too unrealistic, particularly with respect to its fixed-time dependence. In real-time neurons the state of polarization (P) or membrane voltage (V) is used in predicting the time of firing (output) (see Figure 11.6). Some of the delays involved in synaptic transmission and the incremental development of the voltage toward threshold are accounted for in this model. Although this model can

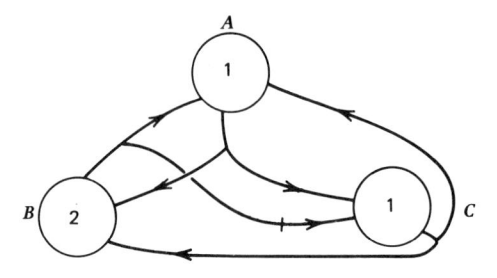

Fig. 11-5 Network of three logical neurons which can be used to mathematically study the operation of the nervous system. (From Griffith, J. S., *Mathematical Neurobiology*, Academic Press, New York, 1971).

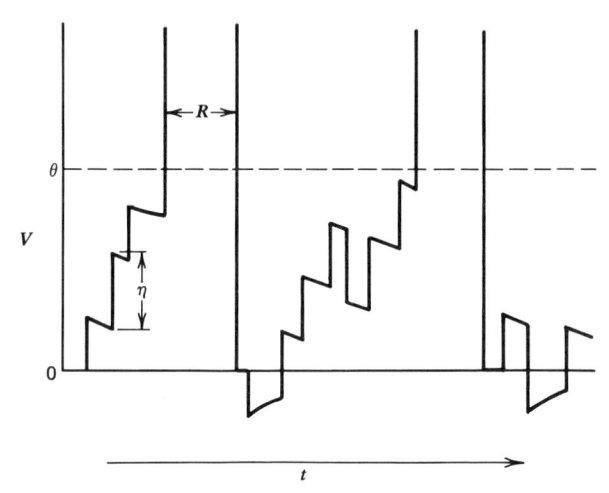

Fig. 11-6 Time course of V for a real-time logical neuron firing two action potentials and assuming reset to zero after each action potential. 0-resting, nonfiring level of neuron voltage. θ-Threshold level for firing an action potential. R-refractroy period during which no action potential can be fired. η-Increment of excitatory postsynaptic potential pushing voltage toward threshold (increments pushing voltage toward and below zero are inhibitory postsynaptic potentials). T-time (from Griffith, J. S., *Mathematical Neurobiology*, Academic Press, New York, 1971).

Fig. 11-7 (*facing page*) Mapping of wave form of postsynaptic potential by cross-correlation histogram. *Above:* Superimposed tracings of intracellular records of visceral ganglion neuron of *Aplysia claifornica*. Stimuli were randomly timed electric shocks to the presynaptic nerve. Records show excitatory postsynaptic potentials that occasionally trigger impulses. *Below:* Cross-correlation histogram between presynaptic events (shocks to presynaptic nerve) and postsynaptic action potentials. Note that most postsynaptic action potentials were produced about 0.125 sec after the presynaptic shock delivered at 0 time. The latency, rise, and fall times of the peak in the histogram are able to map the time course of postsynaptic activity found from direct recordings. Also note the presence of "noisy" background impulse activity in the postsynaptic cell. (From Perkel, D. H. and Bullock, T. H., *Neurosciences Research Program Bulletin*, 6, 1968.)

0.5 sec

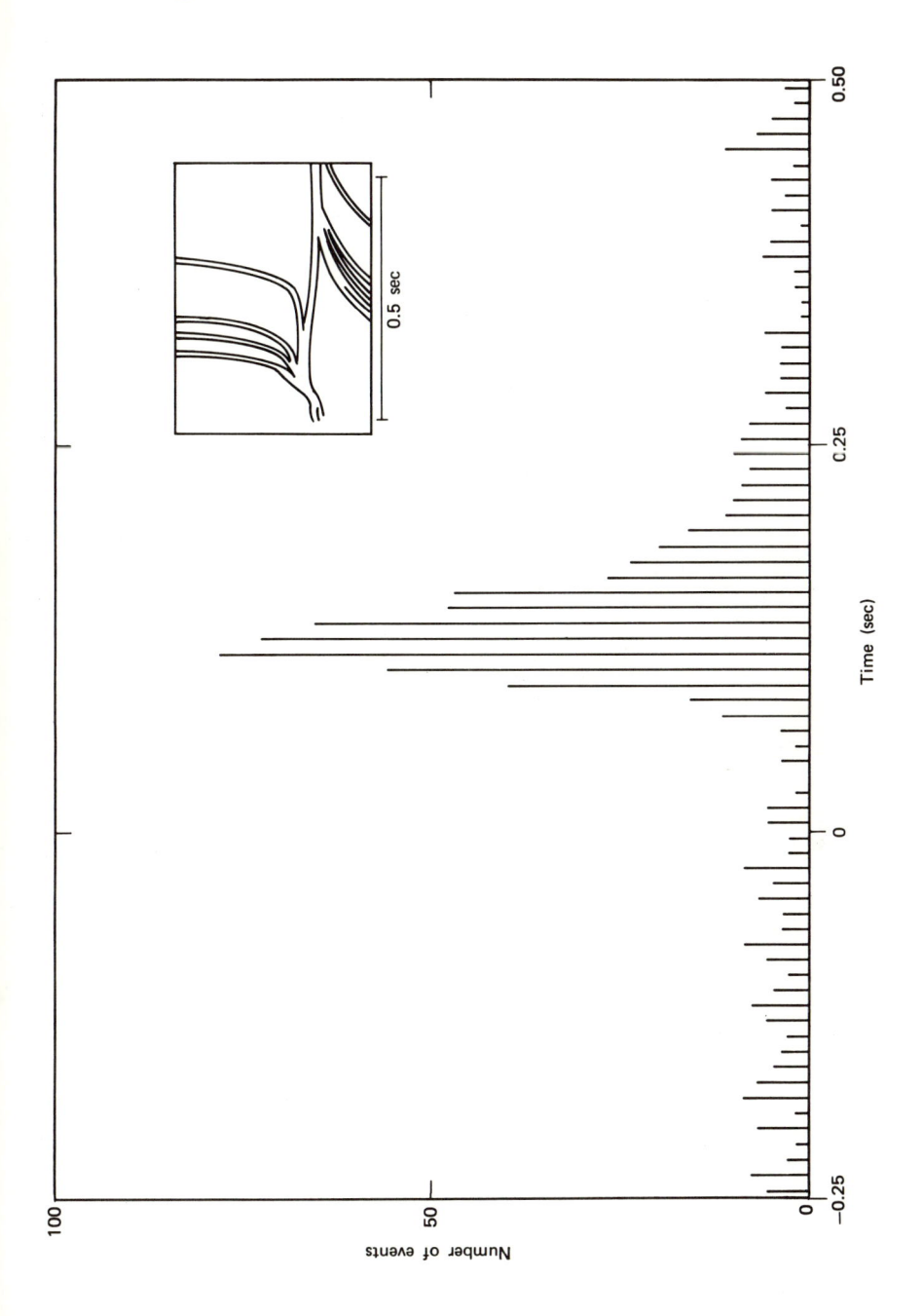

more closely approach the functioning of real neurons in complexity, its use in circuit analysis is still possible by computer.

The nervous system, composed of neurons whose individual behavior can be predicted, must operate as a functioning whole. Is there a neural code or a group of codes by which information is represented and transformed within the nervous system? A large number of possible neural codes are given in Perkel and Bullock,[6] and an example of how the computer can aid in the analysis of simple neural events is shown in Figure 11.7.

At the level of behavior, the investigator is dealing with nerve networks whose intimate operations are unknowable. A functional model is the best that can be constructed, and this model treats the nervous system like a black box or input-output system that has certain characteristics. This constraint does not imply that the model is any less useful toward the eventual understanding of the physiological basis of the behavior. If the properties of the nervous systems have been strongly considered in development of the model and it is rigorously tested with those properties in mind, the model may give great insight into the behavioral process. If the model, however, has little predictive value, whether developed with the use of the computer or not, it can only aid as a framework for organizing the observations and as such will not have advanced understanding to a significant degree.

References

1. Cherry, C.: On Human Communication, Wiley, New York, 1957.
2. Wilson, E. O. and Bossert, W. H.: Chemical Communication Among Animals, *Recent Prog. Hormone Res.*, *19* (1963) 673–716.
3. Hoy, R. R., and Paul, R. C.: Genetic Control of Song Specificity in Crickets, *Science*, **180** (1973) 82–83.
4. Harmon, L. D., and Lewis, E. R.: Neural Modeling, *Physiol. Rev.*, **46** (1966) 513–592.
5. McCulloch, W. S., and Pitts, W. H.: A Logical Calculus of Ideas Immanent in Nervous Activity, *Bull. Math. Biophys.*, **5** (1943) 115–133.
6. Perkel, D. H., and Bullock, T. H.: Neural Coding, *Neurosciences Research Program Bulletin*, **6** (1968) 221–348.
7. Marler, P.: Visual Systems, in *Animal Communication* (T. Sebeok, Ed.) Indiana University Press, Bloomington, 1968.
8. Busnel, R.-G.: Acoustic Communication, in *Animal Communication* (T. Sebeok, Ed.) Indiana University Press, Bloomington, 1968.
9. Mountcastle, V.: Neural Mechanisms in Somesthesia, in *Medical Physiology*, Vol. 1, 13th edition (V. Mountcastle, Ed.), C. V. Mosby Company, St. Louis, 1974.
10. Busnel, R.-G.: *Acoustic Behavior of Animals*, Elsevier, New York, 1963.
11. Konishi, M.: Evolution of Design Features in the Coding of Species-Specificity, *Am. Zoologist*, **10** (1970) 67–72.
12. Griffith, J. S.: *Mathematical Neurobiology*, Academic Press, New York, 1971.

Chapter 12

SOLO, ALTERNATING, AND AGGRESSIVE COMMUNICATION

Introduction and Survey

Alternating and aggressive communication patterns are present in many species. A theory has been developed that explains these two modes of communication. The theory covers both central neural control as well as phonic interaction between two partners. A concrete example (katydid chirping) is shown in detail. Upon receiving acoustical stimuli from the partner a male katydid generates a characteristic response function with three parts, which regulates solo-overlapping chirps, partially delayed chirps, and alternating chirps. Each partner is considered as an element in a closed feedback loop and is described through its response period versus stimulus period curves. It is shown that in alternating mode the communication loop converges toward a stable operating point, whereas in aggressive mode stability is never achieved. Based on the theory, computer models have been designed simulating both deterministic and random components of the communication signal. Computer-simulated chirping sequences are in excellent agreement with field measured data.

In this chapter two functions are introduced for the first time, following Souček[1]: response function and transfer function.

Response function describes the inherent built-in timing program. Upon receiving acoustical stimuli, the animal generates the response function, which determines the phases and the timing of the behavior to follow.

The transfer function describes the input-output relationship, the magnitude of the stimulus versus magnitude of the response. Transfer function is very useful to describe the interaction between two partners or elements in communication loop. Each partner or element could be described through its transfer function. Simple plotting of two transfer functions on the same diagram could be used to explain different patterns of the communication feedback.

12.1 Example of Katydid Chirping

Basic Definitions Adult males of the northern true katydid *Pterophylla camellifolia* (Fab.) (Orthopteria; Tettigoniidae) produce three kinds of sound, termed calling, aggressive, and disturbance signals.

The calling song consists of two to four pulse chirps, Figure 12.1a, and its assumed function is to attract mature females.

If two male katydids are close enough they will respond to each other's songs. One male will be the leader and will increase its chirp rate. The other male, the follower, will try to sing in such a way that the chirps of the two will be in alternation, Figure 12.1b.

The aggressive sounds are made up of multipulsed chirps, Figure 12.1c. The leader by increasing its chirp rate has started an acoustic "war" with the follower. The follower squeezes his chirps in between the leader's, which results in many overlapping chirps. Eventually the acoustically "defeated" katydid might leave the area.

The disturbance sounds consist of very short chirps produced at irregular rates, and are produced whenever male or female katydids are handled, Figure 12.1d.

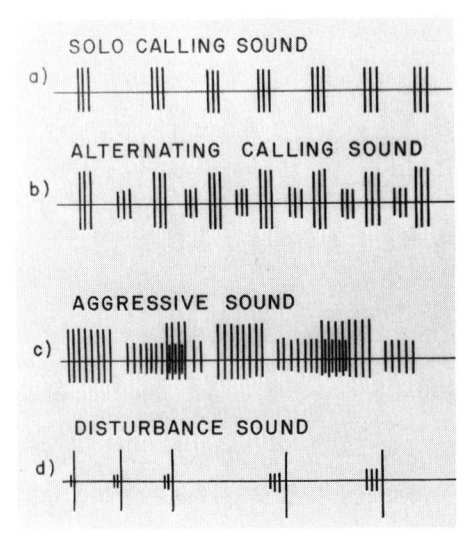

Fig. 12-1 Solo, alternating aggressive and disturbance sounds in northern true katydid *Peterophylla camellifolia* (Fab.) (Orthoptera: Tettigoniidae).

Example of Katydid Chirping

Alternating and aggressive chirping can be explained through some kind of excitatory and inhibitory effects of the sound of one katydid on another, as suggested by Alexander[2] and Jones.[3] Shaw[4] has produced large numbers of measurements, concentrating on the effects of one male's song on another male's song. Huber,[5] Ewing and Hoyle,[6] Jones,[3] Wilson,[7] Shaw,[4] and Heiligenberg[8] have suggested that acoustical output of crickets and katydids is controlled by a neuron oscillator or pacemaker. The pacemaker should be sensitive to sensory inputs, which could turn it on or off.

Recently theories have been developed to explain firefly flashing communication,[9] and also to explain bird dueting.[10] Although there are great differences between katydids, fireflies, and, especially, birds, there are also some similarities in their communication patterns.

In this chapter katydid alternating and aggressive chirping are analyzed. A theory is proposed to explain solo calling, alternating calling, and aggressive sounds. Based on this theory a computer model is designed to simulate katydid chirping sequences. The experimental findings used for the model and the theory are based mostly on the data measured by Shaw.[8] For this reason the definitions of parameters are the same. These definitions are explained in Figure 12.2.

The time measured between beginnings of successive chirps will be called the chirp *period*.

The time measured between the end of one chirp and the beginning of the next will be called a chirp *interval*.

The leader katydid's chirp presents a stimulus to the follower katydid. The follower's chirp reciprocally stimulates the leader.

The time between a stimulus chirp and a responder's chirp is called the *response* period or interval.

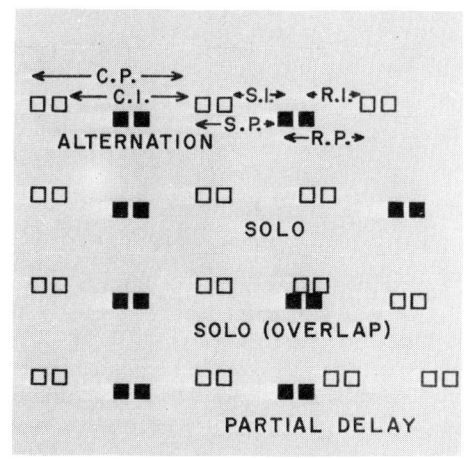

Fig. 12-2 Definition of periods and intervals: CP, chirp period; CI, chirp interval; SP, stimulus period; SI, stimulus interval, RP, response period; RI, response interval. Open box—katydid chirp. Dark box—Stimulus chirp.

The time between a responder's chirp and the next stimulus chirp is called the *stimulus* period or interval.

Review of Experimental Findings The most extensive experimental data dealing with the phonoresponse of the true katydid have been published by Shaw.[8] To find the basis for a theoretical model, it was necessary to carefully examine this experimental data. The basic findings are as follows:

1. A single male produces solo calling song at approximately a constant average rate and with a chirp duration of about 0.2 seconds (temperature 24 to 28°C).

2. Two males produce sounds in alternation. Excitatory and inhibitory effects of the sound of one katydid to another regulate the alternating chirping sequence.

3. The leader's chirp period and response period are functions of the stimulus period produced by the follower. The measured data are presented in Figure 12.3a.

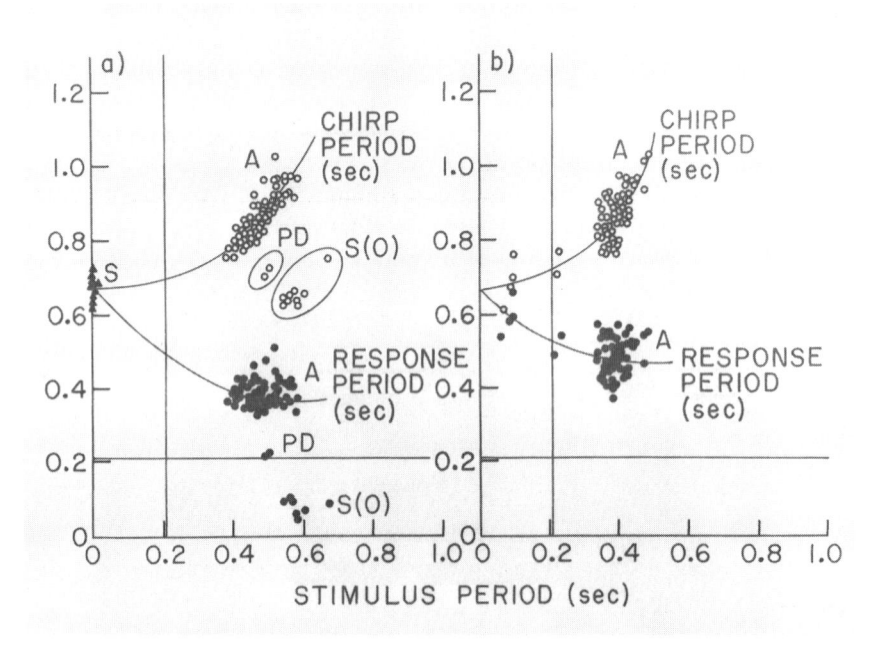

Fig. 12-3 Graphic analysis of acoustical interaction during alternating calling by a pair of *P. camellifolia* males, as measured by Shaw.[4] (a) leader, (b) follower; the letters on the graphs mark the groups of points corresponding to the following types of acoustical responses: A, alternation; S, solo; PD, partial delay; S(O), solo involving overlap of a follower's chirp and a leader's chirp.

Example of Katydid Chirping

4. The follower's chirp period and response period are functions of the stimulus period produced by the leader. The measured data are presented in Figure 12.3b.

5. The mean response interval of the leader is shorter and less variable than the mean response interval of the follower.

6. The response period of the leader in some cases is only slightly longer than the duration of the follower's chirp, which means that the leader in some cases could start chirping only slightly later than the follower. Such cases are called partial delay and are denoted with PD in Figure 12.2 and in Figure 12.3a.

7. No cases of partial delay have been found in the measured data for the follower.

8. The response period of the leader in some cases is shorter than the duration of the follower's chirp. That means that the beginning pulses of the leader's chirp are overlapped with the ending pulses of the follower's chirp. Such cases are called solo overlap, and denoted with S(0) in Figure 12.2 and in Figure 12.3a.

9. Two katydids chirp mostly in alternation. From time to time they lose the antiphony and enter partial delay and solo overlap mode. After one or more overlapping chirps, the alternating antiphony is achieved again.

10. The first solo chirp interval is the shortest interval of all.

11. If the distance between two katydids is shortened (i.e., the sound is greater), they will enter the aggressive mode of chirping. It seems that for entering into aggressive mode, not only acoustical but also other kinds of nonacoustic stimuli are needed. Aggressive chirps have a longer duration (up to 0.6 sec or more).

12. In the aggressive mode the percentage of alternations is decreased, and the percentage of solo overlaps is increased.

13. In aggressive mode the mean chirp interval of solos is about one half of the mean chirp interval of solos in alternating calling.

14. In aggressive mode the response intervals of alternations of the leader are somewhat longer, and of the follower somewhat shorter, than in alternating calling.

15. If the katydid is stimulated with electronically produced chirps, the chirp duration can be varied. There is a shortening of the response interval if the chirp duration is longer.

16. If the stimulus is short, the stimulus intensity has no effect on the response interval.

If the stimulus is long (say 0.6 sec.), the decrease of stimulus intensity to or below 55 Db, shortens the response interval almost to zero. The sixteen effects listed have been used as a basis for the theory and computer modeling.

12.2 Response Function

Alternating Chirping Figure 12.2 presents four possible combinations of the leader-follower acoustical interaction: alternation, solo, solo overlap, and partial delay.

Alternation is the most frequent mode of interaction. The leader's chirps are roughly halfway between the follower's chirps, and vice versa.

In the solo mode the leader chirps one or more times during a very long follower chirp period. In a normal alternating calling song, only a leader produces solo chirps (by definition).

In the solo-overlap mode, the leader starts chirping before the follower's chirp end, and the two chirps partially overlap. Again, as shown in Figure 12.2, the follower's chirp period is greatly extended.

In partial delay the leader chirps almost immediately after the follower. In normal alternation only the leader sings in partially delayed mode.

The model must explain all these modes of acoustical interactions. Three functions are the basis of the model: time function, threshold, and response function, Figure 12.4.

The time function is started every time the male katydid chirps. It could be explained by any integrating process that could be used to measure the time elapsed after the chirp.

The threshold level is compared with the time function. In solo chirping, when the time function reaches the threshold level, a new chirp is produced. As a result the time function is reset and a new process is started. By adjusting the threshold level, the chirp rate can be controlled.

The response function is generated in the male katydid when receiving an acoustical stimulus from another male. The response function is crucial for the explanation of both alternating and aggressive sounds. The response function is added to the threshold, and in this way the crossover point with the time function is changed.

The response function has a spikelike shape, with three parts (see Figure 12.4). First, the negative part of the response function explains the solo-overlapping chirps. Second, the flat part explains partially delayed chirps. Third, the positive part explains the normal alternating chirps.

A typical normal alternating sequence is shown in Figure 12.4. The leading male chirps at time 0 and starts its time function. The follower in this example chirps 0.47 sec later and starts its time function. When the leader hears the follower's chirp, the leader generates the response function. The leader's response function starts at point L and intersects the time function at point L'. At that time (0.86 sec) the leader chirps again, resets its time function to zero, and starts new time function.

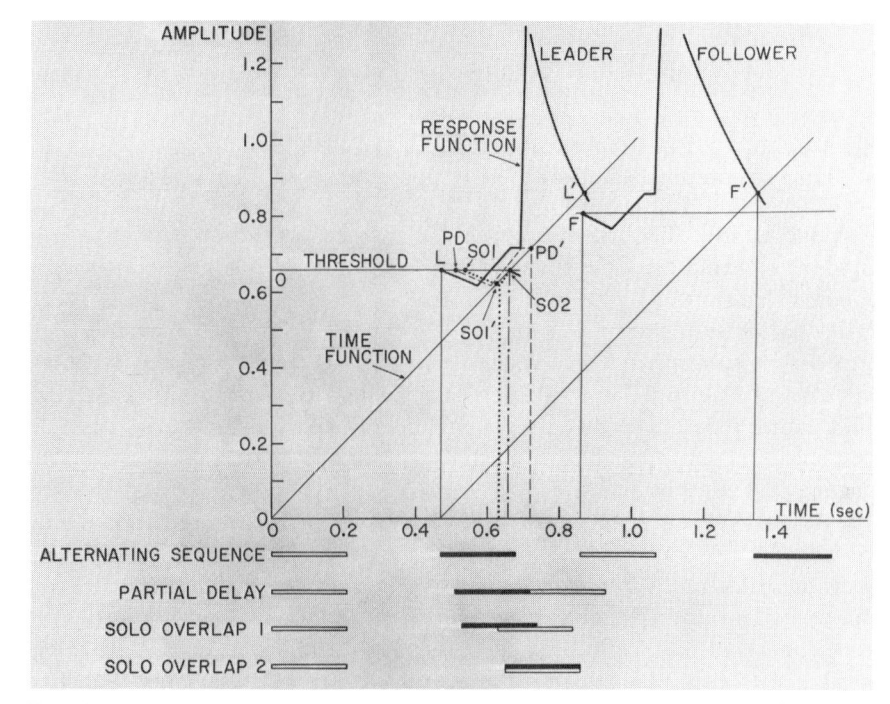

Fig. 12-4 Time function, response function, and threshold. Horizontal scale presents actual time, as observed in experiments. Vertical scale presents relative amplitudes.

The follower hears the leader's chirp and generates its response function. The follower's response function starts at point F and intersects the follower's time function at point F'. As a result the follower now chirps, and the whole sequence is repeated. The leader's response period is $LL' = 0.39$ sec. The follower's response period is $FF' = 0.47$ sec.

The period OL presents the stimulus period for the leader. If this period is approximately zero, the response function exponentially fades out to zero, and the time function intersects the threshold itself, producing the solo chirp period of 0.66 sec. As the stimulus period becomes longer (point L is moved right), the interaction point is on a higher part of the exponential function, and the chirp period is longer. This explains the exponential relationship between chirp period (OL') and stimulus period (OL) and also between response period (LL') and stimulus period (OL), as shown in Figure 12.3.

Without noise and disturbances, the leader and the follower would produce a stable alternating chirping sequence, as dictated by the two response functions arriving at the instants L and F, respectively.

In the computer model, the leader is called katydid #1, and the follower is katydid #2. Each of them has three parameters, which could be adjusted.

- Low level threshold L1 (leader) and L2 (follower).
- Amplitude of the response function A1 and A2.
- Time constants of the tail of the response function, T1 and T2.

Partial Delay Because of noise or disturbance, the standard alternating mode of chirping could be changed into partial delay, solo, or solo-overlap modes of chirping.

Partial delay is shown in Figures 12.2 and 12.3. Note the following: The leader's response period is very short, only slightly longer than the duration of follower's chirp. The resulting leader's chirp period is shorter than in alternating mode. The model explains these facts with the flat part of the response function. One case of partial delay mode of chirping is shown in Figure 12.4 (dashed parts).

The leader has started to chirp at time 0. Without any interaction, the next leader's chirp would occur at 0.66 sec, when the time function intersects the threshold. The follower chirps 0.51 sec after the leader. The leader generates its response function, which starts at point PD and intersects the time function with its flat part, at point PD'. As a result the leader's chirp period is 0.72 sec. This value is larger than 0.66 sec, but smaller than the periods produced in normal alternating mode. The bottom part of Figure 12.4 shows that the leader chirps immediately after the follower.

Note: As the length of the stimulus period OL increases from 0 to 0.5 sec the chirp period OL' becomes longer and longer. However, when the stimulus period OL becomes longer than 0.5 sec, the response function intersects the time function at point PD' rather than at point L'. The resulting chirp period O–PD' is substantially shorter.

Since the probability of partial delay mode of chirping is low, the flat part of the response function is short. In Figure 12.3 there are only two cases of partial delay.

Solo Overlap A case of solo overlap chirping mode is presented in Figure 12.4 (dotted parts). The follower's chirp occurs 0.53 sec after the leader's chirp. The leader generates the response function which starts at point $SO1$ and intersects the time function at point $SO1'$. The resulting chirp period is the shortest period of all, its value being only 0.63 sec. The bottom part of Figure 12.4 shows that the leader's chirp is overlapped with the follower's chirp.

The extreme case of solo overlap is the response function generated at point $SO2$. This function immediately intersects the time function, resulting in a chirp period of 0.66 sec (dash-dotted part of Figure 12.4).

Typical Sequence and Noise

The bottom part of Figure 12.4 shows that the leader's and the follower's chirps are completely overlapped.

The negative part of the response function is longer than the flat part; hence the cases of solo overlapping mode are more probable than the cases of partial delay.

Note: In solo-overlap mode, the response period is shorter than the duration of the follower's chirp. Produced chirp periods are the shortest of all. The first solo overlap period is only 0.63 sec. Other solo overlap chirp periods are between 0.63 and 0.66 sec. The solo-overlap cases are clearly distinguishable in Figure 12.3.

12.3 Typical Sequence and Noise

Table 12.1 shows the regions of interest, as read from Figures 12.3 and 12.4.

Typical observed sequences in alternating calls are as follows: Leader and follower are chirping in alternating mode. Because of noise or disturbance they enter partial delay mode. Partial delay is followed by the solo overlap mode, which is usually followed by the solo chirps of the leader. The next step is establishing again the normal alternating sequence.

The computer model behaves in the same way. Table 12.2 presents one typical sequence produced by the computer model. The disturbance has forced the sequence into the partial delay mode. Table 12.2 shows the standard transition path: partial delay, solo overlap, solo, alternation.

The same sequence is shown in Figure 12.5. The leader chirps at time 0. The follower chirps 0.50 sec later. The leader generates the response function

TABLE 12.1

(all periods in sec)

Mode	Stimulus Period	Response Period	Chirp Period
Alternation	0 to 0.50	0.66 to 0.38	0.66 to 0.88
Partial delay	0.50 to 0.53	0.20 to 0.23	0.70
Solo overlap	0.53 to 0.66	0.10 to 0	0.63 to 0.66

Leader's threshold $L1 = 0.66$
Leader's response function amplitude $A1 = 0.68$
Leader's time constant $T1 = 0.15$ sec.
Follower's threshold $L2 = 0.81$
Follower's response function amplitude $A2 = 3.1$
Follower's time constant $F2 = 0.1$ sec.

TABLE 12.2

Stimulus Period	Response Period	Chirp Period	Chirps
0.5	0.21	0.71	1
0.21	0.61	0.82	2
0.61	0.03	0.64	1
0.61	0.69	0.65	1
0.69	0.32	1.01	2
0.32	0.46	0.78	1
0.46	0.42	0.88	2
0.42	0.41	0.83	1
0.41	0.45	0.86	2
0.45	0.4	0.85	1
0.4	0.46	0.86	2
0.46	0.4	0.86	1
0.4	0.46	0.86	2

$L1$, which intersects the leader's time function 0.21 sec later, and leader chirps in partial delay mode. Now the follower generates the response function $F1$ which intersects the follower's time function 0.61 sec later, and the follower chirps in alternating mode. The leader generates the response function $L2$ which intersects the leader's time function 0.03 sec later, and leader chirps in the solo-overlapping mode. The leader starts a new time function. In the meantime the response function $L2$ exponentially fades out to zero, and the leader's time function intersects the threshold level 0.66 sec later. This results in the solo chirp by the leader. The follower hears the solo chirp and generates the response function $F2$ which intersects the follower's time function 0.32 sec later, and the follower chirps in alternating mode. The leader now generates the response function $L3$ which intersects the leader's time function 0.46 sec later, and leader chirps in alternation. Both leader and follower are now in alternating mode, and this mode will be repeated a number of times. A new disturbance could start the whole transient sequence again.

The sequence presented in Figure 12.5 and in Table 12.2 is generated by the computer model in the case of a "noiseless" environment. In actual biological systems there is always noise present. In the model the noise is added to the threshold level. Sequences produced by noisy computer models are presented in sorted form in Figure 12.6. A very good agreement between measured results, Figure 12.3, and model produced results, Figure 12.6, is obvious. In both cases one could clearly distinguish the regions of alteration,

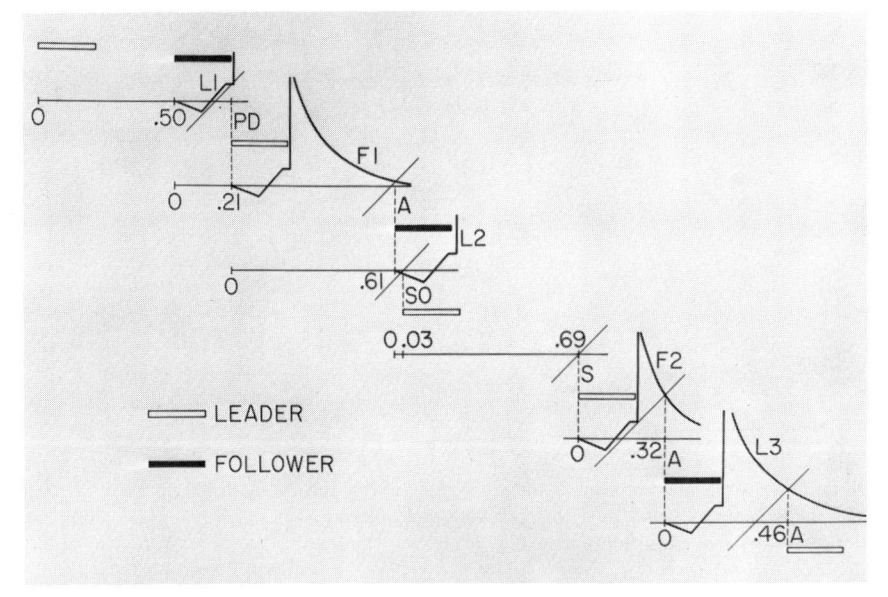

Fig. 12-5 Transition path: alternation; partial delay; solo overlap; solo; alternation.

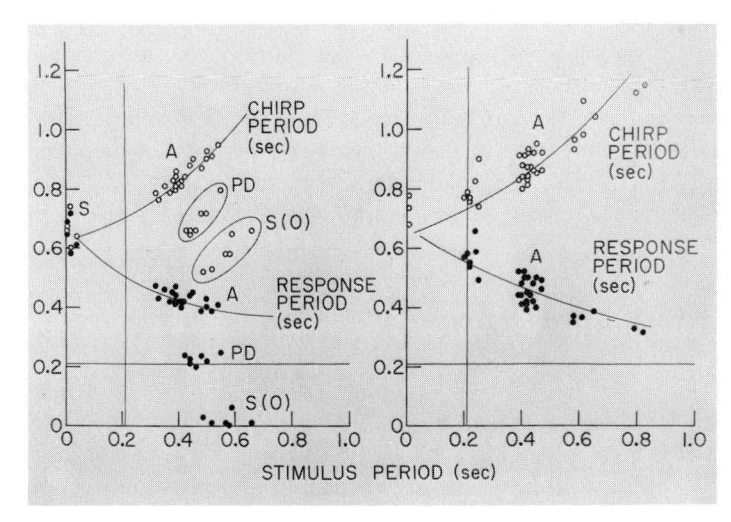

Fig. 12-6 Computer produced data stimulating acoustical interaction in alternating calling. Marked areas are identical to those found in experimental data presented in Fig. 12.3.

213

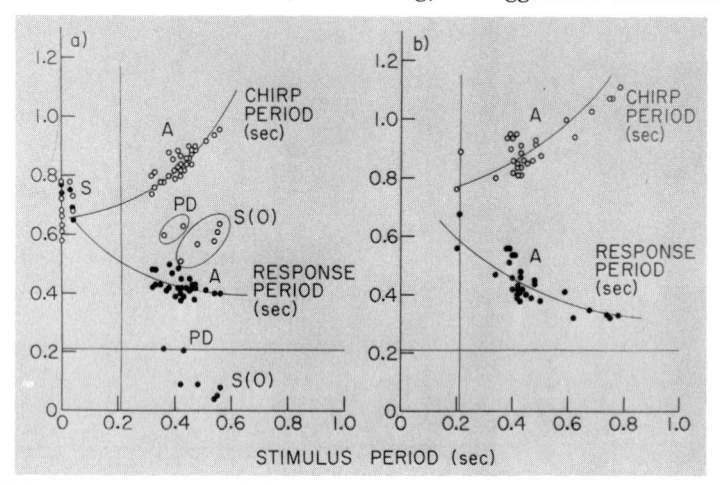

Fig. 12-7 Computer-produced data. Response function is modified in such a way that it emphasizes the solo overlap mode over partial delay mode. Compare this theoretical data with experimental findings in Fig. 12.3

partial delay, and solo overlap in the leader. Both measured and computer-model-produced results show that the follower maintains the alternating mode of chirping only. In Figure 12.3b and 12.6b one could distinguish those follower's chirps that are involved in the leader's partial delay mode. Their stimulus period is around 0.2 sec (duration of the chirp).

Figure 12.7 presents the results of another pair of computer-modeled katydids. In this model the response function is slightly modified. The flat part responsible for partial delay is shortened. The negative part responsible for solo overlaps is extended. As a result, there are only two cases of partial delay in Figure 12.7. The theoretical results presented in Figure 12.7 are in even better agreement with the experimental data in Figure 12.3.

12.4 Stable Sequence, Sliding Sequence, and Transfer Function

The basis for alternating and aggressive chirping is acoustical interaction between two male katydids. Each katydid could be considered as an element in a closed feedback loop, having its input driving force (stimulus period) and responding to it with a measurable output (chirp period). One of the ways to study this feedback loop is by determining the input-output transfer functions for each element. Noiseless computer models have been used for this task.

Stable Sequence, Sliding Sequence, and Transfer Function 215

Obtained response period versus stimulus period curves are plotted in Figure 12.8, for both leader and the follower.

From Figure 12.4 it is clear that the response period of the leader $R1$ is identical with the stimulus period of the follower $S2$. Also, the response period of the follower $R2$ is identical with the stimulus period of the leader $S1$. In Figure 12.8 the curve marked leader presents $R1$ as a function of $S1$. Similarly, the curve marked follower presents $R2$ as a function of $S2$.

The leader curve has three distinguishable parts: alternation, partial delay, and solo overlap. The shape of this curve is, in a sense, the same as the response period curves in Figures 12.3, 12.6, and 12.7. The follower's curve is, in fact, the same shape, except for the fact that partial delay and solo overlap parts are high up out of the frame of Figure 12.8.

The two curves have only one common point, H, and this is the stable operating point of the alternating calls. In a noiseless environment the parameters of alternating calls would be exactly as dictated by the operating point H: $S2 = R1 = 0.39$ sec; $S1 = R2 = 0.47$ sec.

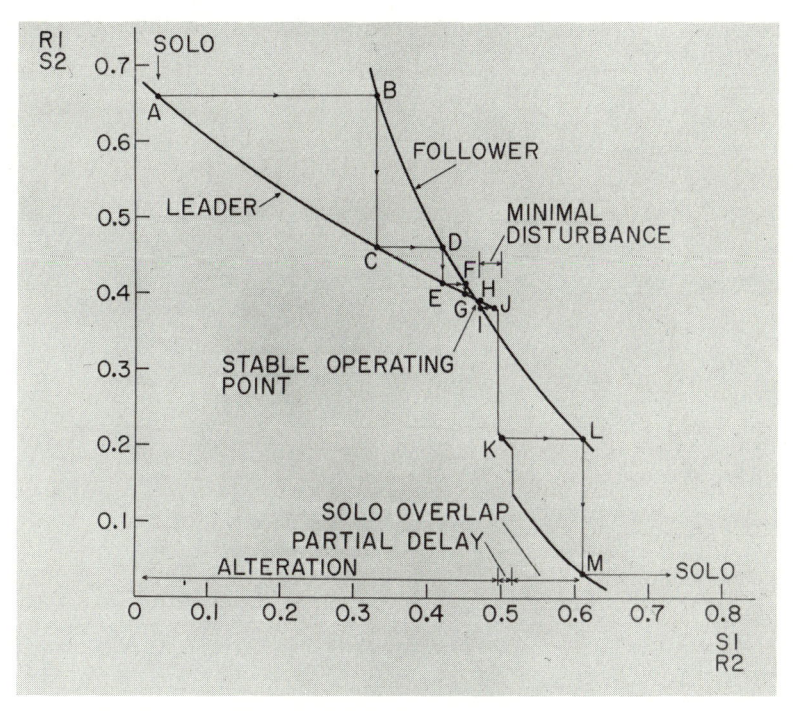

Fig. 12-8 Response period-stimulus period curves for leader and follower. The two curves intersect at the stable operating point H.

216   **Solo, Alternating, and Aggressive Communication**

If for any reason the songs were to start with the parameters of point A, drifting towards the stable point H would occur in the following: leader sings at $A(R1 = 0.67$ sec$)$. As $R1 = S2$, for $S2 = 0.67$ the follower's response is 0.31 (point B). As $R2 = S1$, for $S1 = 0.31$ leader's response is 0.46 (point C), and so forth. The sequence would drift through the path $A\,B\,C\,D\,E\,F\,G\,H$. These eight chirps present the transition. All subsequent chirps will have the parameters of the stable operating point H. Similarly, if the sequence begins at point J, it will drift through the path J, I, H, again ending up in the same operating point H.

An entirely different situation occurs if the sequence starts at point K (partial delay). From point K it will drift into point L (alternation for the follower), and then to point M (solo overlap for the leader).

The next step would lead to a very long response $R2$, which is longer than the chirp period of the leader. As a result, in the next step, the leader produces a solo chirp. The solo chirp period of the leader is 0.66 sec. (see threshold level in Figure 12.4). That means that after point M the next leader's point will be A. The detailed sequence K, L, M, A is shown in Table 12.2

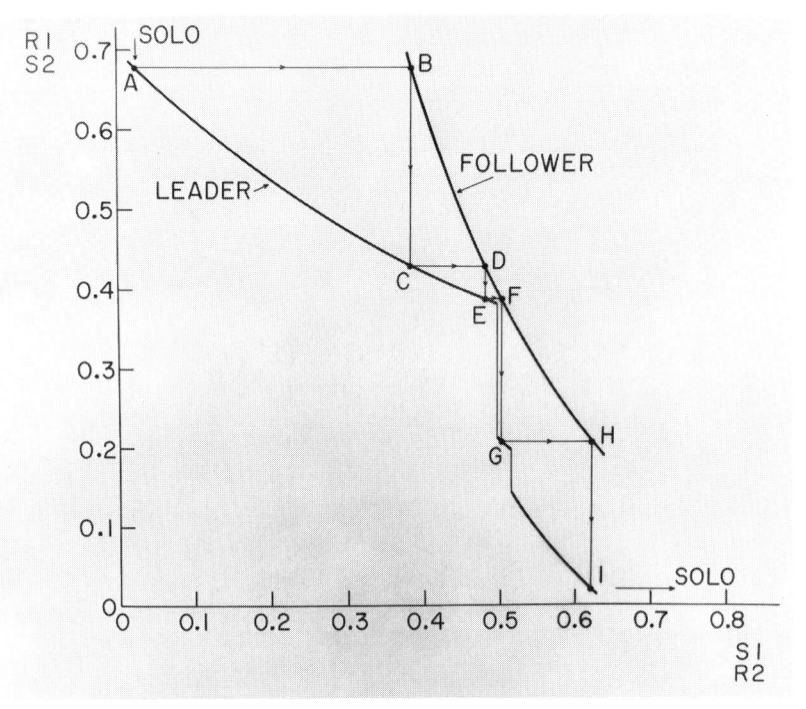

Fig. 12-9 Response period-stimulus period curves, without the stable operating point. Such a combination results in a continuously "sliding" sequence.

Aggression   217

and in Figure 12.5. From point A the sequence drifts to the stable operating point H.

Note that the stable operating point H is distant from the partial delay region of the follower, but very close to the partial delay region of the leader. As a result, noise or disturbance takes the sequence from point H to point K, producing partial delay, solo overlap, and solo parts in the leaders's chirps. Simultaneously, follower is in the alternating mode of chirping.

Another mode of operation is shown in Figure 12.9. The parameters are as follows: $L1 = 0.66$; $A1 = 0.68$; $T1 = 0.15$; $L2 = 0.81$; $A2 = 7$; $T2 = 0.1$. The resulting two curves never intersect. There is no stable point, and the sequence constantly drifts through the points $A\ B\ C\ D\ E\ F\ G\ H\ I$, and back to A. Again, the leader passes through its partial delay (G), solo overlap (I) and solo (A) modes, but not the follower. One detailed sliding sequence is presented in Table 12.3. Results of the sliding sequence with noise are shown in Figure 12.10.

12.5 Aggression

By changing the threshold and the height of the response function (Figure 12.4), both the leader's and the follower's transfer curves can be shifted up and down. Figure 12.11a presents a case in which the follower's solo overlap

TABLE 12.3

Stimulus Period	Response Period	Chirp Period	Chirps
0.47	0.39	0.86	1
0.39	0.5	0.89	2
0.5	0.21	0.71	1
0.21	0.62	0.83	2
0.62	0.02	0.64	1
0.62	0.68	0.65	1
0.68	0.38	1.06	2
0.38	0.43	0.81	1
0.43	0.48	0.91	2
0.48	0.39	0.87	1
0.39	0.5	0.89	2
0.5	0.21	0.71	1
0.21	0.62	0.83	2
0.62	0.02	0.64	1
0.62	0.68	0.65	1
0.68	0.38	1.06	2

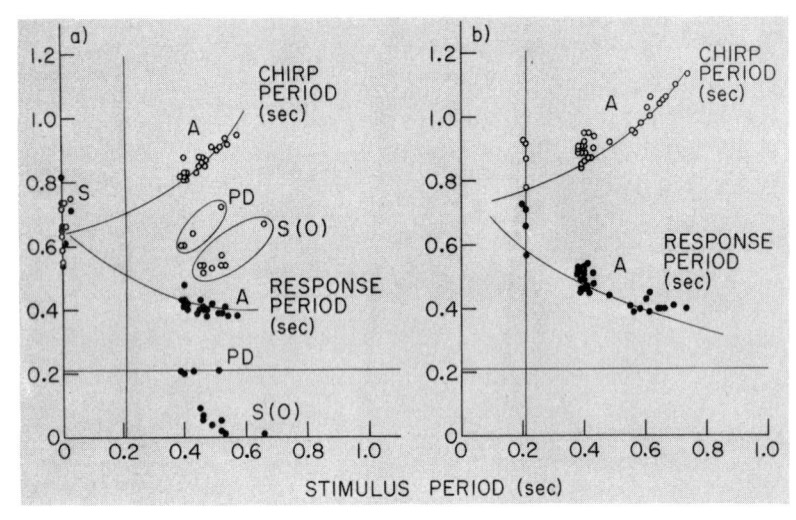

Fig. 12-10 Computer produced data, resulting from sliding sequence and noise.

operating region has a stimulus period $S2$ of similar duration to the leader's solo period. In this case, both the leader and the follower will go through partial-delay and solo-overlap sequences. The parameters are as follows: $L1 = 0.66$; $A1 = 0.68$; $T1 = 0.15$; $L2 = 0.83$; $A2 = 0.50$; $T2 = 0.10$.

If the sequence starts at point A it will follow the full line transition path, including the follower's partial delay. Such a sequence is presented in Table 12.4.

TABLE 12.4

Stimulus Period	Response Period	Chirp Period	Chirps
0.8	0.02	0.82	2
0.02	0.67	0.69	1
0.67	0.21	0.88	2
0.21	0.53	0.74	1
0.53	0.39	0.92	2
0.39	0.43	0.82	1
0.43	0.45	0.88	2
0.45	0.4	0.85	1
0.4	0.47	0.87	2
0.47	0.39	0.86	1
0.39	0.47	0.86	2
0.47	0.39	0.86	1

Aggression

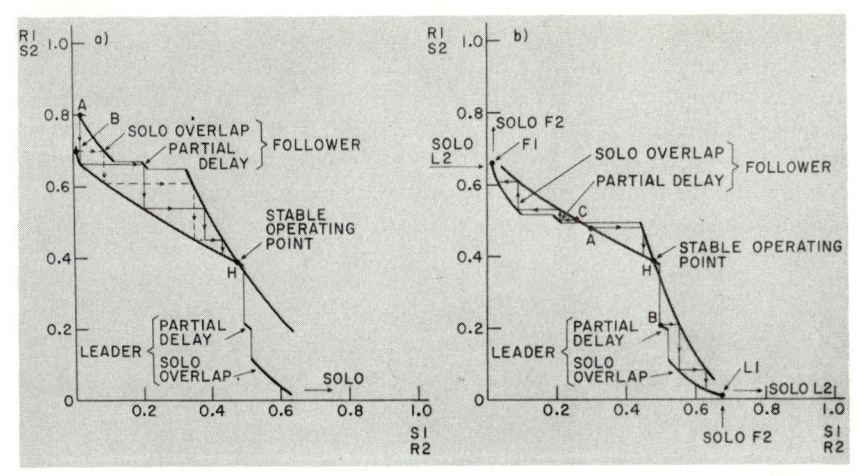

Fig. 12-11 Response period-stimulus period curves. (*a*) Solo overlap and partial delay regions of the follower are close to the solo chirp period of the leader. (*b*) Solo overlap and partial delay regions of the follower are close to the stable operating point.

If the sequence starts at point *B*, it will follow the dashed line transition path, including the follower's solo overlap. Such a sequence is presented in Table 12.5. In both cases the sequence will stabilize in the same operating point *H*.

Of special interest is the case presented in Figure 12.11*b*. The parameters are as follows: $L1 = 0.66$; $A1 = 0.68$; $T1 = 0.15$; $L2 = 0.66$; $A2 = 3$; $T2 = 0.1$.

TABLE 12.5

Stimulus Period	Response Period	Chirp Period	Chirps
0.6	0.04	0.64	1
0.6	0.7	0.65	1
0.7	0.09	0.79	2
0.09	0.61	0.7	1
0.61	0.35	0.96	2
0.35	0.45	0.8	1
0.45	0.43	0.88	2
0.43	0.41	0.84	1
0.41	0.46	0.87	2
0.46	0.4	0.86	1
0.4	0.47	0.87	2
0.47	0.39	0.86	1

In all cases presented up to now the leader's and follower's transfer functions have been substantially different. This time, the two response functions are very much alike. Point H is a stable operating point. If the sequence starts at point A, it will drift toward the stable point H. However, if the sequence starts at point B, it will drift through the leader's partial delay and solo overlap into the leader's solo mode. Also, if the sequence starts at point C, it will drift through the follower's partial delay and solo overlap into the follower's solo mode. Such a case is presented in Table 12.6.

TABLE 12.6

Stimulus Period	Response Period	Chirp Period	Chirps
0.26	0.5	0.76	1
0.5	0.21	0.72	2
0.21	0.53	0.74	1
0.53	0.09	0.63	2
0.09	0.61	0.7	1
0.61	0.03	0.65	2 $F1$
0.61	0.69	0.66	2 $F2$
0.69	0	0.69	1 $L1$
0.69	0.66	0.65	1 $L2$
0.66	0.02	0.69	2
0.66	0.68	0.66	2
0.68	0	0.68	1
0.68	0.66	0.65	1
0.66	0.02	0.69	2
0.66	0.68	0.66	2
0.68	0	0.68	1

Note the follower's chirps: first, $F1$ with the response period of 0.03 sec, and second, $F2$ solo chirp with the chirp period of 0.65 sec. This sequence results in a stimulus period for the leader of 0.68 sec.

Now leader produces two chirps: the first chirp $L1$ is a response to the stimulus $F2$. The second chirp $L2$ is the leader's solo chirp, with a chirp period of 0.65 sec.

Chirp $L2$ presents a stimulus for the follower. In this way the following sequence is established: $F1, F2, L1, L2, F1, F2, L1, L2$ .... Since both leader and follower produce solo chirps, the case presented in Figure 12.11b could be used to explain aggressive chirping.

Noise has been added to the system presented in Figure 12.11b, and a small part of the produced sequence is shown in Table 12.7. Note the many overlapping chirps, as well as solo chirps in the leader and the follower. A com-

TABLE 12.7

Stimulus Period	Response Period	Chirp Period	Chirps
0.6	0.05	0.65	1
0.05	0.66	0.72	2
0.66	0.	0.66	1
0.66	0.66	0.65	1
0.66	0.06	0.73	2
0.06	0.63	0.69	1
0.63	0.03	0.67	2
0.63	0.66	0.63	2
0.66	0.06	0.72	1
0.06	0.62	0.69	2
0.62	0.04	0.66	1
0.04	0.75	0.8	2
0.75	0.	0.75	1
0.75	0.73	0.72	1
0.73	0.	0.74	2
0.73	0.64	0.64	2
0.64	0.09	0.73	1
0.09	0.66	0.76	2
0.66	0.	0.66	1
0.66	0.64	0.63	1

puter-generated sequence in Table 12.7 is very similar to observed aggressive sequences.

As the aggressive mode of chirping has received only minor attention in the past, new experiments are in preparation to concentrate on this mode of operation.

12.6 Model Based on Response Function and Transfer Function

Alternating and aggressive chirping sequences have been analyzed for true katydids. A theory has been developed that explains both central neural control as well as phonic interaction between two partners. It is possible to explain all experimental findings through three functions: time function, threshold, and response function. The response function is generated in a male katydid on receiving acoustical stimuli from another male. The response function has a spikelike shape, with three parts. First, the negative part explains solo-overlapping chirps. Second, the flat part explains partially delayed chirps. Third, the positive part explains the normal alternating chirps.

```
10  REM PARAMETERS
15  T1 = .15
16  T2 = .1
18  L1 = .66
20  L2 = .81
25  A1 = .68
30  A2 = 2
40  S  = 0
45  DIM I1(120), I2(120)
90  K1 = 0
95  PRINT "K2"
100 INPUT K2
102 S2 = .01*K2
160 PRINT
165 PRINT "STM PER", "RSP PER", "CRP PER", "CHIRP",
170 PRINT
```

```
200 REM LEADER 1
210 K1 = K1 + 1
215 C1 = .01*K1
245 X = I1(k1) + L1 + R
250 IF 01 < X GO TO 500
```

```
400 REM CHIRP 1
407 R = S*(RND(1) - .5)
425 FOR J = 1 TO 120
430 I1(J) = 0
435 NEXT j
440 GOSUB 850
455 S2 = .01*K2
460 R1 = S2
461 K1 = 0
465 N = 1
466 C1 = C1 - .01
470 PRINT S1, R1, C1, N,
475 PRINT
```

```
850 REM RESP. FUNC. 2
860 FOR J = 1 TO 7
861 Z = -4.000000E - 03*J
862 I2(K2 + J) = I2(K2 + J) + Z
863 NEXT J
864 FOR J = 8 TO 16
865 Z = z + .01
866 I2(K2 + J) = I2(K2 + J) + Z
867 NEXT J
868 FOR J = 17 TO 19
869 I2(K2 + J) = .09
870 NEXT J
875 FOR J = 1 TO 90
877 I = K2 + J + 18
878 IF I = 120 GO TO 890
880 It(I) = A2*EXP(-.01*J/T2)
885 NEXT J
890 RETURN
```

```
500 REM FOLLOWER 2
510 K2 K2 + 1
515 C2 = .01 * K2
545 Y = I2(K2) + L2 + R
550 IF C2 < Y GO TO 200
```

```
600 REM CHIRP 2
707 R = S*(RND(2) - .5)
725 FOR J = 1 TO 120
730 I2(J) = 0
735 NEXT J
740 GOSUB 950
755 S1 = .01*K1
760 R2 = S1
761 K2 = 0
765 N = 2
766 C2 = C2 - .01
770 PPINT S2, R2, C2, N,
775 PRINT
800 GO TO 200
```

```
950 REM RESP. FUNC. 1
960 FOR J = 1 TO 10
961 Z = -4.000000E - 03*J
962 I1(K1 + J) = I1(k1 + J) + Z
963 NEXT J
964 FOR J = 11 to 19
965 Z = Z + .01
966 I1(K1 + J) = I1(K1 + J) + Z
967 NEXT J
968 FOR J = 20 TO 22
969 I1(K1 + J) = .06
970 NEXT J
975 FOR J = 1 TO 90
977 I = K1 + J + 22
978 IF I = 120 GO TO 990
980 I1(I) = A1*EXP(-.01*J/T1)
985 NEXT J
990 RETURN
```

Each katydid is considered as an element in a closed feedback loop, and could be described through its response period versus stimulus period curve. In an alternating mode the two response curves intersect at a stable operating point, which is near the partial delay region of the leader, but far away from the partial delay region of the follower. Because of disturbance or noise, the communication loop drifts out of the stable point, through the leader's solo chirps, and back to the stable point. In aggressive mode the two response curves are in a position such that the communication loop drifts constantly from solo chirps in the leader to solo chirps in the follower and vice versa. A disturbance moves the communication out of this pattern, which explains the numerous solos by both leader and follower in aggressive mode. The described theory has been used to design a computer model for alternating and aggressive communication. Computer-simulated chirping sequences are in excellent agreement with field measured data.

A katydid's chirping sequences are composed of deterministic and random components. As a result, two programming techniques have been used: continuous system simulation for the deterministic components and Monte Carlo technique for the random components. These techniques have been adopted for application on a laboratory minicomputer, Souček.[11] The complete simulation program in flow chart form is shown in Figure 12.12.

References

1. Souček, B.: Model of Alternating and Aggressive Communication with the Example of Katydid, Chirping, *J. Theor. Biol.* **52** (1975) 399.
2. Alexander, R. F.: *Ann. Rev. Entomol.*, **12** (1967) 195–526.
3. Jones, M. R. R.: *J. Exp. Biol.*, **45** (1966) 15–30.
4. Shaw, K. C.: *Behaviour* **31** (1968) 203–259.
5. Huber, F.: *The Physiology of the Insect Control Nervous System*, Academic Press, New York, 1965.
6. Ewing, A., and Hoyle C.: *J. Exp. Biol.*,**00** (1965) 139–153.
7. Wilson, D. M.: *Symp. Soc. Exp. Biol.*, **20** (1965) 199–228.
8. Heiligenberg, W.: *Z. Vergl. Physiol.*, **65** (1969) 70–97.
9. Souček, B., and Carlson, A. D.: Flash Pattern Recognition in Firefly (in press).
10. Souček, B., and Vencl, F.: *J. Theor. Biol.* **49** (1975) 147–172.
11. Souček, B.: *Minicomputers in Data Processing and Simulation*, Wiley, New York, 1972.

Fig. 12-12 Simulation program. Katydid ($N = 1$) = leader; katydid ($N = 2$) = follower. K_1, K_2, time interval counters; S_1, S_2, stimulus periods; C_1, C_2, chirp periods; R_1, R_2, response periods, $I_1 (K_1)$, $I_2 (K_2)$, response functions; R, random noise; X, comparison level for leader; Y, comparison level for follower.

Chapter 13

COMMUNICATION BASED ON TIMING AND PULSE PATTERN RECOGNITION

Introduction and Survey

A theory has been developed that explains the use of time intervals in an animal communication system. The theory is based upon the observation of female *Photinus macdermotti* firefly responses to artificial male flash sequences. An explanation for time discrimination in a firefly brain is proposed as follows. Upon receiving the flash from the male, the female firefly generates a response function composed of three intervals: initial inhibition or total blocking, acceptance, and long-term inhibition. This response function and discrimination bias generated by the short-term memory, together with background noise, determine the firefly response time window. The female will answer only if the subsequent male's flash is received during this time window.

Short-term memory is used for sensitivity adjustment, whenever sharp change is introduced in the triggering sequence. If the firefly is flashing under favorable conditions, she adjusts herself into a flashing mood and will flash once, even if the triggering changes towards marginal conditions. If the firefly is receiving a wrong input flash sequence, she will enter a nonflashing mood and will not flash on a first good interval. At high frequencies, the firefly blocks her input for a constant period after every flash.

It is shown that the flash sequences predicted by the theory are in very good agreement with actual observed female firefly responses to artificial male flash sequences. In this chapter two principles are introduced for the first time, following Souček and Carlson[1]: pulse pattern recognition and memory for sensitivity adjustment.

Pulse modulation is very frequently used in biological systems for acoustical or visual communication and for neural information transmission. It is shown that the basic feature of such systems is pulse pattern recognition. Similarly, the memory presents the basic element for sensitivity adjustment.

13.1 Example of Firefly Flashing

Courtship communication in fireflies of the *Photinus* genus consists of the exchange of light flashes between flying males and stationary females. As shown by Lloyd[2] much of the basic information transmitted from one member of the species to another member is contained in the interflash interval of male flashes and in the female's flash response latency. When courting, fireflies must exist with sympatric species; error in the information transmission and processing can lead to attraction of a nonproductive partner. Hence precise time discrimination is needed to recognize the species-specific signal and to identify the correct partner.

Lloyd[2] was able to classify *Photinus* species on the basis of their characteristic flash signal patterns. One firefly species can be distinguished from others by the duration of the flash or by the interflash interval and Lloyd[2] has demonstrated that females of sympatric species can indeed make the appropriate choice of males based upon these parameters. In this study the responses of the female firefly *P. macdermotti*[2,3,4,5] are analyzed and used as a basis for a computer model.

The basic experimental findings are as follows. Males of *P. macdermotti* emit rhythmic 200 msec patrolling flashes while searching for an appropriate female, Figure 13.1a. Upon receiving an answering flash of correct latency the males switch to courtship flash pairs of shorter INTRAPAIR interval, Figure 13.1b, and each pair is separated by a longer INTERPAIR interval, Figure 13.1b. Females of the same species normally answer after the second flash of each courtship flash pair. They often answer consecutive patrolling flashes, however.

Double Flash Interval or "Double Interval" Females will readily respond to flashes (pulses) generated by an electronic flasher which allows the investigator to vary interflash intervals at will. When the role of interflash interval is investigated by this means a range of intrapair intervals is found in which the female responds after the second pulse of the pair. This is called the female's double flash interval, and it is between 1.0 and 2.1 sec at 24°C. (see Figures 13.1c and f). The female answers with high probability in the middle of the double interval range (1.5 sec at 24°C.), and her response probability falls off on either side (see Figure 13.1f).

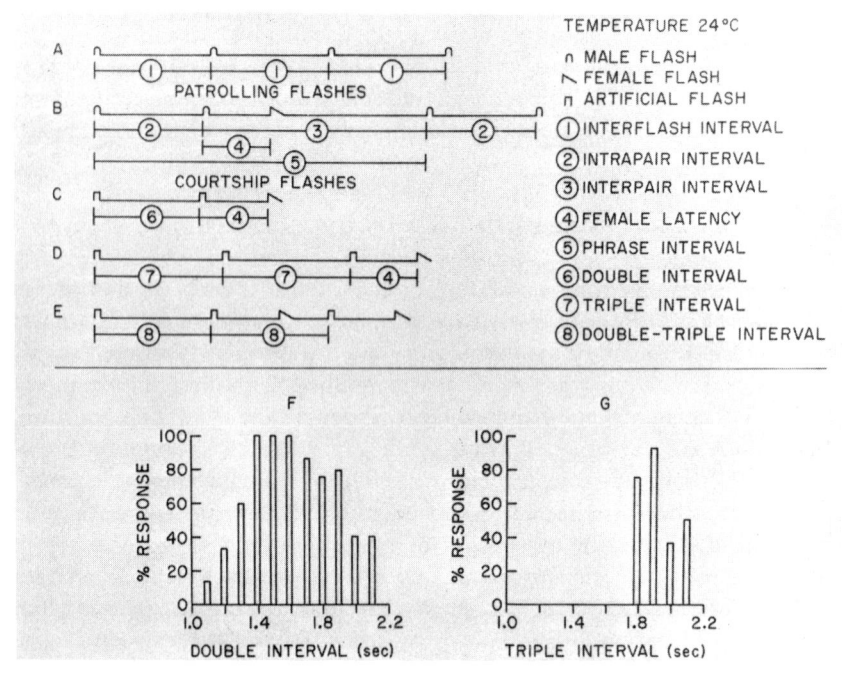

Fig. 13-1 (*A–E*) Basic definitions of time intervals in firefly flashing sequences. (*F*) Statistics of the responses to the double interval excitation. (*G*) Statistics of the responses to the triple interval excitation.

Triple Flash Interval or "Triple Interval"   At long interflash intervals near the upper end of the double interval range females will answer after the second pulse, and after a third pulse and at still longer intervals she may answer after only the third pulse of a series of rhythmically repeated pulses (see Figures 13.1*d* and *e*). The interflash interval range over which she will answer after the third pulse is called the triple flash interval; it is between 1.8 and 2.2 sec at 24°C (see Figure 13.1*g*). Again, the female's response probability to the third pulse varies over the triple interval range.

Short Interpair Interval   If the interval between pulse pairs is sufficiently long (above 2.2 sec at 24°C.) the female will always answer after the second pulse of the pair. If the interpair interval becomes shorter the female answers in a more random fashion, which nevertheless often follows the probabilities of the double and triple interval ranges.

Blocking   If the interpair interval is shorter than 1.8 sec at 24°C the female will respond to the second pulse of the first pair and will then answer every third pulse presented. It appears that the female blocks her input for a period close to 1.8 sec after she produces her flash response.

Memory and Self-Adjusting In switching from the sequence of pulse pairs with effective double intervals (say, 1.5 sec) to the new sequence of pairs with ineffective double intervals (say, 1.1 sec) the female will answer the first ineffective pair. Also, switching from a sequence of ineffective pairs (say, 1.1 sec) to effective pairs (say, 1.9 sec), the female will not answer after the first effective pair but will answer after the second. She apparently remembers the previous conditions and adjusts herself for it.

Based on the experimental findings of Carlson et al.[5,6] a theory has been developed and a computer model has been designed to explain the basic signal processing elements in the female *P. macdermotti* firefly. The model also explains message exchange between the firefly sexes. The model covers all experimental findings: double interval, triple interval, short interpair interval, blocking, memory, self-adjustment, and random fluctuations in firefly flash sequences.

The basic components of the model, explaining the deterministic part of firefly sequences are described in Section 13.2. The components that explain the random part of firefly sequences, as well as memorizing effects, are described in Sections 13.3 and 13.4.

13.2 Response Function

Upon receiving a flash from the male it is assumed that the female firefly generates a neural response function that alternates through a cycle of inhibition–excitation–inhibition. This cycle is diagrammed in Figure 13.2

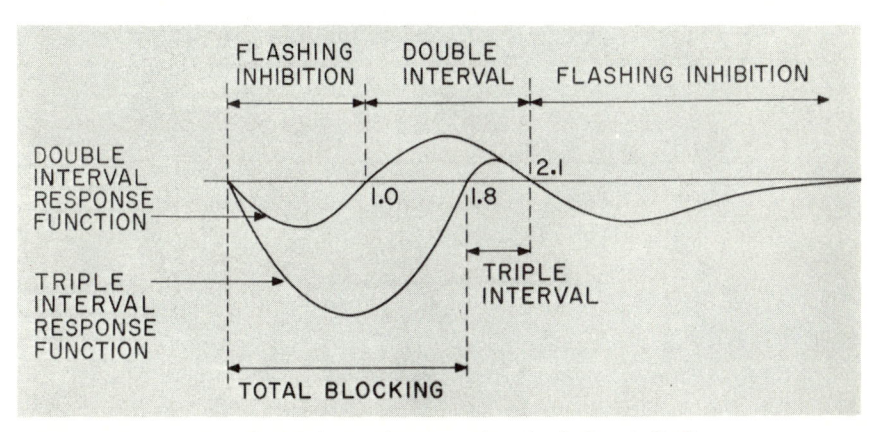

Fig. 13-2 Double and triple interval response function in female firefly.

and labeled double interval response function. It is proposed that subsequent male flashes arriving after generation of the response function act in the following manner:

1. Male flashes arriving during the early and late inhibitory (negative) phases serve to generate a new response function.
2. Male flashes arriving during the excitatory (positive) phase act to generate a female answer.
3. Male flashes which induce a female answer generate a different response function with different time parameters labeled triple interval function in Figure 13.2.

The response function can be generated in many ways. Such a wave form can be easily generated in an electrical or mechanical feedback circuit with damping. The wave will have a few oscillations, and then it will exponentially fade out.

The response function will be used to explain the deterministic components in firefly sequences. Statistical fluctuations, memory, and self-adjustment will be explained after that.

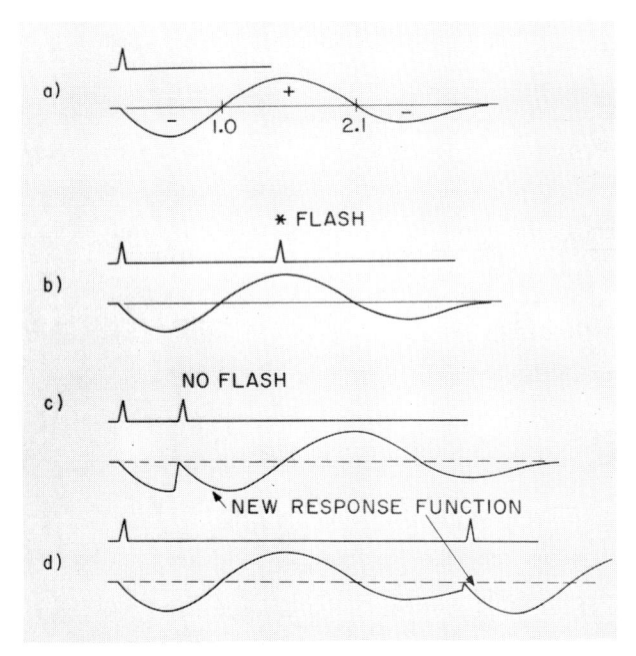

Fig. 13-3 Double interval response function. (*a*) Generation of the response function; (*b*) Accepted second male's flash; (*c*) and (*d*) rejected male's flashes.

Response Function

Double Interval A double interval sequence is presented in Figure 13.3. The first male flash starts a new double interval response function, with positive period of $1.0 \leq t < 2.1$ sec. If the second male's flash arrives during this positive period, the female will produce the answer, Figure 13.3*b*.

If the second male's flash arrives during the negative parts of the response function, the female will not produce the flash. In this case she will reset the response function and start a new one, Figures 13.3*c* and *d*. Note that the double interval response function is initiated by the male flashes that do not result in a female flash.

Triple Interval Sequences A triple interval sequence is presented in Figure 13.4. The first male flash starts a new double interval response function, with a positive period of $1.0 \leq t < 2.1$ sec. The second male flash arrives after 2 sec. It is not in the middle of the positive period, but it is still inside the margins, and it may produce the female flash. The second male flash produces the female flash and also initiates a new, triple interval response function, with a positive period of $1.9 \leq t < 2.1$ sec. The third male flash arrives during this positive period, and it may produce another female flash.

Because of noise and statistical fluctuations, only a small fraction of second and third pulses are actually arriving during the positive period. These facts are explained later.

Blocking Figure 13.5 presents the double interval case, in which the interpair interval is shorter than 1.8 sec. The first input results in a double interval response function. The second input arrives during the positive period of the response function and it will result in a female flash. It also starts a triple interval response function. The third input arrives during the

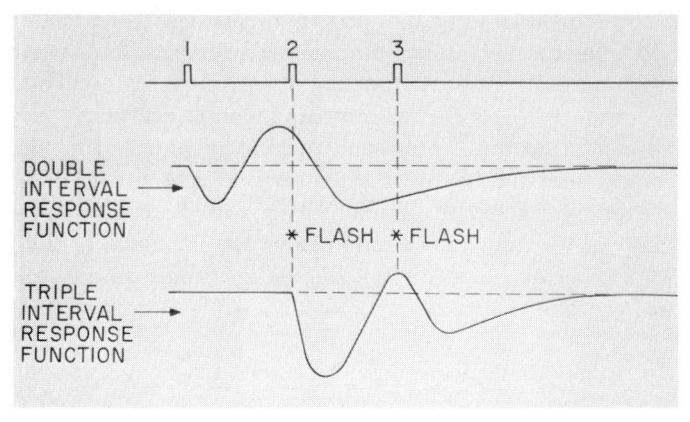

Fig. 13-4 Triple interval pattern.

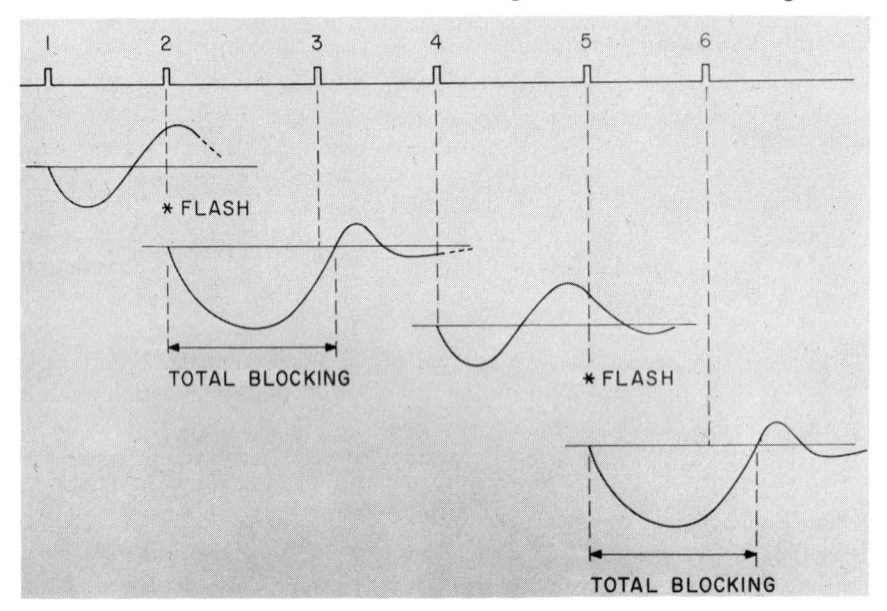

Fig. 13-5 Blocking due to the short intrapair interval.

total blocking period of the triple interval response function. As a result, the third input is rejected, leaving no trace on the female firefly. The fourth input is accepted under the same conditions as the first one, and the sequence is repeated all over again.

Inserted Flash Sequences Figure 13.6 presents the case of three male flashes at constant interval with an inserted male flash. The first male flash initiates a double interval response function. The inserted flash arrives during the negative part of the response function. Hence the female will not produce an answer, but it will reset the old response function, and start a new double interval response function. The second flash arrives during the positive part of the response function. Hence the female generates the flash, and also a new, triple interval response function. The third flash arrives during the positive part of the response function, and the female generates another answer. Again because of statistical fluctuations, only a small number of female answers will be actually generated.

Noise In realistic biological preparations one must expect fluctuations or noise. As a result, the response function voltage is not compared with ideal zero level, but with a noisy reference level, Figure 13.7.

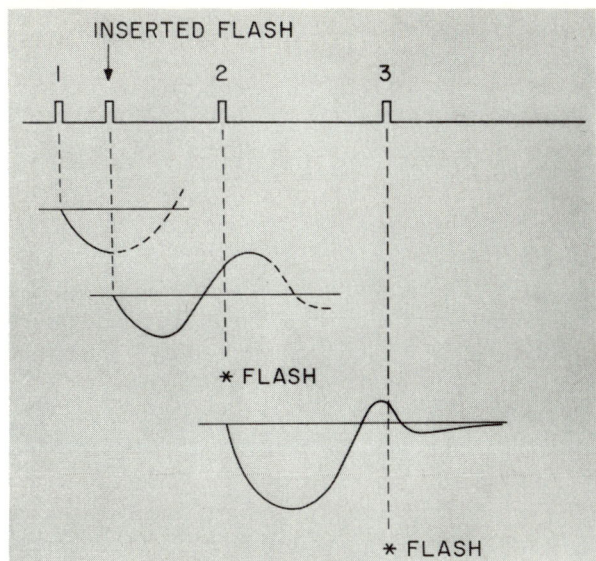

Fig. 13-6 Effect of the inserted male's flash.

If the male flash arrives during the middle part of the positive period it will be answered with 100% probability. If the male flash arrives at the beginning or at the end of the positive period, its acceptance will be determined by the noise. Because of fluctuations the discrimination window will be wider or narrower. Cases 1 and 2 in Figure 13.7, show opposite effects on the input flash. Noise in case 1 narrows the window, and the input flash is rejected. Noise in case 2 widens the window, and the input flash is accepted.

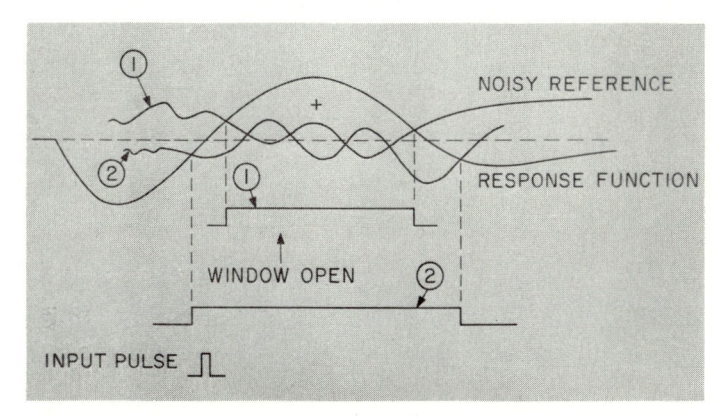

Fig. 13-7 The effect of the random noise on the time window width.

Figures 13.1*f* and *g* present the experimental data showing statistical fluctuations in the window width.

13.3 Sensitivity Adjustment and Memory

When a sequence of male flash pairs of effective intrapair interval is switched to a sequence of ineffective intrapair intervals the female often answers the first ineffective pair. Sensitivity can be changed by adjusting the discriminating reference level, Figure 13.8. The wave of the response function is compared with the discriminating reference level. If the response function wave is more positive than the reference level, the window is open, and the second flash of the male pair will produce the female answer.

Whenever a sharp change is introduced in triggering sequence, the female makes a self-adjustment of her sensitivity. If the female is receiving flash pairs of effective intrapair intervals, she adjusts herself into a flashing mood and will flash once, even if the males intrapair interval changes towards the marginal ends of her double interval. In the model, the female increases her sensitivity by lowering the reference level, Figure 13.9. If the firefly is re-

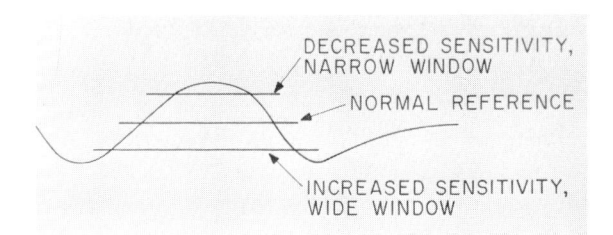

Fig. 13-8 Sensitivity adjustment through discrimination voltage control.

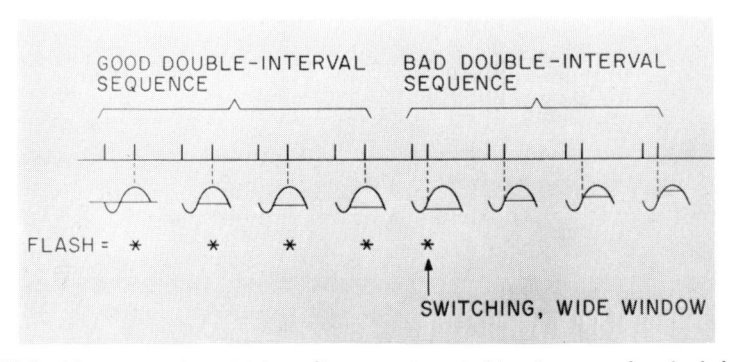

Fig. 13-9 Memory and sensitivity adjustment in switching from good to bad double-interval sequence.

Sensitivity Adjustment and Memory 233

ceiving flash pairs of ineffective intrapair interval, she will be in a nonflashing mood and will not answer the first flash pair of effective intrapair interval. In the model the female decreases her sensitivity by raising the reference level, Figure 13.10.

Note that self-adjustment is applied only at switching points (effective sequence into ineffective, or vice versa). Hence only the first pair of a new sequence is affected.

Memory To control the self-adjustment, a memory is needed. The memory must store two results, old and new. Each result can be either "effective" (female flashes) or "ineffective" (female does not flash). The old result describes the status of the female just before the new pair arrives. In Figure 13.9 it is old effective, whereas in Figure 13.10 it is old ineffective at a switching point. The new result describes the output produced by the pair that has just arrived, and it will be used for motor control in the analysis of the next pair. In Figure 13.9 it is new ineffective, whereas in Figure 13.10 it is new effective for a first pair after switching the conditions.

The memory elements, new and old, are used to control reference level, Figure 13.8. The response function wave is compared with the reference level. If old = new, the reference level is zero, and sensitivity is at normal level. If old = effective; the discrimination level is lowered, and the sensitivity is increased, (Figure 13.9). If old = ineffective, the discrimination level is raised, and the sensitivity is decreased, Figure 13.10.

The memory can be explained with a network of neurons or with an analog level. The memory must have two independent states, new and old, Figure 13.11. The first stage of memory is reserved for a new result M, which is produced in the sensory discriminator, with zero bias. The second stage of the

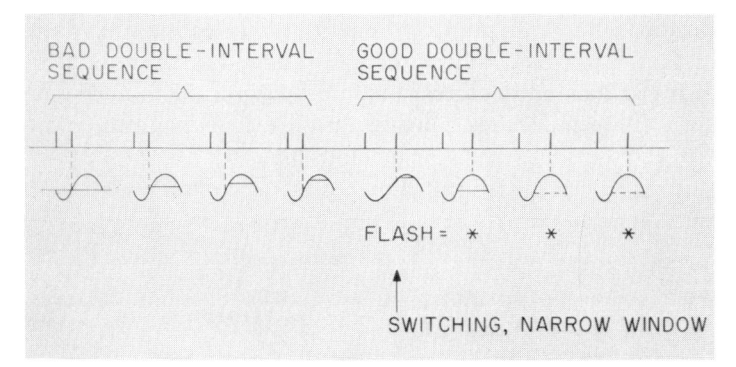

Fig. 13-10 Memory and sensitivity adjustment in switching from bad to good double-interval sequence.

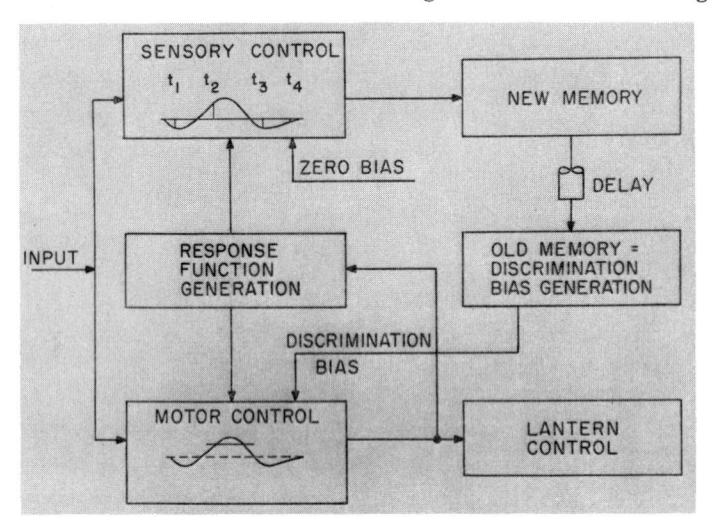

Fig. 13-11 Functional model of flash pattern recognition in female firefly.

memory keeps the old discriminator value D, which describes the condition of the past.

In fact, the measurement of each interval is performed in two steps (see Figure 13.11):

1. Sensory control. The discrimination level is not applied and the reference level is zero. The window has a normal width. The incoming male's flash samples the instantaneous value of the response function, and the obtained sample is stored in the memory. This sampling is shown in sensory control block in Figure 13.11. If the male's flash arrives, for example, at the instant t_1 or t_3, a negative result is deposited in memory. If the male's flash arrives at the instant t_2, a positive result is stored. If the male's flash arrives at the instant t_4, a practically zero value is stored in the new memory M. This result will be used to correct the sensitivity for the following pair.

2. Motor control. The discriminator level D is now applied, and the window width and sensitivity are adjusted, following the trend of the past history. The result of this control is applied on the motor circuit to control the flash. This result also determines the new response function.

13.4 Locked-In Sequences

The new discrimination value of D is a function of the old value of D and last value of M. Although the experimental results are not precise enough

Locked-In Sequences

to be used for exact mathematical expression for M and D, one could predict the general form of the relation

$$D_{new} = T_1 \cdot (T_2 \cdot D_{old} - T_3 \cdot M) \qquad (13.1)$$

If $T_1 = 1$, $T_2 = 0$, equation 13.1 changes into

$$D_{new} = -T_3 \cdot M \qquad (13.2)$$

This presents a short-term memory, where discrimination level is determined only by the pair previous to the one just measured.

If $T_1 = 1$, $T_2 = 1$, equation 13.1 changes into

$$D_{new} = D_{old} - T_3 \cdot M \qquad (13.3)$$

This equation is used for computer simulation, with the following limits

Saturation: $D_{max} = +1$

$\qquad\qquad D_{min} = -1$

Resetting: $D = 0 \quad$ for $(D - M) = 0$

An example is shown in Figure 13.12.

The male flashes repeat the sequence: effective, effective, ineffective .... Each effective input produces positive M, and ineffective input produces negative M. Because of the lack of precise measurement results, all positive values will be approximated with $+1$, all negative values with -1, and no response will be presented with 0 stored in new memory. As a result, the discrimination level follows an up and down pattern. However, as there are two effective inputs for every one ineffective, the discrimination level gradually approaches the value -1.

As long as the discrimination level is smaller than the correction level, the sensitivity adjustments are small, and normal female sequence is produced.

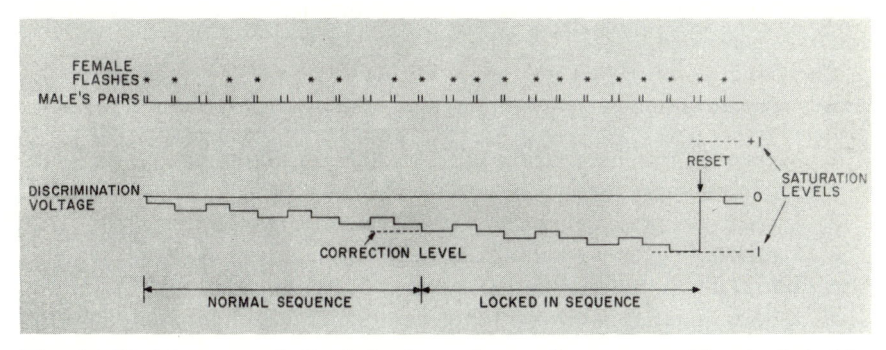

Fig. 13-12 Memory and discrimination voltage adjustment resulting in locked-in sequences.

When the discrimination level falls below the correction level, the sensitivity is changed enough to widen the time window. Now even the ineffective input pairs are accepted, and the female sequence is locked, producing flashes all the time. Eventually the discrimination reaches the saturation level. The first ineffective input will now reset the discrimination level to zero, and the whole pattern repeats.

In this example the sequence with mostly effective pairs has been analyzed, resulting in saturation level $D = -1$. If the sequence with mostly ineffective pairs is analyzed, the discrimination level would gradually rise towards the value of $+1$, following virtually the same pattern. The female is locked in a no flashing mood.

13.5 Preparation for Model

The firefly communication signal consists of a series of light flashes in which the basic information to be transmitted from one member of the species to the other is contained in time intervals. Precise time discrimination is needed to recognize the signal and to identify the partners as belonging to the right species.

Upon receiving the male flash, the female firefly generates the response function. It is shown that the response function must have three periods: (1) initial inhibition or total blocking, (2) acceptance, and (3) long-term inhibition. Only the male flashes received during the acceptance period might result in a female answer.

If the response function is regarded as an analog voltage, it could be generated in an electromechanical tuned circuit. Critically tuned, such a circuit produces two oscillations, after which the output voltage exponentially fades out towards zero. Such a process is very simple and could be easily generated, and yet its intrinsic characteristic is the stability and accuracy of the time intervals generated.

The basic neural signal processing elements, necessary to recognize the male's sequences and to generate the female's answer, are the following:

1. A time discriminator, controlled by the response function and discrimination reference level.
2. A generator producing a three-phase response function.
3. A memory element recalling the results of the previous excitation, and contributing to the discrimination level.
4. A discrimination level control, generating the discrimination level for the time discriminator.
5. A lantern motor control.

The developed theory covers all the experimental findings including double interval, triple interval, short interpair interval, blocking, memory, self-adjustment, and random fluctuations in firefly sequences.

Based on the theory, a computer model has been designed and used to produce artificial firefly random sequences. Computer simulated sequences and their deterministic and stochastic properties, are in very good agreement with actual measured firefly sequences. A comparative study between measured and computer simulated sequences is presented in the following chapter.

References

1. Souček, B., and Carlson, A. D. Flash Pattern Recognition in Fireflies (in press).
2. Lloyd, J. E.: *Misc. Publ. Mus Zool.*, Univ. Michigan, No. 130 (1966) 1–95.
3. Lloyd, J. E.: *Coleopt. Bull.*, 23 (1969) 29–40.
4. Carlson, A. D., Copeland, J., Raderman, R., and Bullock, A. E. M. Role of Interflash Intervals in a Firefly Courtship (*Photinus macdermotti*).
5. Carlson A. D., Copeland, J., Raderman, R., and Bulloch, A. E. M. Response Patterns of Female *Photinus macdermotti* Firefly to Artificial Flashes. (in press).
6. Carlson, A. D., and Souček, B. Computer Simulation of Firefly Flash Sequences. (in press).

Chapter 14

COMPUTER SIMULATION OF FIREFLY FLASH SEQUENCES

Introduction and Survey

Experiments involving artificial courtship of female *Photinus macdermotti* fireflies have been used to develop a functional model of flash analysis for firefly communication. Based on this theory a computer model was designed that simulates the behavior of the female firefly. The model explains both deterministic and stochastic components present in the flashing sequences. The model's output is a stochastic process whose statistical averages are equivalent to the statistical averages of a real firefly sequence. The model recognizes double and triple intervals regardless of their position in the sequence. A comparative study between measured and computer simulated sequences is presented. It is shown that computer generated flash-answer sequences are in very good agreement with observed flash-answer sequences.

In this chapter the computer modeling of a biological communication system is described in detail, and its significance is discussed. Modeling has helped in many ways. In the concrete case of firefly flashing, modeling and the response function describe the system with one single function. This function replaces about 100 tables describing different experimentally observed flashing sequences. Also, modeling has helped to predict the behavior, and then the experimenter has actually found such a behavior in his experimental data. It is shown that modeling is very useful also in designing and planning new experiments. Modeling presents the major step toward theoretical explanation of complex biological systems.

14.1 Firefly Flash Sequences

Basic Findings In *Photinus* firefly flash communication interflash interval is an important information parameter, Lloyd.[2] Male fireflies flash in rhythmic patterns, producing group of flashes. The female examines the interflash intervals between the flashes within the group and between the groups. Only if the pattern and interflash intervals fall within certain limits will the female answer.[1,2] Precise measurements of firefly flashing sequences in males and females of *P. macdermotti* have been performed by Carlson et al.[3,4] Over 100 experiments and field measurements have been performed on the female firefly. Only basic results of the experiments will be shown here and compared with computer-simulated sequences.

The limits of interflash intervals accepted by females are temperature dependent.[3] As the temperature decreases, the upper and lower male interflash intervals accepted shift toward longer periods. The results from experiments performed at temperatures of 25, 24, and 22°C are shown here. The percentage of responses of female fireflies to double and triple flashes are shown in Figure 14.1 for the three groups of experiments. The following

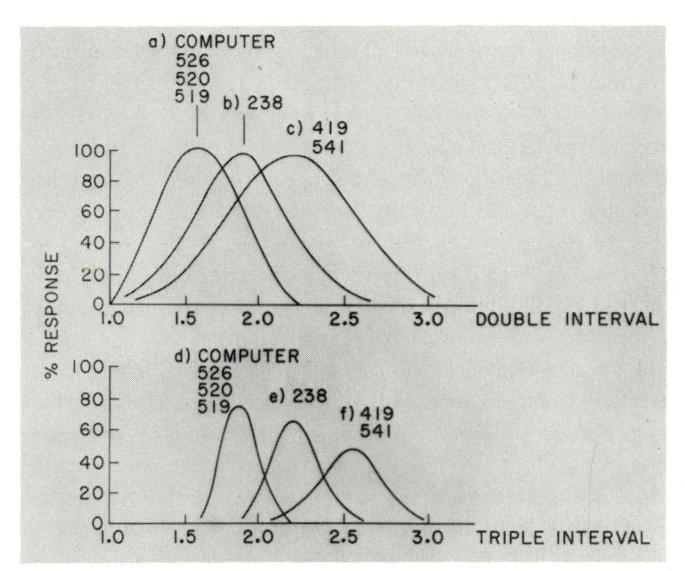

Fig. 14-1 Above: Percent response of female firefly to male double flash intervals in seconds. Response curves of particular experiments and computer at temperatures of (a) 26°C (b) 24°C and (c) 22°C.

Below: Percent response of female firefly to male triple flash intervals in seconds. Response curves of particular experiments and computer at same temperatures as in upper graph.

experiments were performed at the temperatures indicated: #519, #520, and #526 at 26°C; #238 at 24°C; and #419 and #451 at 22°C.

There are some variations in experimental results because of the individual variations in the large number of fireflies used for those field measurements. Nevertheless, the same basic findings have been observed in the majority of experiments.[3,5]

The results of the experiments have been used to develop a functional model for female flash analysis. This analysis is performed through time discrimination, sensitivity adjustment, and accumulative memory. It is proposed that male flash generates a response function in the female firefly with three intervals: initial inhibition of total blocking, acceptance, and long-term inhibition. The developed theory is presented in Souček and Carlson.[5]

The developed theory has been used to design the computer model, which explains neural control in the female firefly. The model simulates the behavior of the female firefly and generates random firefly flashing sequences. The theory and the computer model explain both deterministic and stochastic components found in firefly sequences.

In this chapter the statistical computer model for firefly sequence generation is described. A comparative study between measured and computer-simulated sequences is presented. The theory of firefly communication and the generation of firefly flashing sequences is explained in the previous chapter. Thus the measured and computer-generated sequences are compared here, without repeating the explanations of basic processes.

The computer simulations presented here are based on experimental data at 26°C, as shown in the left-hand side curves a and d in Figure 14.1.

Standard Double and Triple Interval Sequences The male generates the pair of flashes. The female will answer only if the intrapair interval is a right one for a given ambient temperature.

Table 14.1 shows one of the results of experiments #520. The summary of this experiment is shown in curve a in Figure 14.1. The interflash interval of 1.15 sec is a marginal one. The female will respond in average roughly on 50% of the pairs. This is evident from the sequence shown in Table 14.1.

Table 14.1 also presents a small part of a computer-generated sequence for a similar intrapair interval (1.3 sec). One could see that the sequence is random, and also that the simulated female responds in average roughly on 50% of the pairs.

The next experiment is with triple interval. The male generates three flashes. The female examines the interflash interval and answers on either the second or the third flash, or sometimes on both. Table 14.2 shows the sequence of triplets in the experiment #419. This is a low-temperature experiment; its summary is presented on curves c and f in Figure 14.1. The

TABLE 14.1 Standard Double Interval Sequence

Experiment #520

Male:	1.15	6	1.15	6	1.15	6	1.15	6	1.15	6	1.15	6	1.15	6	1.15	6
Female:	1	0	0	0	1	0	0	0	0	0	0	0	1	0	1	0

Computer Experiment

Male:	1.3	6	1.3	6	1.3	6	1.3	6	1.3	6	1.3	6	1.3	6	1.3
Female:	0	0	1	0	1	0	1	0	1	0	0	0	0	0	1

TABLE 14.2 Standard Triple Interval Sequence

Experiment #419

Male:	2.7	6	2.7	6	2.7	6	2.7	6	2.7	6	2.7	6	2.7	6	2.7	2.7
Female:	1	0	0	1	0	1	0	1	0	1	0	0	0	1	1	1

Computer Experiment

Male:	2	2	6	2	2	6	2	2	6	2	2	6	2	2	6	2	2
Female:	0	0	0	1	0	0	1	0	0	1	0	0	0	0	0	1	1

interval of 2.7 sec is suitable for both pairs and triplets. Answers on the second, third, and on both the second and third flash occur in the sequences. Also, in some cases the female fails to answer.

Table 14.2 also presents a computer-generated sequence with triple interval of 2 sec. The sequence is random, and again all combinations of female answers are present. It has been checked on longer computer-generated sequences that the average number of responses is in agreement with curve *d* in Figure 14.1.

Blocking, Short Interpair Interval and Extra Flash If the interval between the group of male's flashes is shortened, eventually the female cannot distinguish one group from another.

Table 14.3 shows the sequence of pairs in experiment #419. The intrapair interval is 2 sec, and the interpair interval is also 2 sec. Note that the female answers every fourth flash. One could see on Figure 14.1*f* that 2 sec corresponds with the beginning of triple interval range for this experiment.

Table 14.3 also presents a computer-generated sequence for similar conditions. One could see from Figure 14.1*d* that the triple interval range starts from 1.6 sec. Hence this interval has been chosen in the computer-simulated experiment for both intrapair and interpair interval. Note that the simulated female answers exactly in the same way as a real one, that is, on every fourth flash.

In the next experiment, #419, the interpair interval has been increased to 2.6 sec, Table 14.4. This interval is longer than the blocking interval, and yet it is not long enough for clear distinction between the pairs. As a result the female sometimes missed two intervals in a row, and another time she missed only one interval. Note in Figure 14.1*f* that the interval of 2.6 sec is in the middle of the triplets range.

In Table 14.4 the computer-simulated sequence shows the same feature. The interpair interval of 2 sec is in the middle of the triplets range, Figure 14.1*d*. Note that the female misses sometimes two intervals in a row, and another time she misses only one interval.

Table 14.5 presents a triple interval experiment #541. An extra male's flash is inserted between first and second normal flash. The normal triple interval of 2.35 sec is broken into two intervals: 0.55 and 1.7 sec. The female answers sometimes to the 1.7-sec interval, because it belongs to her double range. She also answers sometimes to the 2.35-sec interval, because it belongs to her triple range. However, as both of these intervals are marginal, in many cases the female does not answer.

Table 14.5 also presents the computer simulated triplets with an inserted extra flash. The triple interval is 2.1 sec. The extra flash breaks this interval into two parts: 0.8 and 1.3 sec. Again the female answers sometimes to the

TABLE 14.3 Blocking

Experiment #419

Male:	2	2	2	2	2	2	2	2	2	2	2	2	2	2	2	2	2	2	2	2
Female:	0	0	0	1	0	0	1	0	0	1	0	0	1	0	0	1	0	0	0	1

Computer Experiment

Male:	1.6	1.6	1.6	1.6	1.6	1.6	1.6	1.6	1.6	1.6	1.6	1.6	1.6	1.6	1.6	1.6	1.6	1.6	1.6	1.6
Female:	1	0	0	1	0	0	1	0	0	1	0	0	1	0	0	1	0	0	1	0

TABLE 14.4 Short Interpair Interval

Experiment #419

Male:	2	2.6	2	2.6	2	2.6	2	2.6	2	2.6	2	2.6	2	2.6	2	2.6	2	2.6	2	2.6	2
Female:	0	1	1	0	1	1	0	0	1	1	0	1	1	0	1	1	0	1	0	0	1

Computer Experiment

Male:	1.5	2	1.5	2	1.5	2	1.5	2	1.5	2	1.5	2	1.5	2	1.5	2	1.5	2	1.5	2	1.5	2
Female:	1	0	1	1	0	1	1	1	0	0	1	1	0	1	1	0	1	0	1	0	1	1

TABLE 14.5 Extra Flash Triple Interval Sequence

Experiment #541

Male:	0.55	1.7	2.35	6	0.55	1.7	2.35	6	0.55	1.7	2.35	6	0.55	1.7	2.35	6
Female:	0	1	0	0	0	1	0	0	0	0	1	0	0	0	1	0

Computer Experiment

Male:	.8	1.3	2.1	6	.8	1.3	2.1	6	.8	1.3	2.1	6	.8	1.3	2.1
Female:	0	1	0	0	0	0	1	0	0	0	0	1	0	0	1

1.3-sec interval, because it belongs to her double range. She also answers to the 2.1-sec interval, because it belongs to her triple range. Again, as both of these intervals are marginal, in many cases the simulated female does not answer.

14.2 Switching the Conditions

If the firefly is flashing under favorable conditions, she adjusts herself into a flashing mood and will flash once, even if the triggering changes towards marginal conditions.

Table 14.6 shows an example of switching the conditions. In the first half of the sequence, the double interval is 1.4 sec. For the experiment #520, one could see from Figure 14.1a that 1.4 sec is a very effective interval, and the firefly flashes every time. In the second half of the sequence, the double interval is 1 sec. This is a rather ineffective double interval and the firefly does not respond to it. However, at the switching point, when the interval of 1 sec is introduced for the first time, the firefly is still in the answer mood, and she flashes on a first ineffective interval.

Table 14.6 also shows computer-generated sequence, switching from an effective interval of 1.5 sec, to a rather ineffective interval of 1.2 sec. Note the same behavior as in a real experiment: the firefly answers the first ineffective interval, and then she stops flashing.

Table 14.7 shows an extreme example of switching conditions in experiment #526. From an effective interval of 1.4 sec the sequence is switched to the extremely ineffective interval of 0.8 sec. Although the firefly is in the answering mood, she will not flash on the first, extremely ineffective interval.

Table 14.7 also shows the computer-simulated sequence for the switch, effective–extremely ineffective. Again the first extremely ineffective pair is not answered.

Now the opposite effect will be described. If the firefly is receiving the ineffective input sequence, she will enter a nonflashing mood and will not flash on the first effective interval. An example provided by experiment #419 is shown in Table 14.8. The double interval of 1.25 sec is ineffective, whereas the double interval of 1.7 sec is rather effective. Note that the firefly fails to answer on the first effective interval. The computer simulation of a similar sequence is also shown in Table 14.8. Again, the firefly fails to answer on the first effective interval.

Table 14.9 shows an extreme example of switching conditions, from experiment #526. From an ineffective interval of 1 sec, the sequence is switched to the very effective interval of 1.4 sec. Although the firefly is in a nonanswering mood, she immediately flashes on the first effective interval. The computer simulation in Table 14.9 mimics this behavior of the female firefly.

TABLE 14.6 Switching from Effective to Ineffective Sequence

Experiment #520

Male: 1.4 6 1.4 6 1.4 6 1.4 6 1.4 6 1.0 6 1.0 6 1.0 6 1.0
Female: 1   0 1   0 1   1 1   0 1   0 1   0 0   0 0   0 0

Computer Experiment

Male: 1.5 6 1.5 6 1.5 6 1.5 5 1.2 6 1.2 6 1.2 6 1.2 6 1.2
Female: 1   0 1   0 1   0 1   0 1   0 0   0 0   0 0   0 0

TABLE 14.7 Switching from Good to Very Bad Sequence

Experiment #526

Male: 1.4 6 1.4 6 1.4 6 1.4 6 1.4 6 0.8 6 0.8 6 0.8 6 0.8
Female: 1   0 1   0 1   0 1   0 1   0 0   0 0   0 0   0 0

Computer Experiment

Male: 1.6 6 1.6 6 1.6 6 1.6 6 1.6 6 0.8 3 0.8 3 0.8 3 0.8
Female: 1   0 1   0 1   0 1   0 1   0 0   0 0   0 0   0 0

TABLE 14.8 Switching from Bad to Good Sequence

Experiment #419

Male:	1.25	6	1.25	6	1.25	6	1.7	6	1.7	6	1.7	6	1.7
Female:	0	0	0	0	0	0	0	0	1	0	1	0	1

Computer Experiment

Male:	1	6	1.8	6	1.25	6	1.8	6	1.8	6	1.8	6	1.8
Female:	0	0	0	0	0	0	1	0	0	0	1	0	1

TABLE 14.9 Switching from Bad to Perfect Sequence

Experiment #526

Male:	1	6	1	6	1	6	1.4	6	1.4	6	1.4	6	1.4
Female:	0	0	0	0	0	0	1	0	1	0	1	0	1

Computer Experiment

Male:	1	6	1	6	1	6	1.5	6	1.5	6	1.5	6	1.5
Female:	0	0	0	0	0	0	1	0	1	0	1	0	1

14.3 Memory and Locked-In Sequences

If the female firefly receives the sequence of male's pairs composed of fairly effective double intervals, she will begin to adapt such that she eventually answers after every pair. Her sequence is now locked to a continuous stream of answers.

An example of a locked in sequence is shown in Table 14.10, taken from experiment #419. The double interval of 2 sec results in approximately 80% responses, Figure 14.1c. However, Table 14.10 shows the probability to respond gradually approaches 100% as the experiment continues. Toward the end the female answers to every pair. The computer-simulated sequence with the locking feature is also shown in Table 14.10. The double interval of 1.8 sec results in approximately 80% responses, Figure 14.1a. However, Table 14.10 shows that female responds irregularly at the beginning of the sequence, and she responds very regularly in the more advanced phase of a sequence.

If the female firefly receives the sequence of male's pairs with relatively ineffective double intervals, she will go through the opposite procedure of adaptation. At the beginning, she will answer occasional pairs. However, as the time passes, she gradually enters the nonanswering mood, and she will stop answering altogether. An example of this type of locked-in sequence is shown in Table 14.11 from experiment #519. The double interval of 1.9 sec results in approximately 40% responses, Figure 14.1a. The probability to respond, however, gradually decreases towards 0% as the experiment continues. Toward the end the female firefly stops answering altogether.

An example of a computer generated-sequence which mimics this behavior is also presented in Table 14.11. The interval of 1.2 sec results in approximately 20% responses, Figure 14.1a. The female responds a few times, and then she stops responding altogether. She is now locked in the no-flashing sequence.

14.4 Computer Model

Different computer models have been designed recently to simulate animal communication. Here are a few examples: cricket chirping,[6] bird duetting,[7] and katydid communication.[8]

The computer model described here, is based on the theory presented in the previous chapter. Because the firefly sequences present, in fact, the random process, Monte Carlo simulation technique has been employed.[9] The actual program is written in BASIC language on the minicomputer PDP-8. The same minicomputer is equipped with an analog to digital convertor and is used also for spike and latency on-line measurements and histogram analysis. The program is divided into two parts: male's sequence

TABLE 14.10 Locking in toward Flashing Sequence

Experiment #419

Male:	2	3.2	2	3.2	2	3.2	2	3.2	2	3.2	2	3.2	2	3.2	2
Female:	0	0	1	0	0	0	0	1	0	1	0	1	0	1	1

Computer Experiment

Male:	1.8	3	1.8	3	1.8	3	1.8	3	1.8	3	1.8	3	1.8	3	1.8
Female:	0	0	1	0	0	0	0	1	0	1	0	1	0	1	1

TABLE 14.11 Locking in toward No Flashing Sequence

Experiment #519

Male:	1.9	6	1.9	6	1.9	6	1.9	6	1.9	6	1.9	6	1.9	6	1.9
Female:	0	0	1	0	0	0	0	0	0	0	0	0	0	0	0

Computer Experiment

Male:	1.2	3	1.2	3	1.2	3	1.2	3	1.2	3	1.2	3	1.2	3	1.2
Female:	1	0	1	0	1	0	0	0	0	0	0	0	0	0	0

Computer Model

generation and female simulation. Here only the female simulation part is described.

The flow chart of the program is presented in Figure 14.2.

Step 1. Presents setting up initial condition in the female firefly. The female is prepared to receive the first male's flash under the following conditions.

- Presently female is not flashing, $F = 0$ (in the future when the female will flash, this will be notified as $F = 1$).
- Female's input is not blocked, $B = 0$ (in the future when the female's input gets blocked, it will be blocked for 1.8 sec, hence $B = 1.8$).
- Female's double interval response function is in progress, $V = 1$. That assumes that she has received one male's flash, just before the simulation starts (in the future, when response function will not be in progress, this will be notified as $V = 0$).
- Memory scaling factor $T_3 = 0.15$ and discrimination scaling factor $E = 0.15$ are introduced. These two parameters describe the contribution of one male's flash to the female's memory and discrimination bias. By choosing the proper values for T_3 and E, the computer-simulated sequences are adjusted to match in the best way the experimentally observed sequences.
- Upper limit of time window is set; $L = 2.3$. The male's interval $I > L$ is ineffective, Figure 14.1a.

Step 2. Is the first step in the actual analysis of one male's flash. The male firefly generates an interval of the length I.

Step 3. The female firefly examines the male's interval and checks if her input is still blocked, $I < B$. If the input is blocked, the male's interval I is rejected, leaving no trace on female.

Step 4. The female firefly checks if her response function is in progress, $V = I$. If there is no response function in progress, the input is the first one of the group. It will produce no answer, but it will start a new response function.

Step 5. This step examines if the response function which is in progress has passed the upper limit point, $I > L$. If yes, the response function has entered the long-term negative inhibitory phase. The male's flash will not produce an answer, but instead it will reset the existing response function, and start a new one.

If the input is not blocked, if there is the response function, and the response function did not enter the long term negative inhibitory phase, the male's flash will be accepted for examination.

Step 6. Presents the discrimination bias generation. The new discrimination bias D is

$$D_{\text{new}} = D_{\text{old}} - T_3 \cdot M$$

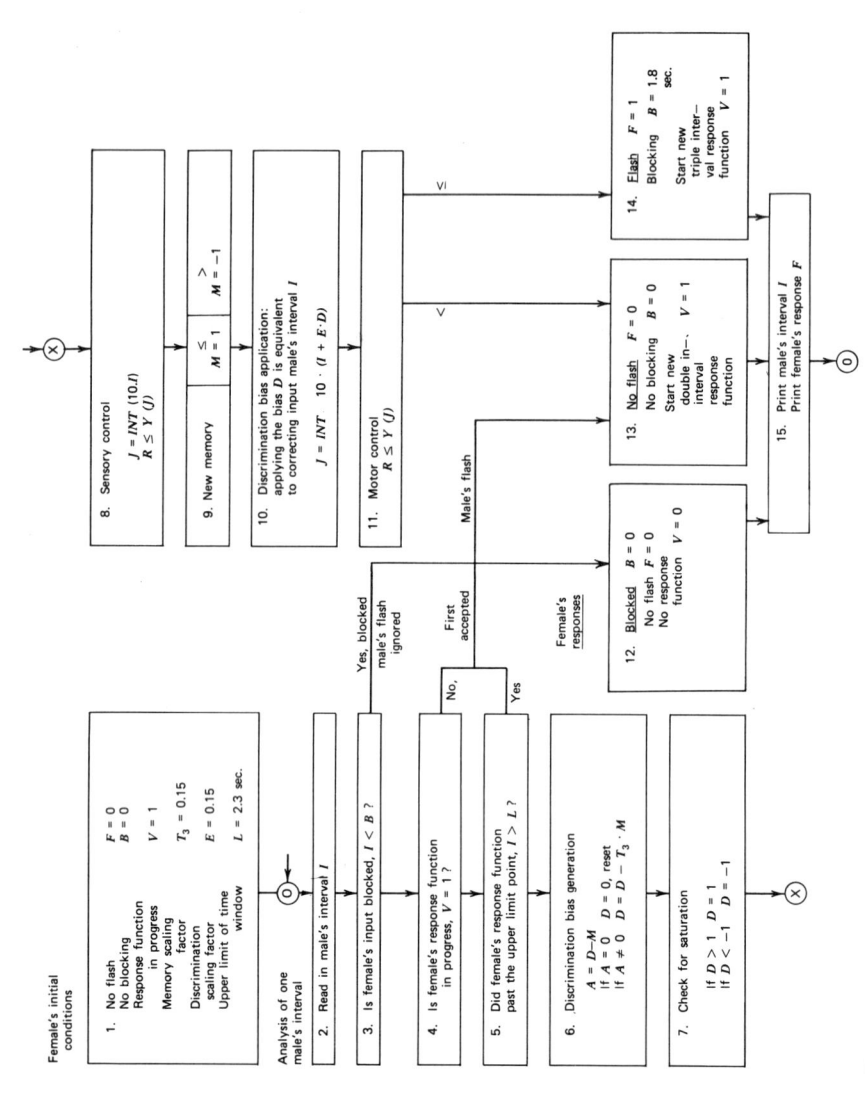

Fig. 14-2 Computer program of female flash response to simulated male flashes of varying interflash interval.

TABLE 14.12

```
1 REM DOUBLE INTERVAL DISTRIBUTION
2 DIM P2(6Ø)
7 RESTORE
8 FOR K=1Ø TO 25
1Ø READ X
11 P2(K)=X
12 NEXT K
22 GO TO 7Ø
25 REM
3Ø REM EXPERIMENTAL DISTRIBUTION
4Ø DATA Ø,.1,.3,.6,1,1,1,.9,.8,.7,.4,.3,Ø,Ø,Ø,Ø
41 RESTORE
42 REM
43 REM
44 REM MALE INTERVAL SEQUENCE GENERATION
46 "ENTER 4-INTERVAL BASIC MALE SEQUENCE"
47 PRINT "FIRST="
49 INPUT Z1
51 PRINT "SECOND="
52 INPUT Z2
54 PRINT "THIRD="
56 INPUT Z3
57 PRINT "FOURTH="
58 INPUT Z4
6Ø FOR K=K1 TO K2
61 N=4*K-3
64 I1(N+3)=Z4
65 I1(N)=Z1
66 I1(N+1)=X2
56 I1(N+2)=Z3
68 NEXT K
69 RETURN
7Ø DIM I1(2ØØ)
71 PRINT "FIRST 1Ø BASIC MALE SEQUENCES"
72 K1=1
73 K2=1Ø
74 GOSUB 47
75 PRINT "SECOND 1Ø BASIC MALE SEQUENCES"
76 K1=11
77 K2=2Ø
78 GOSUB 47
```

(Continued)

TABLE 14.12 (*continued*)

```
         ⎧ 81 PRINT "ENTER 1,-1,0 FOR FLASH,NO FLASH,NEW SEQUENCE"
         ⎪ 82 INPUT M
         ⎪ 83 D=-M
         ⎪ 84 F=0
         ⎪ 85 B=0
         ⎪ 86 V=1
   1.    ⎨ 90 T3=.15
         ⎪ 92 E=.15
         ⎪ 93 L=2.3
         ⎪ 96 PRINT
         ⎪ 97 PRINT "MALE","FEMALE","BIAS",
         ⎩ 98 PRINT
           100 FOR K=1 TO 200
         ⎧ 105 REM MALE INTERVAL IS I
   2.    ⎨ 110 I=I1(K)
         ⎩ 115 IF I=0GO TO 428
   3.    { 120 IF I-BGO TO 290
   4.    ⎰ 121 IF V=0GO TO 200
         ⎱ 122 IF I<1GO TO 200
         ⎧ 124 IF I>LGO TO 200
   5.    ⎨ 125 GO TO 430
         ⎩ 200 GO TO 300
         ⎧ 280 REM BLOCKED INPUT
         ⎪ 290 V=0
  12.    ⎨ 295 F=0
         ⎪ 296 B=0
         ⎪ 297 GO TO 400
         ⎩ 299 GO TO 428
         ⎧ 300 REM NO FLASH,NO BIAS
         ⎪ 310 F=0
  13.    ⎨ 315 B=0
         ⎪ 316 V=1
         ⎩ 320 GO TO 400
         ⎧ 350 REM FLASH AND BIAS
         ⎪ 355 F=1
  14.    ⎨ 360 B=1.8
         ⎪ 365 V=1
         ⎩ 370 GO TO 400
         ⎧ 400 REM PRINT OUTPUT
  15.    ⎨ 424 PRINT
         ⎩ 426 PRINT I,F,D,
```

(*Continued*)

TABLE 14.12 (*continued*)

```
     428 NEXT K
     429 GO TO 999
    ┌ 430 REM DISCRIMINATION BIAS GENERATION
    │ 432 A=D-M
    │ 435 IF A=ØGO TO 445
 6. ┤ 440 D=D-T3*M
    │ 442 GO TO 450
    └ 445 D=Ø
    ┌ 450 IF D>1GO TO 460
    │ 455 IF D<-1GO TO 470
    │ 458 GO TO 500
 7. ┤ 460 D=1
    │ 465 GO TO 500
    └ 470 D=-1
    ┌ 500 REM SENSORY CONTROL
    │ 502 REM NOISE IS R
    │ 505 R=RND(1)
 8. ┤ 510 X=I-1.6
    │ 515 J=INT(16+10*X+.5)
    │ 520 Y=P2(J)
    └ 525 IF R<YGO TO 550
    ┌ 530 M=-1
 9. ┤ 540 GO TO 560
    └ 550 M=1
    ┌ 555 REM DISCRIMINATION BIAS APPLICATION
    │ 560 IF X<=ØGO TO 580
    │ 565 X=X+D*E
    │ 570 IF X>ØGO TO 590
    │ 571 IF B=ØGO TO 574
    │ 572 X=B-1.6
    │ 573 GO TO 590
10. ┤ 574 X=Ø
    │ 578 GO TO 590
    │ 580 X=X-D*E
    │ 585 IF X<=ØGO TO 590
    │ 588 X=0
    │ 590 J=INT(16+10*X+.5)
    └ 620 Y=P2(J)
    ┌ 622 REM MOTOR CONTROL
11. ┤ 625 IF R<YGO TO 350
    └ 630 GO TO 300
     999 END
```

where M is the result stored in the memory during the examination of previous interval. If D reaches the saturation level, it is reset to zero.

Step 7. Checks for saturation.

Step 8. Presents the sensory control. The male's interval is checked against the time window. It will be accepted only if it is within the time window. The upper and lower limits of the time window vary in random fashion, according to the curve *a* in Figure 14.1. This curve, $Y(J)$ is stored in memory as an array. To find the probability of acceptance, the random number generator is used.

For a given male's interval I, the program calculates the index value $J = \text{INTEGER } (10 \cdot I)$, and reads the acceptance probability $Y(J)$ from the curve *a* in Figure 14.1. The random number R is compared with $Y(J)$.

Step 9. Determines the value for a new memory M.

$$\text{If} \quad R \leq Y(J) \quad \text{accepted}, \quad M = +1$$
$$R > Y(J) \quad \text{rejected}, \quad M = -1$$

The sensory control is used only to store the result in the memory M. This result will be used in the future to correct the discrimination level, for the next interval to be analyzed, step 6.

Step 10. is discrimination bias application for motor control. Discrimination bias D is used to adjust the female's sensitivity. In the program this is done by correcting the male's interval by the value $D \cdot E$, where E is discrimination scale factor. For a given male's interval I, the program calculates the index value $J = \text{INTEGER } [10(I + D \cdot E)]$, and reads the acceptance probability $Y(J)$, from the curve *a* in Figure 14.1.

The same curve is shown also in Figure 14.3. Note the following: Negative discrimination bias, $D < 1$, increases the sensitivity, that means it increases the probability of acceptance of a given male's interval I. Positive discrimination bias, $D > 1$, decreases the sensitivity, that means it decreases the probability of acceptance of a given male's interval I.

Step 11. is actual motor control. The discrimination bias has been applied to the time discriminator, and the window width has been modified accordingly in the step 10. Now the random number generator is used, to find out if the male's interval is inside the modified time window:

$$\text{If} \quad R \leq Y(J) \quad \text{female produces the flash}$$
$$\text{If} \quad R > Y(J) \quad \text{female will not flash}$$

Steps 12, 13, and 14 present possible female's responses.

Step 12. Female's input is blocked. As a result male's flash is rejected.

Step 13. Female's input is open, but the male's flash is outside of the time window, or response function has not been started. Start new response function.

Significance of the Computer Modeling

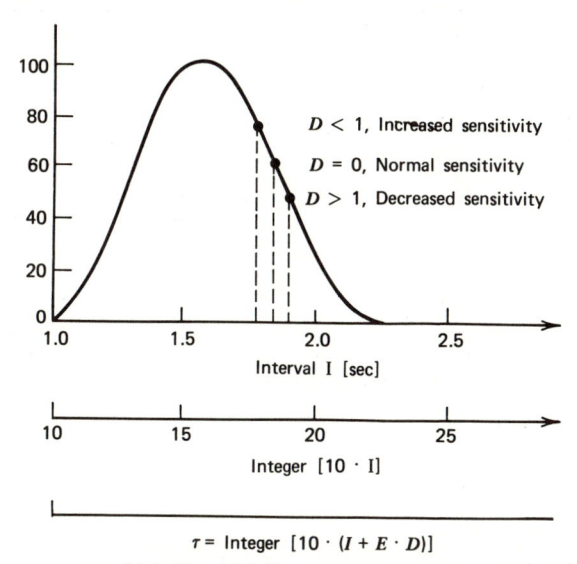

[Graph showing percent response curve peaking near I = 1.5 sec, with markers indicating D < 1 Increased sensitivity, D = 0 Normal sensitivity, D > 1 Decreased sensitivity. Interval I [sec] axis from 1.0 to 2.5; Integer [10 · I] axis from 10 to 25.]

$\tau = \text{Integer}\,[10 \cdot (I + E \cdot D)]$

Fig. 14-3 Same as curve (a) in Fig. 14.1. Percent response of female firefly to male double flash intervals in seconds. Response curve of computer and experiment at 26°C, D is discrimination bias, I is male flash interval, E is discrimination scale factor, and τ is index value.

Step 14. Male's flash is inside the time window, female produces the flash, blocks her input for 1.8 sec, and starts new triple interval response function.

Note that in this program the triple interval acceptance probability curve, Figure 14.1d, is replaced with the double interval curve, Figure 14.1a: for the values of $I < 1.8$ sec probability is kept at zero. For the values of $I > 1.8$ sec curve a is used. For the purpose of this simulation, the tail of the curve a is close enough approximation of the curve d.

Step 15. Prints male's intervals and corresponding female's responses.

This is the end of the analysis of one male's interval. Now the program goes back to point 0, new male's interval is generated and the whole procedure is repeated.

14.5 Significance of the Computer Modeling

The observations of responses of female *Photinus macdermotti* fireflies to artificial male flashes of various interflash intervals have been used to develop a functional computer that can predict female response patterns. With the model the computer can be programmed to simulate the behavior of the female firefly to male flashes.

A comparative study between measured and computer simulated sequences has been performed. Special attention has been given to the study of the following phenomena:

1. Standard double interval sequences and their random fluctuations.
2. Standard triple interval sequences and their random fluctuations.
3. Blocking, short phrase sequences, and extra flash sequences.
4. Switching the conditions from effective intervals to ineffective and vice versa.
5. Memory effects and locked in sequences.

The theory and the designed computer model have enabled us to recognize the above phenomena in the large amount of data produced in experiments. Without the model it would be especially difficult to recognize some of the rare events, because of the strong random component present in firefly flashing sequences.

It is shown that the theory and the computer-generated sequences are in very good agreement with experimental findings. The computer model has generated correct sequences for all the conditions that have been investigated. It is shown that the simple, three phase response function can explain all the deterministic components in the firefly sequences. Statistical fluctuations in time window width result from random noise, which is always present in actual biological structures. The theory and the model explain both, deterministic and random components present in the sequences.

It is shown that short-term memory, discrimination bias, and sensitivity adjustments are needed to explain the behavior in switching from effective to ineffective conditions and vice versa. Additional experiments would be needed to find out the contribution of each flash to the memory. Also, new measurements may be performed to find the range and the amount of sensitivity adjustments.

The memory also explains the effect of locking in the sequences. If the female firefly receives the sequence of male's flashes with fairly effective double intervals, she will go through the procedure of adaptation. At the beginning she will not answer on all pairs. However, as the time passes by, she gradually enters into the answering mood, and she will answer virtually on all pairs. Her sequence is not locked to a continuous stream of answers. To measure precisely the locking in phenomena, longer sequences will be needed than those available in present experiments.

The computer model is a description of the female flash response functions and their interrelations. It does not predict how the process is carried out in terms of neural networks. However, the separation of the flash recognition and response process into separate functions facilitate the design of new experiments. Some experiments that may further refine the model are as follows:

1. Contribution of each male flash to female memory. What are the short-term and long-term time constants involved in enhancement and reduction of females response probability?

2. Numerical parameters of sensory control. What is a unit increase or decrease of window width? What are the maximal and minimal window widths?

3. Does reset go completely to zero when an incorrect flash pair is presented?

4. How does the positive bias increase in the female over long periods of male flash deprivation?

References

1. Lloyd, J. E.: *Misc. Publ. Mus. Zool.*, Univ. Michigan, No. **130** (1966) 1–95.
2. Buck, J., and Buck, E.: *Biol. Bull.*, **142** (1972) 195–205.
3. Carlson, A. D., Copeland, J., Raderman, R., and Bulloch, A. G. M. Role of Interflash Intervals in a Firefly Courtship (*Photinus macdermotti*).
4. Carlson, A. D., Copeland, J., Raderman, R., Bulloch, A. G. M. Response Patterns of Female *Photinus macdermotti* Firefly to Artificial Flashes.
5. Souček, B., and Carlson, A. D. Flash Pattern Recognition in Fireflies. (in press).
6. Heiligenberg, W.: *Z. Vergl. Physiol.*, **65** (1969) 70–97.
7. Souček, B., and Vencl, F.: *J. Theor. Biol.*, **49** (1975) 147–172.
8. Souček, B. Model of Alternating and Aggressive Communication with the Example of Katydid Chirping. *J. Theor. Biol.* **52** (1975) 399.
9. Souček, B.: *Minicomputers in Data Processing and Simulation*, Wiley, New York, 1972.

Chapter 15

NEURAL, COMMUNICATION, AND BEHAVIORAL SEQUENTIAL PATTERNS

Introduction and Survey

Sequential behavioral patterns present the basis for animal communication. For precise study of such patterns large amounts of data must be analyzed. Statistical analysis of the motor patterns could be best accomplished using programmed digital computer. Developed methods and computer programs could be used in general to analyze any coded behavioral pattern, or neural spike pattern.

The pattern could be a sequence of chirps or sounds used in acoustical communication. It could be also a sequence of flashes used in firefly communication, a sequence of displays describing animal behavior, or the like. Here the methods are used to study organization of bird songs and animal communication in bird duetting (white-crested jay thrush). The duetting songs are composed of 4 syllables for female and up to 20 syllables for male. Transition probabilities between syllables, pairs, triplets, and subsequences have been measured. Programmed associative memory is used to efficiently store and display hundreds of songs in the form of one single tree. The tree pattern is a new method to present both the communication process and behavioral control. As the songs are displayed in a sorted and comparative way, one can read from the tree the basic message units, decision making points, and variations in songs, frequencies, and probabilities of transitions.

This chapter describes both sequential pattern analyzing methods and concrete computer programs. The programs are tested on the example of bird song analysis,[1] but they can be used to analyze other sequential patterns as well.

15.1 Basic Elements or Syllables

Example of the Bird Song Study of bird songs provide much information about the sequential organization and control of a behavior which functions as communication.[2-6] Recently, theories have been developed to explain firefly flashing communication,[7] and also to explain katydid chirping communication.[8] Although there are great differences between katydids, fireflies, and especially the birds, there are also some similarities in their communication pattern. The present chapter gives an account of song organization and duetting in the white-crested jay thrush (family Timaliidae, *Garrulax leucolophus patkaicus*).

Bird singing is expressed as a series of elements of varied frequency lying between 1.5 and 8 kHz. The individual elements of the song are called syllables. Syllables result from waveform analysis of acoustical records through continuous sampling and Fourier transformation. Four different syllables are found in the female song, coded 1 to 4, and twenty syllables are found in the male song, coded 5 to 25.[9]

A syllable can be recognized as a continuous trace on the sonogram which is separated from other traces of the same individual by 75 msec or more. Two adjacent syllables in the song form a pair, and three syllables form a triplet. Syllable, pair, triplet, and song definitions are outlined in Figure 15.1.

One of the basic questions in sequential behavioral analysis is How strong is the influence of a given syllable, pair, or triplet, on the element (syllable), which will appear N time positions later in the sequence? The presented

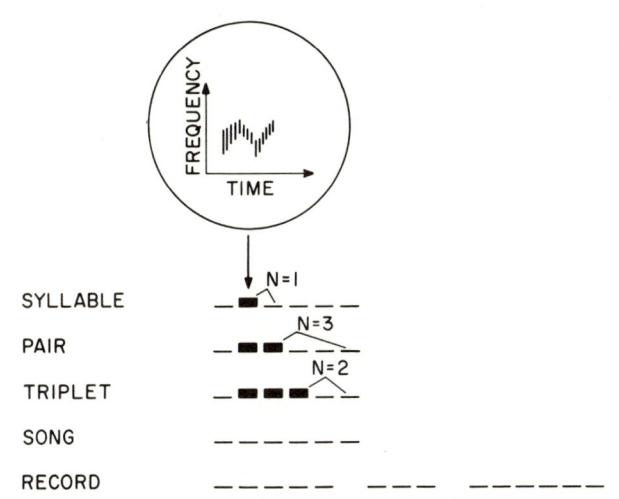

Fig. 15-1 Definitions of syllables, pairs, triplets, songs, and records.

computer programs are written in such a way that the experimenter can select any value for the distance N. However in reality only the distances 1, 2, or 3 are of interest.

The analyzed record is composed of 141 duetting songs and contains approximately 3000 syllables. By coding the female syllables with codes 1 to 4 and male syllables with codes 5 to 25, a digital record has been produced as the input to the computer.

By introducing digital computer in bird communication study, it was possible to provide many classes of analysis with high statistical accuracy. Special attention is given to the study of motor patterns of each individual and to the study of message switching in between individuals during duetting.

The study is based on recordings of one pair of birds in summer 1973. Songs were recorded at 9.5 cm/sec, on a UHER-4200 Report stereo tape recorder using a UHER M517 microphone. Frequency spectra defined as syllables were produced on a Kay electric sonagraph. Coded sonagrams were preanalyzed on a minicomputer PDP-8, in BASIC language. Final analysis was performed in a PDP-10 computer in FORTRAN.

Analysis of Syllables The computer program for analysis of syllables is presented in Figure 15.2. It reads the record and produces the transition matrix one syllable one syllable. Table 15.1 shows the matrix for the dis-

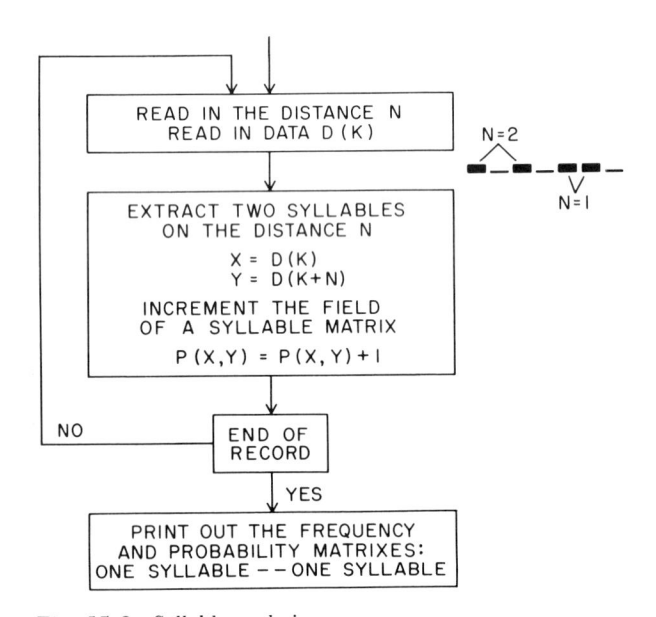

Fig. 15-2 Syllable analysis program.

TABLE 15.1

```
N = 1
ONE SYLABLE - ONE SYLABLE
FREQUENCIES
                NEXT ELEMENT
1ST  2ND  3RD  SUM    1    2    3    4    5    6    7    8    9   10   11   12   13   14   15   16   17   18   19   20   21   22   23   24   25
 1    0    0   463                        1   86        2    3                                                 23         266   26   32
 2    0    0   138         3    3    6    4   15                                                                     1              78   20
 3    0    0   316         4  118    6   10    3   26    2    3    2              14   21    9    1                  6    7    1    9   79    2
 4    0    0   585         2  147   43   14   60   16   36   38              11   18   44         1   20   10   19   1    1   93   10
 5    0    0    57        11   42                                   1
 6    0    0   110   91    3       18
 7    0    0   100        12   41   46
 8    0    0    18              4   14
 9    0    0    40              5   34         1
10    0    0    44              4   34              1                              1    5
11    0    0     0
12    0    0     0
13    0    0    33             16   17                                        1
14    0    0    50   18    3    8   19                        1                              1                                               1
15    0    0    57             7   48
16    0    0     1                                                                                        1
17    0    0    16   15                                                                                   1    1
18    0    0     1                  1
19    0    0    20                 20
20    0    0    40         1   24   14
21    0    0    27              7   20
22    0    0   356  353         1                        1
23    0    0   155  132   22                                                                                                       1
24    0    0   225    1    7  179   38
25    0    0    14        1   12    1
SUM =     2866  478  161  341  648   59  105  101   18   41   45    0    0   33   50   59    1   15    1   20   41   27  268  115  225   14
```

261

tance between syllables N = 1. The program reads the data and analyses two syllables at a time D(K) and D(K + N). The values of those syllables X = D(K) and Y = D(K + N), are used to address the matrix P(X,Y). The addressed location is used as a counter. Whenever the analyzed two syllables produce the values X and Y, one is added to the location P(X,Y). This operation is repeated, until the end of record is reached. The program then prints, the frequency tables, such as table 15.1. The table displays the transition frequencies from syllable X (first element, vertical) into syllable Y (next element, horizontal). It also displays row and column sums as well as a total sum. The total sum is used to normalize the frequency matrix, producing a new, probability matrix, as shown in Table 15.2. Note that in the probability matrix only the transitions with the probabilities higher than 1% are displayed.

From the frequency Table 15.1 we see that the most frequent transition is from syllable 22 (male) into syllable 1 (female). This transition occurs 353 times. From probability Table 15.2 one can read that this transition presents 12% of all analyzed transitions.

Note that transitions from male syllables (5 to 25) into female syllables (1 to 4) are very frequent. Also, the reverse is true. On the other hand, male-male transitions are almost nonexistant and could be treated as spurious events. At the same time female-female transitions are very frequent: syllable F3 is followed by F4 for 118 times; syllable F4 is followed by F4 for 147 times. Hence we can conclude that the male will not sing without an immediate female response; the female will sing without an immediate male response. Also, the female will repeat the same syllable, namely F4.

Table 15.3 shows the frequencies of syllable transitions for N = 2. Note very frequent transitions: F1 into F1 (with another syllable in between) 361 times; F4 into F4, 311 times, M22 into M22, 262 times. Also one is able to see the diagonal over the whole matrix, showing the fact of very frequent repetition of the same syllable (with another in between).

From the tables presented one could always read conditional probabilities: given an element, what are the relative frequencies of the elements to follow? For example, from Table 15.1, one could read: given syllable 23, the probability that syllable 2 will follow is 132/155.

15.2 Pairs

A computer program for analysis of pairs is presented in Figure 15.3. Here the patterns are analyzed with a repertoire of up to 25 different syllables. Hence they could form 25 × 25 = 625 different pairs. Without computer the analysis of pairs would be very limited. The computer program is written

TABLE 15.2

```
N = 1
ONE SYLABLE - ONE SYLABLE
PROBABILITIES *100
```

NEXT ELEMENT

1ST	2ND	3RD	SUM	1	2	3	4	5	6	7	8	9	10	11	12	13	14	15	16	17	18	19	20	21	22	23	24	25
1	0	0	16																						9		1	
2	0	0	4																							2		
3	0	0	11				4																				2	
4	0	0	20				5	1		2		1															3	
5	0	0	0				1						1															
6	0	0	3	3																								
7	0	0	3			1																						
8	0	0	0																									
9	0	0	1				1																					
10	0	0	1				1																					
11	0	0	0																									
12	0	0	1																									
13	0	0	1																									
14	0	0	1																									
15	0	0	1				1																					
16	0	0	0																									
17	0	0	0																									
18	0	0	0																									
19	0	0	0																									
20	0	0	1																									
21	0	0	0																									
22	0	0	12	12																								
23	0	0	5		4																							
24	0	0	7			6	1																					
25	0	0	0																									
SUM =			100	16	5	11	22	2	3	3	0	1	1	0	0	1	1	2	0	0	0	0	1	0	9	4	7	0

TABLE 15.3

N = 1
ONE PAIR - ONE SYLABLE
FREQUENCIES

NEXT ELEMENT

1ST	2ND	3RD	SUM	1	2	3	4	5	6	7	8	9	10	11	12	13	14	15	16	17	18	19	20	21	22	23	24	25
1	1	0	61	61																								
1	6	0	212	210	1																		*8				
1	22	0	22	1	5	16																					
1	24	0	71		61	10																				1	
2	23	0	70				3	2	10		5	2				1	4	1	1	1	1			7			34	
3	4	0	50			39	11			12	7	3	4			2	6	1				2	1	6			15	2
3	24	0	72				7	3																				
4	4	0	32			1	29																					
4	5	0	40			9	30											1										
4	7	0	29			2	26	1				1																
4	9	0	26			1	19											5										
4	17	0	27			1	26																					
4	15	0	70			54	16																					
4	24	0	29				8	6	1	1		9	1			1		2										1
5	4	0	76			1		1	59	11			2				2											
6	1	0	20				6			6	1																	
7	3	0	29				14	2	9		1																	
7	4	0	21				6	5	1	2		12	4					2										
9	4	0	26				3	2	1	1			1				1	21										
10	4	0	32				1	3	17											5	1	2	6					2
15	4	0	309																						245			
22	1	0	98				4			8						1								1		10	1	
23	2	0	123		2	3	57	1	6	6						7	2	2		5				3		66	26	
24	3	0	26			2	24			1														1		45	11	
24	4	0	0																								6	
0	0	0	0																									
SUM =			1586	272	68	140	288	25	88	58	14	27	12	0	0	12	15	34	1	5	1	2	7	18	247	86	142	7

264

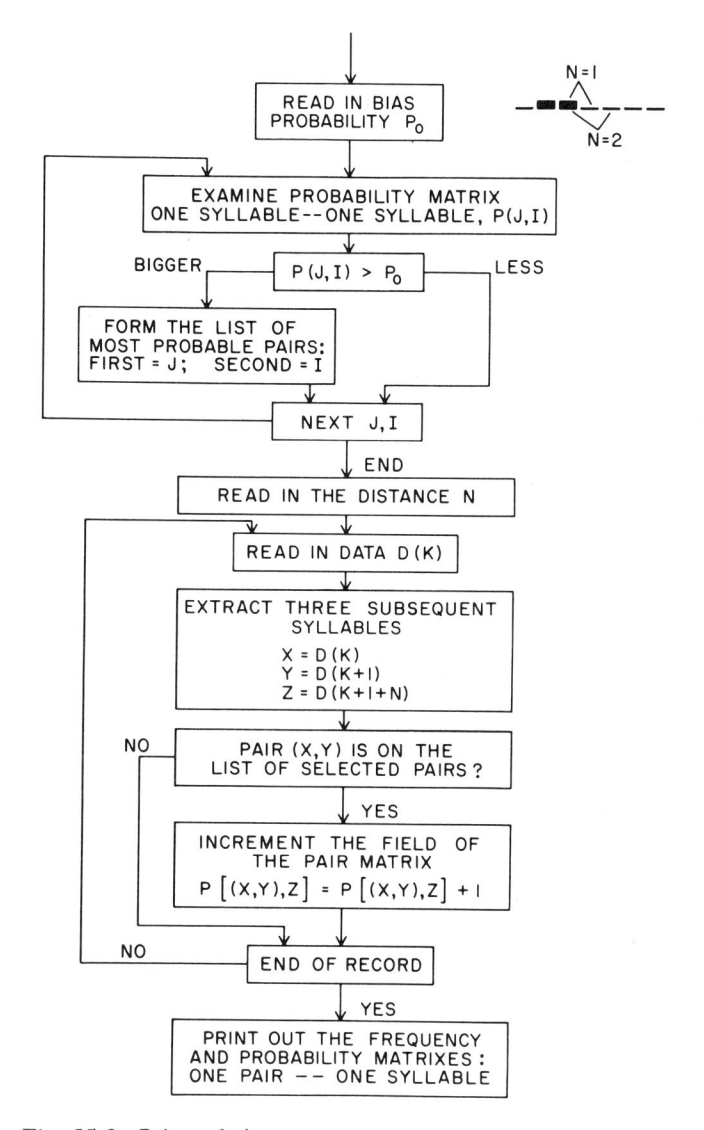

Fig. 15-3 Pair analysis program.

in such a way that it selects out of all possible pairs only the most probable pairs. The program, in fact, investigates the probability matrix one syllable-one syllable, Table 15.2. It compares the probability for each transition with bias probability Po. The experimenter can choose any value for the bias probability Po. If the probability of a transition $P(J, I) \geq P_o$, the pair J, I is inserted into a list of most probable pairs. For example, for Po = 1%, one could see from Table 15.2 that the list of pairs will look as follows:

$$
\begin{aligned}
&\text{1st pair} = 1, 6 \quad \text{(probability 3)} \\
&\text{2nd pair} = 1, 22 \quad \text{(probability 9)} \\
&\text{3rd pair} = 1, 24 \quad \text{(probability 1)} \\
&\text{4th pair} = 2, 23 \quad \text{(probability 2)} \\
&\text{and so forth}
\end{aligned}
$$

The program now prepares the space for a new matrix = one pair-one syllable, as shown in Table 15.4. The vertical column presents the list of most probable pairs, horizontal column presents the syllable following after the pair.

The transition frequency for a given pair-syllable combination is obtained in the following way (Figure 15.3).

The program reads three subsequent data $D(K)$, $D(K + 1)$, and $D(K + 1 + N)$. The values $X = D(K)$ and $Y = D(K + 1)$ are treated as a pair. The value $Z = D(K + 1 + N)$ is the syllable following the pair at a distance N.

The new pair XY is compared with all pairs on the list of most probably pairs. If such pair is on the list, one is added to the matrix element $P[(X,Y),Z]$.

After the whole record is analyzed, the frequency matrix one-pair-one syllable is printed, as shown in Table 15.4.

Dividing all values of Table 15.4 with the total number of analyzed transitions one finds transition probabilities, as shown in Table 15.5. Again, only probabilities above 1% are displayed. The most frequent pair-syllable transitions are as follows:

- Pair (22, 1) is followed by syllable 22, 15% of all transitions.
- Pair (1, 22) is followed by syllable 1, 13% of all transitions.
- Pair (24, 3) is followed by syllable 4, 3% of all transitions.
  and so on.

Pair-syllable matrixes will prove very useful in explaining the basic sequences of songs.

TABLE 15.4

```
N = 1
ONE PAIR - ONE SYLABLE
PROBABILITIES *100
```

					NEXT ELEMENT																						
1ST	2ND	3ND	SUM	1	2	3	4	5	6	7	8	9	10	11	12	13	14	15	16	17	18	19	20	21	22	23	24	25
1	6	0	3	3																								
1	22	0	13	13																								
1	24	0	1			1																						
2	23	0	4		3																							
3	4	0	4																								2	
3	24	0	3			2																						
4	4	0	4																									
4	5	0	1				1																					
4	7	0	2				1																					
4	9	0	1				1																					
4	10	0	1				1																					
4	15	0	1				1																					
4	24	0	4			3	1																					
5	4	0	1																									
6	1	0	4						3																			
7	3	0	1																									
7	4	0	1				1																					
9	4	0	1																									
10	4	0	1																									
15	4	0	2																1									
22	1	0	19						1																15		1	
23	2	0	6																							4		
24	3	0	7			3																					2	
24	4	0	1				1																					
0	0	0	0																									
SUM =			100	17	4	8	18	1	5	3	3	1	0	0	0	0	0	2	0	0	0	0	0	1	15	5	8	0

267

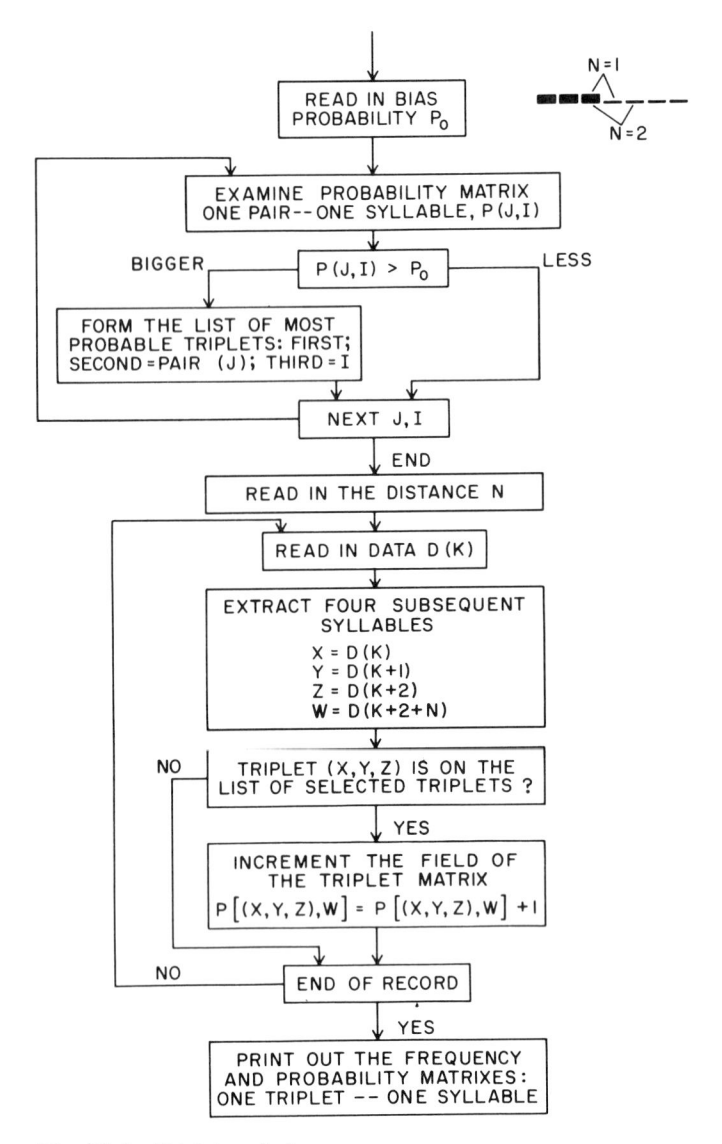

```
                    ┌─────────────────────┐
                    │   READ IN BIAS      │
                    │   PROBABILITY P₀    │
                    └─────────────────────┘
                              │
                    ┌─────────────────────────────────┐
            ┌──────▶│ EXAMINE PROBABILITY MATRIX      │
            │       │ ONE PAIR -- ONE SYLLABLE, P(J,I)│
            │       └─────────────────────────────────┘
            │                 │
            │   BIGGER  ┌───────────────┐  LESS
            │  ┌────────│  P(J,I) > P₀  │────────┐
            │  │        └───────────────┘        │
            │  ▼                                 │
            │ ┌──────────────────────────────┐   │
            │ │ FORM THE LIST OF MOST        │   │
            │ │ PROBABLE TRIPLETS: FIRST;    │   │
            │ │ SECOND = PAIR (J); THIRD = I │   │
            │ └──────────────────────────────┘   │
            │             │                      │
            │             ▼                      │
            │       ┌──────────────┐             │
            └───────│  NEXT J, I   │◀────────────┘
                    └──────────────┘
                          │ END
                          ▼
                  ┌──────────────────────┐
                  │ READ IN THE DISTANCE N│
                  └──────────────────────┘
                          │
            ┌─────────────▼──────────────┐
            │      ┌──────────────────┐   │
            │      │ READ IN DATA D(K)│   │
            │      └──────────────────┘   │
            │            │                │
            │   ┌────────────────────────┐│
            │   │ EXTRACT FOUR SUBSEQUENT││
            │   │        SYLLABLES       ││
            │   │      X = D(K)          ││
            │   │      Y = D(K+1)        ││
            │   │      Z = D(K+2)        ││
            │   │      W = D(K+2+N)      ││
            │   └────────────────────────┘│
            │            │                │
            │   NO  ┌──────────────────────┐
            │  ┌────│TRIPLET (X,Y,Z) IS ON │
            │  │    │LIST OF SELECTED      │
            │  │    │TRIPLETS ?            │
            │  │    └──────────────────────┘
            │  │         │ YES
            │  │         ▼
            │  │  ┌──────────────────────────┐
            │  │  │ INCREMENT THE FIELD OF   │
            │  │  │   THE TRIPLET MATRIX     │
            │  │  │P[(X,Y,Z),W]=P[(X,Y,Z),W]+1│
            │  │  └──────────────────────────┘
            │  │         │
            │  │   NO    ▼
            └──┴─────┌──────────────┐
                     │END OF RECORD │
                     └──────────────┘
                            │ YES
                            ▼
                  ┌────────────────────────────┐
                  │ PRINT OUT THE FREQUENCY    │
                  │ AND PROBABILITY MATRIXES:  │
                  │ ONE TRIPLET -- ONE SYLLABLE│
                  └────────────────────────────┘
```

Fig. 15-4 Triplet analysis program.

15.3 Triplets

Computer program for analysis of triplets is presented in Figure 15.4. Here the patterns are analyzed with repertoire of 25 syllables, and the number of possible triplets is so big that it would be impossible to make any analysis without the help of a computer. Again, out of all possible triplets, the computer selects only the most probable triplets. The program investigates the probability matrix one pair-one syllable, Table 15.5, and compares each value with the bias probability Po. If the probability of a transition $P(J, I) \geq$ Po, the triplet is inserted into the list of most probably triplets. For example, for Po = 1%, one could see from Table 15.5 that the list of triplets will look as follows:

1st triplet: pair (1, 6), syllable 1 = 1, 6, 1 (probability 3)
2nd triplet: pair (1, 22), syllable 1 = 1, 22, 1 (probability 13)
3rd triplet: pair (1, 24), syllable 3 = 1, 24, 3 (probability 1)
4th triplet: pair (2, 23), syllable 2 = 2, 23, 2 (probability 3)
and so on

The program now prepares the space for a new matrix: one triplet-one syllable, as shown in Table 15.6. The vertical column presents the list of most probably triplets; the horizontal column presents the syllable following after a triplet.

The transition frequencies for a given triplet-syllable combination are obtained in the following way (Figure 15.4).

The program reads four subsequent data $D(K)$, $D(K + 1)$ $D(K + 2)$, and $D(K + 2 + N)$. The values $X = D(K)$, $Y = D(K + 1)$ and $Z = D(K + 1)$ are treated as a triplet. The value $W = D(K + 2 + N)$ is the syllable following the triplet at a distance N.

The new triplet X, Y, Z is compared with all triplets on the list of most probable triplets. If such a triplet is found, one is added to the matrix element $P[(X, Y, Z), W]$.

Table 15.7 presents probability matrix for $N = 1$. The most frequent triplet-syllable transitions are as follows: (22, 1, 22)1 (20% of all transitions; (1, 22, 1)22 (14%); (23, 2, 23)2 (5%), and so on. Note the strong tendency of the male to repeat his syllable, and of the female to repeat her syllable.

15.4 Trees

Although transition matrixes are useful for many discussions, for a more detailed picture the time information will be needed. For this reason a computer program was written that superimposes the songs. The song is split

TABLE 15.5

N = 1
ONE TRIPLET - ONE SYLABLE
FREQUENCIES

NEXT ELEMENT

1ST	2ND	3RD	SUM	1	2	3	4	5	6	7	8	9	10	11	12	13	14	15	16	17	18	19	20	21	22	23	24	25
1	6	1	61			1		1	47				1				2								1			
1	22	1	200						10														5		154	8	24	
1	24	3	16																							7	16	
2	23	2	42		2		3			7														1		27	2	
3	4	24	32			25	7																					
3	24	3	34				13	1	3			7				4		1									12		
4	5	4	24				8	4	1	1			1			1		2											1
4	7	4	24				14	2	6		1																		
4	9	4	18				6	3	8		1																		
4	10	4	18				1	2		1		9	4					1											
4	15	4	23					1		1			1				1	17											
4	24	3	42				30			3						2							2				1	5	1
4	24	4	16				11																	1					4	
6	1	6	45	45																									
7	4	4	11				2			5		1	1															1	1
15	4	15	16				16																						
22	1	6	16	16																									
22	1	22	211	209		1																						1	
22	1	24	20	1	5	14																							
23	2	23	64		55	9																							
24	3	4	49				3			7	1	1				1								3			33	
24	3	24	30			23	7																						
24	4	4	15							2	1																	12	
0	0	0	0																									
0	0	0	0																									
SUM =			1028	271	62	73	121	14	65	36	4	18	8	0	0	8	3	21	0	0	0	0	5	7	155	42	111	3

270

TABLE 15.6

N = 1
ONE TRIPLET - ONE SYLABLE
PROBABILITIES *100

NEXT ELEMENT

1ST	2ND	3RD	SUM	1	2	3	4	5	6	7	8	9	10	11	12	13	14	15	16	17	18	19	20	21	22	23	24	25
1	6	1	5						4																			
1	22	1	19																						14			
1	24	3	1																									1
2	23	2	4																							2		
3	4	24	3			2																					1	
3	24	3	3				1																					
4	5	4	2																									
4	7	4	1				1																					
4	9	4	1																									
4	10	4	1																									
4	15	4	2															1										
4	24	3	4				2																					
4	24	4	1				1																					
6	1	6	4	4																								
7	4	4	1																									
15	4	15	1				1																					
22	1	6	1																									
22	1	22	20	1																								
22	1	24	1	20																								
23	2	23	6			1																						
24	3	4	4		5																							
24	3	24	2			2																						
24	4	4	1																									
0	0	0	0																									
0	0	0	0																									
SUM =			100	26	6	7	11	1	6	3	0	1	0	0	0	0	0	2	0	0	0	0	0	0	15	4	10	0

271

TABLE 15.7

```
N = 2
ONE SYLABLE - ONE SYLABLE
FREQUENCIES
                NEXT ELEMENT
1ST 2ND 3RD SUM  1   2   3   4   5   6   7   8   9  10  11  12  13  14  15  16  17  18  19  20  21  22  23  24  25
 1   0   0  459 361  32  58   4                                                                              1
 2   0   0  135   3  79  42   4                                                                                  3
 3   0   0  309  12   8 113  63   1  21   1              8       3   5   3                           1           7
 4   0   0  547   5 111 311  16   6   1  22   7   7   1                   3   8  13           1       9   1       1  55
 5   0   0   55              14       1   2   1  13  20       3   1   3              2   8           1   2       2  25
 6   0   0  109           1   1   1  68      25          1                                   1                  10       4
 7   0   0   86              6       1       1   1       4                                                       9       3
 8   0   0   13               3         1        1   1                                                                2
 9   0   0   35             11  10   2       1       6                   1
10   0   0   41           1   8   2   2                              2
11   0   0    0
12   0   0    0
13   0   0   30             16                       1               3   8   2                                   2
14   0   0   48           1   3                          2          13   4  27  1           22                   2
15   0   0   52           1   8   5   2   1                              2
16   0   0    1
17   0   0   16                   1                                                      9           6
18   0   0    1
19   0   0   18                                                                      1  15
20   0   0   38           1  15       2   2                          16                      1   3
21   0   0   26              17      18                                       4
22   0   0  342   1       1      10                                       2   3   6          6   2 262  16  32
23   0   0  132       4   3  11       9                                       7   3              2  72  23
24   0   0  214       4 112           4                                           0              4   1   1  72
24   0   0   14               3   1                                                                              1

SUM = 2721 382 123 338 647  59 104 101  17  41  43   0   0  32  50  59   1  15   1  20  41  27 268 113 225  14
```

into time intervals T1, T2 ··· T50. One time interval belongs to one syllable. The program overlays the songs in such a way that the time interval T1 of all the songs is synchronized. The program then counts the frequency of occurrence of a syllable at a given time interval. To show the dependence on the preceding syllables, overlaying technique produces the branches, whenever a new song differs from those songs already forming overlying tree. For example:

$$\text{Song 1: } 1, 1, 2, 3, 4, 4$$
$$\text{Song 2: } 1, 1, 3, 4, 4$$

The overlay will look as follows:

$$\frac{\text{Syllable}}{\text{frequency}} \quad \frac{1}{2} \to \frac{1}{2} \to \frac{2}{1} \to \frac{3}{1} \to \frac{4}{1} \to \frac{4}{1}$$
$$\searrow \frac{3}{1} \to \frac{4}{1} \to \frac{4}{1}$$

Note that for time T1 and T2 both songs follow the same pattern, but after T2 the songs are different, producing two separate branches in the overlaying tree. As a result the frequencies (lower number) for the first and second node are 2, whereas for higher nodes the frequencies are 1. The tree program is shown in Figures 15.5 and 15.6. The program reads in one song at a time. It compares first syllable of the song $F(1)$, with the root syllable R. One can choose any root syllable. If the song starts with the selected root syllable, it will be analyzed; otherwise, the program reads in a new song sequence.

The next step is the overlaying procedure. The new song series is compared with the tree structure formed up to this moment. The analysis starts for time interval T1 and is proceeded until the end of a song is reached. For each time interval the syllable of the song is compared with the tree structure for this time interval. An example of the tree produced using this technique is shown in Figure 15.7. The program allows up to five branches to go out from each node of the tree. The new syllable is treated as a descriptor for searching the tree. The syllable investigates five branches, with the following possible outcomes:

- Branch is empty (B = 0). In this case new descriptor takes this branch from now on. Each branch keeps the following information:

    address A—pointing to the origin of the branch
    descriptor D—syllable
    time interval of the branch T
    frequency counter C

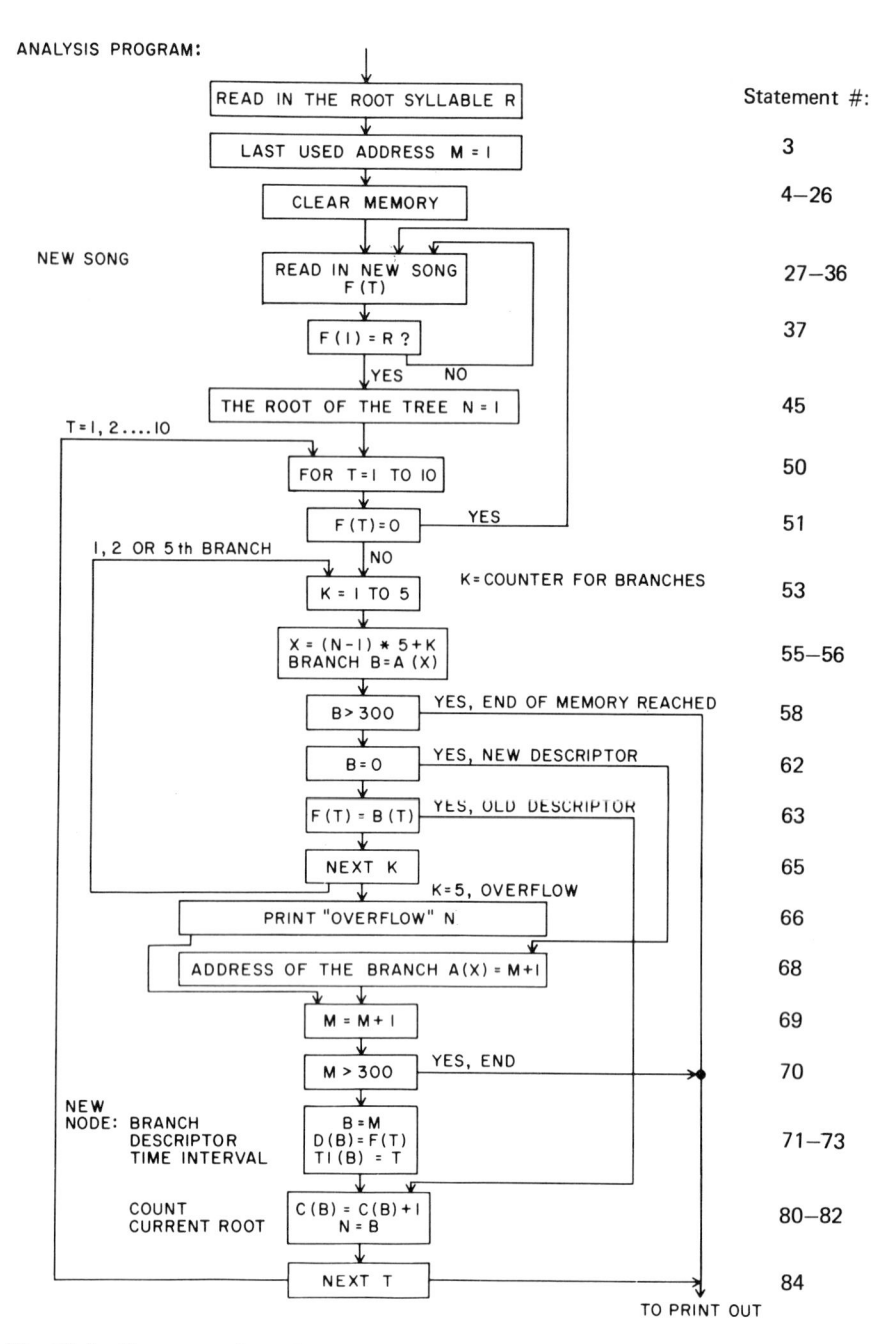

Fig. 15-5 Tree generating using programmed associative memory. For program statement labels see Table 15.10.

Trees

```
PRINT OUT:
```

Statement #:

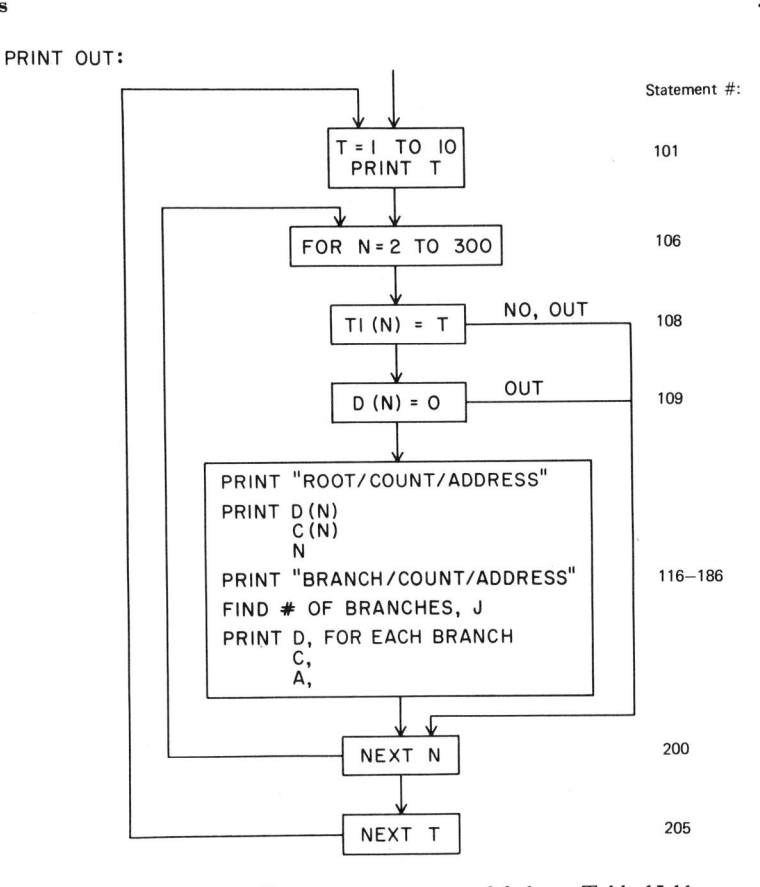

- T = 1 TO 10, PRINT T — 101
- FOR N = 2 TO 300 — 106
- TI (N) = T — NO, OUT — 108
- D (N) = 0 — OUT — 109
- PRINT "ROOT/COUNT/ADDRESS"
  PRINT D (N)
        C (N)
        N
  PRINT "BRANCH/COUNT/ADDRESS"
  FIND # OF BRANCHES, J
  PRINT D, FOR EACH BRANCH
        C,
        A, — 116–186
- NEXT N — 200
- NEXT T — 205

Fig. 15-6 Print-out program. For program statement labels see Table 15.11.

- Branch is occupied with the descriptor that is identical to the one being analyzed. In this case frequency counter C of this descriptor is incremented by one.
- Branch is occupied with a different descriptor. In this case analysis proceeds on the next branch.

Note that computer memory could not be allocated to the branches in advance. The branches take the memory locations from a common space as they are added to the tree. In fact, the tree is organized as a programmed associative memory; the search is based on the content of a cell, rather than on the address of the cell.[10]

After the analysis the printout program produces the results. This program is shown in Figure 15.6. The program prints out all the branches for T = 1,

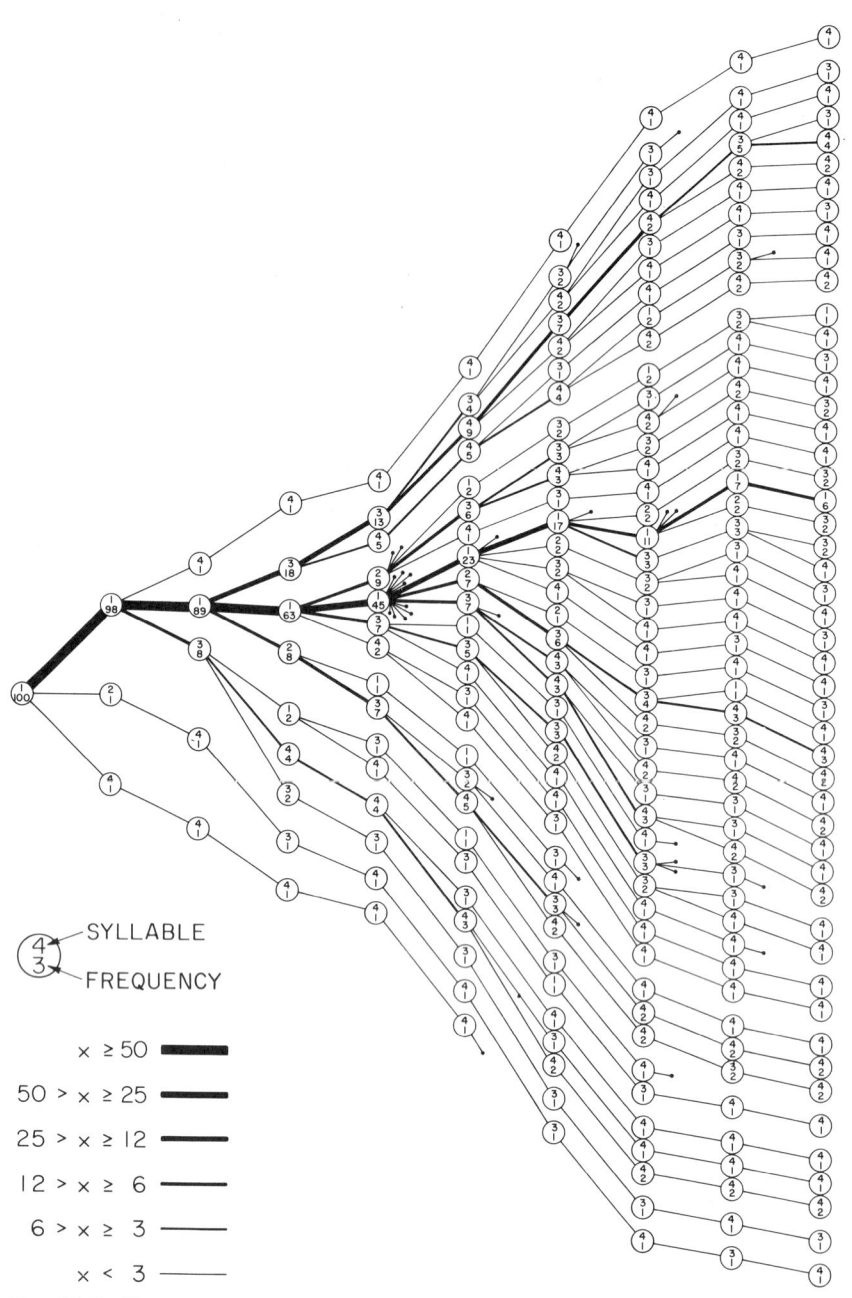

Fig. 15-7 Female transition tree. Tree stores only female portion of sequences beginning with F_1. Tree contains 100 duets (frequenceis are percentages). Tree is formed by the summation of common syllable types at each node in a time period. Reading the tree

Trees

then for T = 2 and so on. Each branch is described with the following information: descriptor = syllable, frequency, connectors to the preceding nodes, connectors to the following nodes.

Figure 15.7 presents a female tree. Figure 15.8 presents a duet tree. One is able to identify the structure of the songs that are repeated many times (heavy lines), as well as many rare variations, especially toward the end of the song.

Note that the tree keeps the songs of the whole experiment. Each individual song can be recognized on the tree. The tree method presents in fact the most efficient way to store hundreds of behavioral sequences. In the same time the tree compares and sorts the patterns. One could read directly from the tree the transition frequencies between individual syllables, including the time dependence. The tree also displays pairs, triplets, and longer subsequencies, and their time dependence.

One can conclude that the tree presents an actual model of the behavior under analysis. The tree not only keeps all the information from the experiment, but also displays all the characteristic features of the patterns, subpatterns, probabilities of transitions, branching points, and time dependence.

Here are few examples how to read the information from the female tree in Figure 15.7. The female songs are produced by extracting the female syllables from duets.

Whenever a song ends at a time T, a pin is inserted in the tree node at the time level T.

The left-hand node 1/100 shows that the tree is composed of the songs starting with syllable 1, and that 100 songs are analyzed.

The top branch presents a rare song = 1, 1, 4, 4. This song has occurred only 1 time.

The second branch presents a song: 1, 1, 1, 3, 3, 3, 3, 3.

The seventh node on this branch (3/2), reads: 3 = syllable; 2 = frequency, i.e., two songs. One song is ended at this node (see the pin), whereas the second song proceeds for one more time interval (node 3/1).

Next are few examples dealing with duet tree, Figure 15.8.

The first node 22/89 shows that songs starting with the syllable 22 are displayed, and that 89 songs are stored in the tree.

The heavy branch presents the most frequent sequence: 22, 1, 22, 1, 22, 1. At the time T = 6, on this branch one finds the node 1/64, with few outgoing branches. The first top branch shows that one song ends here. The second branch points to the node 20/2, showing that in two songs next

---

from left to right, 10 time periods are given, $T = 1$ to 10. If a new syllable type is uttered, a branch is formed. Pins symbolize the termination of a sequence. Repetition of F_1 gives a strong major pathway or branch.

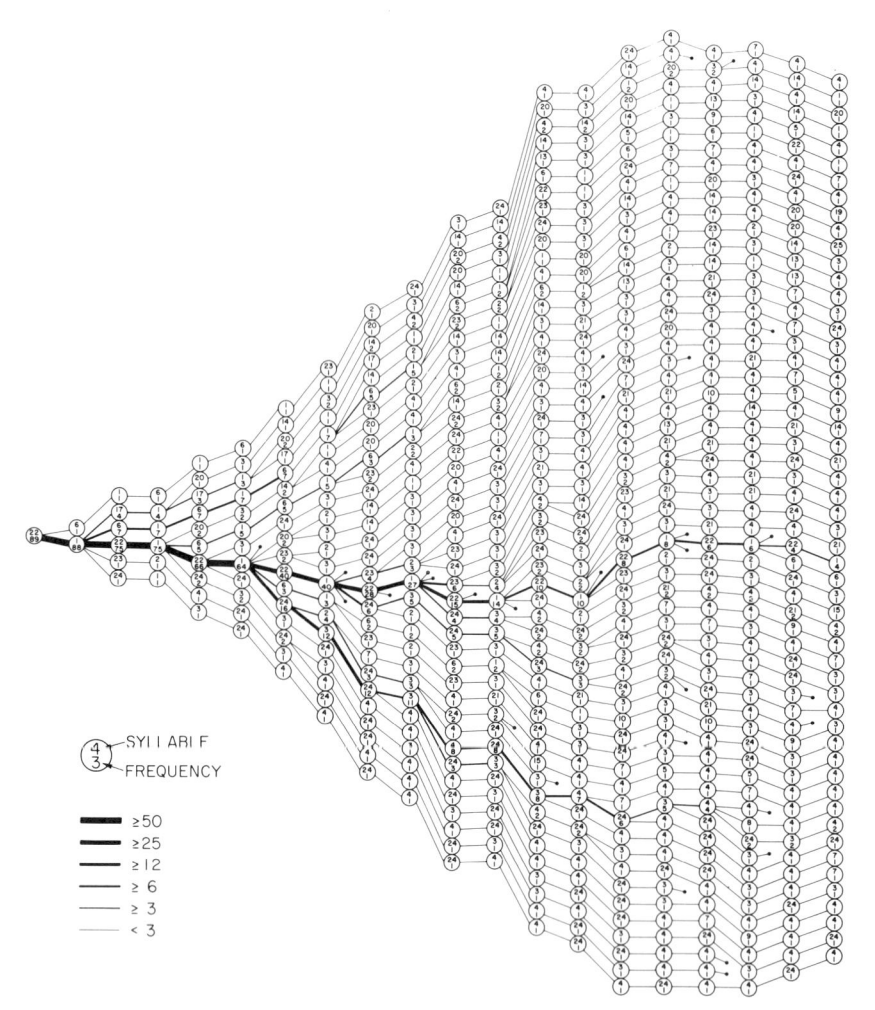

Fig. 15-8 Duet transition tree. Tree graphically stores all sequences from the study beginning with M_{22}. The tree contains 89 such sequences. The tree is formed by overlapping of common syllables at nodes. Nodes occur at each time period (from left to right $T = 1$ to 20). Branching occurs if the sequence is different at any node. Syllable type is given as the top value. Bottom value is the frequency of occurrence at that node. Pins symbolize the termination of sequences. Lines connecting nodes are weighted. Overall branching pattern reveals pictorially the decision points and major sequence pathways. Many sequences are formed from an alternation of M_{22} and F_1 and produce the thick trunk of the tree. At $T = 10$–11 and $T = 12$–13 major branching occurs indicating decision points. At $T = 6$, M_{24} initiates a major side branch.

syllable was 20. The third branch points to the node 23/2, showing that in two songs next syllable was 23. The fourth branch points to the node 22/40, showing that in forty songs next syllable was 22, and so on.

Note that after the time $T = 6$, the two heavy branches exist:

$$22, 1, 22, 1, 22, 1$$
$$24, 3, 24, 3, 24, 3$$

In forty songs, the male has responded at time $T = 7$ with usual 22, and the standard sequence 22, 1, 22, 1 continues.

In sixteen songs however, the male has responded with 24, and a new subsequence is developed: 24, 3, 24, 3.

This case shows definite interdependence between syllables produced by the male and those produced by the female.

In a similar way one could examine different portions of the tree, and follow subsequences and transitions.

15.5 Significance of the Tree Method

In contrast to the transition matrices, the tree stores all observed duet sequences preserving the relative time dependent relationships between syllables. Each duet sequence is compared, counted, and displayed for analysis.

Female Songs Figure 15.7 represents the transition tree for the female part of the duets beginning with F_1 only. F_2 was less frequent (43%) as the females first call, whereas F_3 and F_4 were almost never given first. Most frequently recurring sequences are identified by line weightings and counts at nodes. Rarer variations appear later in the duet. Sequences shorter than 10 syllables terminate as periods between nodes. Dominant features such as bound groups, subpatterns, transitional probabilities, and switch points are summarized as follows:

1. From the 100 sequences stored in the tree, 45% reach the time $T = 5$ as a repetition of F_1 (heavy branch).

2. Distributions of syllables relative to sequence time are given for $T = 1$, $T = 5$, $T = 10$ in Figure 15.9. F_1 is the dominant syllable in the beginning of the duet. At time $T = 5$, F_3 totals 29%. At time $T = 10$, F_4 composes 49% of the cases, whereas F_1 is quite rare (7%).

3. Figure 15.7 shows that many sequences (11%) terminate at time $T = 5$. $T = 5$ appears to be a decision point. Those sequences which continue past time $T = 5$ are likely to be composed of new units (44%), whereas only a few revert to previously given units (4%).

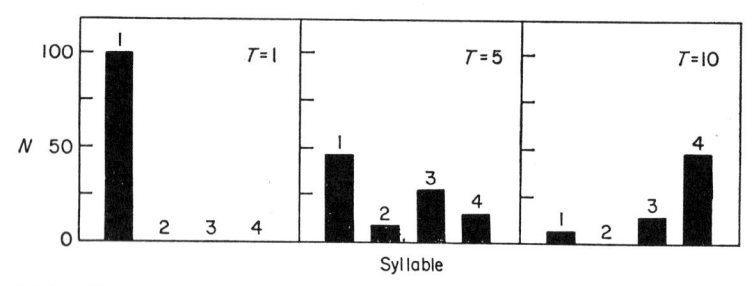

Fig. 15-9 Histograms showing distribution of syllables in female tree. At the beginning of the song ($T = 1$), syllable F_1 is dominant. In the middle of the song ($T = 5$), all syllable are possible. At the end of the song ($T = 10$), syllable F_4 is dominant.

4. Fifty-two percent (52%) of the sequences continue past $T = 5$ as a repetition of the same syllable type.

5. Linear dependencies exist. For example, F_1 and F_2 precede F_3 and F_4 98% of the time. F_1 follows F_2 less than 3% of the time.

6. Some syllables commonly occur together; they may be said to be coupled. They are expressed in Table 15.8 as a percentage of the total syllables at time periods $T = 1-2$; $T = 5-6$; and $T = 9-10$.

Duet Songs Figure 15.8 represents a duetting transition tree composed of only those sequences beginning with M_{22}. Forty-eight duets, not sampled by this tree, begin with M_{23}; four begin with M_{24}; and none with any other male note. The base node indicates that 89 sequences were initiated by the male with M_{22}. By the time $T = 20$ the main pathways have dwindled and the appearances of rarer syllables are most frequent. Distribution data for

TABLE 15.8

Message Unit and Time Period

Message Unit	Time period		
	$T = 1-2$	$T = 5-6$	$T = 9-10$
(1 → 1)	98	24	6
(3 → 4)	0	18	15
(4 → 4)	0	9	12
(3 → 3)	0	12	6
(1 → 2)	1	9	0
(1 → 3)	0	9	1
(1 → 4)	1	3	1

Significance of the Tree Method

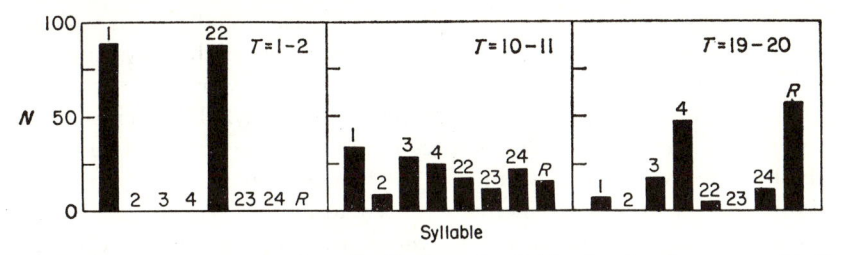

Fig. 15-10 Histograms showing distribution of syllables in the duet tree. At the beginning of the duet ($T = 1$–2), syllables F_1 and M_{22} are dominant. In the middle of the song ($T = 10$–11) many combinations are possible. At the end of the song ($T = 19$–20), syllable F_4 and rare male syllables become dominant.

the beginning, middle, and end of the tree is summarized in Figure 15.10. All rare events are denoted by R. The most common message units and their times of appearance are given in Table 15.9.

By introducing digital computer into the study of animal communication, it was possible to provide many classes of analysis with a high statistical accuracy. Special attention is given to the study of motor patterns of each individual, and to the study of message switching between individuals during bird duetting.

It is found that the individual pattern of female usually starts with syllable code 1 and exhibits a tendency of shifting toward syllables with higher code values. As a result, the female song usually ends with the syllable F_4. Also, the female frequently produces two syllables in a row, without waiting for the male's answer.

TABLE 15.9

Message Units and Their Times of Appearance

Message Unit	$T = 1$–2	$T = 5$–6	$T = 10$–11	$T = 15$–16	$T = 19$–20
$(4 \to 4)$	0	0	4	3	8
$(4 \to 3)$	0	0	12	9	6
$(22 \to 1)$	88	82	22	10	6
$(24 \to 3)$	0	3	2	15	5
$(24 \to 4)$	0	4	3	10	6
$(23 \to 2)$	0	1	2	1	0
Terminations	0	0	2	0	2
New branches	0	0	2	0	
Rare transitions	0	5	14	20	22

It was found that individual pattern of the male usually starts with syllable code M_{22}, and has a tendency to shift toward syllable M_{23} and M_{24}, and then it spreads over a variety of syllables with lower frequencies. The male initiates the song, and after that generates one syllable following each answer from the female side. Almost never does the male produces two syllables in a row, without waiting for the female answer.

Programmed associative memory is used to efficiently store and display hundreds of songs in the form of one single tree. The tree pattern is a new method to present both the communication process and behavioral control. Since the songs are displayed in a sorted and comparative way, one can read from the tree the basic message units, decision-making points, variations of songs, frequencies, and probabilities of transitions. Each individual song can be recognized in the tree. The tree method presents in fact the most efficient way to store hundreds of behavioral sequences.

Through inspection of the duet tree, it is found that the first message unit in most of the songs has a pattern 22, 1, 22, 1, 22, 1. At the time T = 6 major branching occurs, initiated by the male responding with a syllable which is different from 22.

Large numbers of songs will continue with the pattern 22, 1, but also other new message units are initiated. If the male has responded with 24, a new message unit starts, with the pattern 24, 3, 24, 3, 24. This is an obvious case of peripheral influence. It shows the interdependence of syllables produced by male and those produced by female.

In a similar way one could examine the whole tree and find other message units and patterns, as well as probabilities of transitions.

The actual tree program is shown in Tables 15.10 and 15.11. The statement labels from the Table 15.10 are also displayed next to the corresponding boxes on the flowchart in Fig. 15.5. The statement labels from the Table 15.11 are displayed in the same manner on the flowchart in Fig. 15.6. Note the following details in the program: code 98 is used to identify the end of one song; code 99 is used to identify the end of record (and of the experiment); data are first stored in the disk file and then used by this program.

TABLE 15.10

1	REM TREE PATTERN
2	DIM G(5)
3	M=1
4	FILES DUET
5	DIM F(10)
6	DIM D(300)
8	DIM C (300)

(Continued)

TABLE 15.10 (*continued*)

```
10  DIM T1(300)
12  DIM A(1500)
15  FOR I=1 TO 300
16  D(I)=0
17  C(I)=0
18  T1(I)=0
20  FORK=1 TO 5
22  X=(I-1)*5+K
24  A(K)=0
25  NEXT K
26  NEXT I
27  PRINT "ROOT,FILE ?"
29  INPUT R,Z
30  FOR J=1 TO 10
31  INPUT #1,F(J)
32  IF F (J)=99 GO TO 92
35  IF F(J)=98 GO TO 37
36  NEXT J
37  IF F (1)<>RGO TO 30
45  N=1
50  FOR T=1 TO 10
51  IF F(T)=0GO TO 30
52  IF F(T)=98 GO TO 30
53  FOR K=1 TO 5
55  X=(N-1)*5+K
56  B=A(X)
58  IF B>=300 GO TO 99
62  IF B=0GO TO 68
63  IF F(T)=D(B)GO TO 80
65  NEXT K
66  PRINT "OVFL.ROOT=",N
67  GO TO 69
68  A(X)=M+1
69  M=M+1
70  IF M>=300 GO TO 99
71  B=M
72  D(B)=F(T)
73  T1(B)=T
80  C(B)=C(B)+1
82  N=B
84  NEXT T
90  GO TO 30
92  PRINT"OUT=0,MORE=2"
93  INPUT W
94  IF W=2 GO TO 27
95  GO TO 100
99  PRINTS"300 NODES"
```

TABLE 15.11

```
100  PRINT
101  FOR T=1 TO 10
102  PRINT
103  PRINT "TIME POSITION=",T
106  FOR N=2 TO 300
108  IF T1(N)<>TGO TO 200
109  IF D(N)=0GO TO 200
115  PRINT
116  PRINT "ROOT/CO/ADR"
117  PRINT D(N)
118  PRINT C(N)
120  PRINT N
126  PRINT
130  PRINT "BR/CO/ADR"
131  J=0
150  FOR K=1 TO 5
152  X=(N-1)*5+K
153  IF A(X)=0GO TO 160
154  J=J+1
155  G(K)=A(X)
160  NEXT K
165  FOR K=1 TO J
168  Z=G(K)
170  PRINT D(Z),
171  NEXT K
172  PRINT
173  FOR K=1 TO J
174  Z=G(K)
175  PRINT C(Z),
177  NEXT K
178  PRINT
180  FOR K=1 TO J
184  PRINT G(K),
185  NEXT K
186  PRINT
200  NEXT N
205  NEXT T
300  END
```

It was common practice in the past to analyze bird songs as a stationary sequence of events, assuming that the probabilities of different outcomes do not change with time. The present analysis clearly shows, that duetting songs are highly nonstationary process: the probabilities of outcomes are very much different at the beginning of the song from those in the middle or at the end of the song. In contrast to the transition matrices, the tree method stores all observed duet sequences preserving the relative time dependent relationships between syllables.

References

1. Souček, B., and Vencl, F.: Bird Communication Study Using Digital Computer, *J. Theoret. Biol.*, **49** (1975), 147–172.
2. Beer, C. G.: On the Responses of Laughing Gull Chicks to the Calls of Aduts. I. Recognition of the Voices of Parents, *Anim. Behav.*, **18** (1970).
3. Bertram, B.: The Vocal Behaviour of the Indian Hill, Mynah *Gracula religiosa*, *Anim. Behav. Monog.*, **3** (1970) 2.
4. Evans, R.: Imprinting and Mobility in Young Ring-Billed Gulls, *Larus delawarensis*, *Anim. Behav. Monog.*, **3**, Part 3 (1970) 193–248.
5. Lemon, R., and Chatfield, C.: Organization of Song in Cardinals, *Anim. Behav.*, **19** (1971) 1–17.
6. Thorpe, W. H.: Bird-Song, Cambridge University Press, London (1961).
7. Souček, B., and Carlson, A. D.: (in press). Flash Pattern Recognition in Firefly.
8. Souček, B.: Model of Alternating and Aggressive Communication with the Example of Katydid Chirping. *J. Theoret. Biol.* **52** (1975) 399.
9. Vencl, F., and Souček, B.: Structure and Control of Duet Singing in the White-Crested Jay Thrush (in press).
10. Souček, B.: *Minicomputers in Data Processing and Simulation*, Wiley, New York, 1972.

Chapter 16

COMMUNICATION BASED ON FREQUENCY PATTERN RECOGNITION

Introduction and Survey

Acoustical communication based on the sound wave modulation is used in human speech, in bird singing, and in a terestial and acquatic animal communication. Insects are frequently using a pulse modulation of the sound wave producing short chirps. Timing and pulse pattern recognition suffice to identify and to decode such a signal in the receiving animal. On the other side, mammals and birds use amplitude, frequency, and phase modulation of the sound wave, producing complex communication signals. The frequency pattern is basic carrier of information in such a complex signal. Hence the research in this area is concentrated on the frequency pattern recognition study.

The problem could be divided into three parts:

- Identification, definition, and classification of basic elements or syllables the sound signal is composed of.
- Recognition of messages composed of fixed sequences of syllables.
- Identification of internal and external dependences controlling the song sequences.

This chapter gives the general descriptions of the problems and methods, and after that shows concrete examples in detail. The examples of bird duetting analysis are shown. The models are developed, explaining deterministic and random components in song sequences. It is shown that many bird songs present nonstationary, time-dependent process and that this fact is crucial in selecting the right analyzing method. The analysis of complex communication signals was possible only through the usage of digital computer. Concrete experimental data have been used to check the computer algorithms, Vencl and Souček.*

16.1 Example of the Bird Duetting

Duetting, the combined song vocalizations of two birds, demonstrates the complexity and refinement of patterned behavior.[1-8] The functional aspects of duetting have been examined by Thorpe and North,[9] North and Thorpe,[10] Todt,[11] Thorpe et al.,[12] and others. According to these workers, duetting in some species appears to maintain social bonds and isolate species, whereas in others it aids in recognition, maintains territories, or synchronizes the reproductive state of a pair of birds.[13-25]

This study is aimed at understanding the form of duet songs in the laughing thrush with the hope of being able to isolate factors that control their composition. This chapter presents data in the form of transition matrices and probability tables. For the first time stereophonic recordings is used to define precisely each bird's vocabulary. A method of sequence analysis is presented that combines information from several matrices into one diagram, called a control program. Control programs summarize transitional data in terms of probabilities. They predict points of interaction between birds, endogenous patterning, and when syllable types are likely to appear in the sequence. Several models are discussed that might explain duet sequencing. Eventually we hope to learn exactly what information is coded in duets.

16.2 Definitions and Classification of Syllables

One of the most difficult tasks in the sound-wave analysis is the definition and recognition of basic elements or syllables. In animal communication study, the recognition process is done by the researcher, who looks at the frequency patterns or "sonagrams" and recognizes the basic elements. This procedure is slow and limited to a small amount of data. Engineers have developed computer systems for automatic recognition of speech.[26] Such systems are used not only for the research in the area of human communication but also in industries. Some of the applications in which human voice input systems have already become operational are

1. Quality control and inspection
2. Automated material handling.

*For detailed report on experimental data see *Behavior* **57** (1976).

3. Parts programming for numerically controlled machine tools.
4. Direct voice input to the computer.[27]

The first voice input systems to be used by industry in these various applications were installed in late 1973 and early 1974.

All automatic speech recognition systems can be classified under two categories: continuous or connected speech systems and isolated or discrete speech systems. The differences between these two types of systems can become obscure and overlapping when attempting to classify a particular approach as either isolated or continuous. Isolated speech systems can be defined as those systems that require a short pause before and after utterances that are to be recognized on entities. The minimum duration of a pause that separate independent utterances is on the order of 100 msec.

A number of techniques have been developed for automatic recognition of isolated, single-word-length utterances. In fact, until recently the vast majority of recognition work in the field has been at the presegmented level, and only recently has there been widespread interest in the continuous speech problem.

It is customary to capture, frequency convert, and quantize the acoustic speech signal into a form that is usable by subsequent processing. There are at least three major candidates for this task, filter-bank analysis, discrete Fourier transformation, and linear predictive coding. It is not clear at this point which of these is superior in application to the speech-analysis problem.

A practical, limited-vocabulary system achieves recognition processing by comparing an unknown utterance with a set of stored samples of the vocabulary words obtained from the user of the system. These reference data must be stable over long periods of time for practical applications. Once the reference data have been obtained the operator should be able to use the voice input system with little or no retraining overtime.

Frequency-domain representation of the speech signal is particularly advantageous for two reasons:

1. It is known that the human auditory system performs a crude frequency analysis at the periphery of auditory sensation (preprocessing of the signal).

2. An exact description of the speech sound can be obtained with a natural frequency concept model of speech production.

Time-domain representation of the speech signal has been also used recently for the speech analysis based on the linear predictability of speech waveforms.

Each pattern recognition system must define the key processing function as the feature extractor. The feature-extraction processes frequently used are the spectral shape of the speech signal and the time derivative of the spectral envelope function. The spectral shape and its changes with time are

Definitions and Classification of Syllables

continuously sampled over the frequency range of interest. Combinations and sequences of these samples are processed to produce a set of significant acoustic features. This segmentation process is performed using both special hardware and computer software. The feature extraction system has two modes of operation: training and recognition.

During the training mode the system automatically extracts a time-normalized feature for each repetition of a given speech signal (word). A consistent matrix of feature occurrences is required before the features are stored in the reference pattern memory.

During the recognition mode each speech signal (word) entering the system is processed: the features are extracted, digitized, and time normalized. The result is then compared digitally to each stored reference matrix. The stored reference producing the highest overall correlation is selected as the test signal.

Today limited-vocabulary systems exist and are used in both the research laboratories and industries. Advances in microcomputer technology promises that by 1985 practical systems capable of handling 1000–5000 word vocabularies will become available. In the animal communication study even today's systems with limited vocabulary could be successfully used for feature extraction and recognition.

Here we show a concrete example of the analysis of sound sequences in bird duetting. The study is based on 140 duets, which are composed of 2866 discrete elements we have called syllables. Each syllable of a duet was coded onto the punch tape according to its category type and sequence position. The duets were statistically analyzed by computer for syllable interrelationships using transition matrices (see Chapter 15). The male syllables are coded M_1 to M_{24}. The female syllables are coded F_1 to F_4.

Preliminary analysis on the duetting of three pairs of birds indicates that a few syllables are shared and that most are specific to each pair. Female vocabularies are quite similar. Males share only syllables M_{22}, M_{23}, and M_{24}. The remainder of male vocabularies vary in extent and show many variations. Syllables F_1, F_2, F_3, F_4, and M_{22}, M_{23}, and M_{24} comprise 75% of the total vocalizations recorded. The majority of male syllables are seldom heard.

In the beginning of this study, the three classes of syllables, M_{17}, M_{22}, and M_6, were observed and assigned discrete codes. It turned out that the behavior of the female to these three types was indistinguishable. Moreover, the transitions from these to other male elements were identical. Finally, the linear interrelationships between M_6 and M_{22} suggested that both were members of a graded continuum. Since the female apparently failed to discriminate between them, both are assigned to one group, M_6/M_{22}. M_{17} is also a member of this group, but because of its rarity, it is not included in the group code.

16.3 Analysis of Syllable Sequences

The ultimate goal of the interactive behavior analysis is to produce the theoretical model that will follow the same set of roles as observed in the experimental data. In the case of bird duetting, the model should simulate both male and female calls and their interactions.

Two questions are fundamental in the analysis of interactive behavior patterns like bird duetting. First, it is important to be able to describe statistically the various alternative sequences. Some sequence may be rare, whereas some may be quite common. Second, the factors that determine what syllables are to be used must be isolated. This last point can be posed as follows: Is the selection of a syllable to be sung dependent on the partner's previous syllable (exogenous)? Or, is a syllable uttered according to some internal template or program (endogenous)? The essential idea is that the occurrence of a previous event may influence the probability that certain other events will succeed it in time. The basis of this influence may lie within the individual or may be the consequence of interaction with another individual, in this case the bird's mate.

Transition matrices have frequently been employed to analyze sequencing of events. A matrix is constructed by plotting successive syllables against all remaining syllables. Each syllable is taken as a preceding event and the next syllable is plotted as a following event. Tables 16.1, 16.2, 16.3, and 16.4 illustrate the result of this procedure. The number of preceding or following events can be varied. The sequence interval between the two events can also be varied. Even though matrices have been used often in animal behavior, they are by themselves limited in their usefulness. An individual matrix provides no information about the unsampled portions of long sequences or about the probable distribution of types of sequences, choice points within sequences, or the appearance of events relative to the total sequence. How can the information about syllables in matrices looking at different sequence intervals be arranged into longer, continuous summaries of all the behavior observed? Can evidence for larger patterns obscured by random variations be revealed by looking at many matrices together? An extension of the information in individual matrices is provided by a correlation technique. This method utilizes several kinds of matrices in concert. The correlation of matrices can reconstruct sequences of events, which will provide the following advantages: (1) a summary of common and rare sequences based on probabilities, (2) a description of cases of endogenous or exogenous determination of events, (3) when in the sequencing controls do operate, (4) when in the sequence certain events are likely to appear.

Tables 16.1 to 16.3 are used to build the communication model. Table 16.3 lists the most frequent pairs of syllables observed in the duetting. The

TABLE 16.1

```
N = 1
ONE SYLLABLE-ONE SYLLABLE
PROBABILITIES *100
```

				NEXT ELEMENT																							
1ST	2ND	3RD	SUM	1	2	3	4	5	6	7	8	9	10	11	12	13	14	15	16	17	18	19	20	21	22	23	24	25
1	0	0	16						3																9		1	
2	0	0	4																							2	2	
3	0	0	11				4																				3	
4	0	0	20				5	1		2																		
5	0	0	1				1																					
6	0	0	3	3																								
7	0	0	3			1																						
8	0	0	0																									
9	0	0	1				1					1																
10	0	0	1				1					1	1															
11	0	0	0																									
12	0	0	0																									
13	0	0	1															1										
14	0	0	1																									
15	0	0	1				1																					
16	0	0	0																									
17	0	0	0																									
18	0	0	0																									
19	0	0	0																									
20	0	0	1																									
21	0	0	0																									
22	0	0	12	12																								
23	0	0	5		4																							
24	0	0	7			6	1																					
25	0	0	0																									
SUM =			100	16	5	11	22	2	3	3	0	1	1	0	0	1	1	2	0	0	0	0	1	0	9	4	7	0

Matrix shows only most probable syllable to syllable transitions. The probability raw and column sums were obtained by dividing values from Table 16.2 with total sum (2866) and rounding-off quotients to lower integer. The probabilities less than 1% are excluded.

TABLE 16.2

```
N = 1
ONE SYLLABLE - ONE SYLLABLE
FREQUENCIES
                    NEXT ELEMENT
1ST 2ND 3RD  SUM    1    2    3    4    5    6    7    8    9   10   11   12   13   14   15   16   17   18   19   20   21   22   23   24   25
 1   0   0   463              2    2    1   86                             3                                                      23   1  266   26  32  25
 2   0   0   138         3    3    3         1   15                        6    3                                                      1   78   20
 3   0   0   316         4    4  118    6    4    3   26    2    3    2   10        1    7    9                  15                 6    7    9   79   2
 4   0   0   585              4  147   10   43   14   60   16   36   38   14   21   44                 1         1   20        10   19    1    1   93  10
 5   0   0    57         3    3   11    2                                  11   18
 6   0   0   110    91        3   11                                                                        1
 7   0   0   100             12   41   46                1                                       1
 8   0   0    18              4   14   34                                                                                                                 1
 9   0   0    40              5    4                          1
10   0   0    44              4    4        1                            5
11   0   0     0
12   0   0     0
13   0   0    33             16   19   17
14   0   0    50   18    3    8   48                                                    1
15   0   0    57              7
16   0   0     1                                                                                                                  1
17   0   0    16   15                                                                                                             1
18   0   0     1
19   0   0    20              1   20
20   0   0    40         1   24   14
21   0   0    27              7   20
22   0   0   356  353    1                                                                  1
23   0   0   155        132   22                              1
24   0   0   225    1   7 179   38                                                                                                                   1
25   0   0    14              1   12    1
SUM = 2866        478 161 341 648   59  105 101   18   41   45    0    0   33   50   59    1   15    1   20   41   27  268  115  225  14
```

The matrix gives frequencies of transitions (first order) from preceding syllables (column) into following syllables (row). Only successively uttered syllables are tabulated ($N = 1$). Syllables are coded along both axes: 1–4 for female and 5–25 for the male.

TABLE 16.3

N = 1
ONE PAIR – ONE SYLLABLE
FREQUENCIES

1ST	2ND	3RD	SUM	NEXT ELEMENT 1	2	3	4	5	6	7	8	9	10	11	12	13	14	15	16	17	18	19	20	21	22	23	24	25
1	6	Ø	61	61																								
1	22	Ø	212	210		1																					1	
1	24	Ø	22	1	5	16																						
2	23	Ø	71	61		10																						
3	4	Ø	70				3	2		10	5	2				1	4			1				7			34	1
3	24	Ø	50		39		11																					
4	4	Ø	72				7	3		12	7	3	4			2	6	1	1			2	1	6			15	2
4	5	Ø	30			1	29																					
4	7	Ø	40			9	30																					
4	9	Ø	29			2	26	1																				
4	10	Ø	26			1	19					1						5										
4	15	Ø	27			1	26																					
4	24	Ø	70			54	16																					1
5	4	Ø	29				8	6	1	1		9	1			1		2										
6	1	Ø	76			1		1	59				2				2								2	8	1	
7	3	Ø	20							11																2		1	
7	4	Ø	29				6	2		6	1																	1	
9	4	Ø	21				18	5	9	1	1																		
10	4	Ø	26				6	2	1	2		12	4					2											
15	4	Ø	32				3	1	3	1			1				1	21											2
22	1	Ø	309			1			17											5			6		245				
23	2	Ø	98	2		3	4			8						1	2							1		10	1	
24	3	Ø	123		2	57	1			6						7		2						3		66	11	
24	4	Ø	26			18				1														1			45	
24	4	Ø	Ø																									6	
SUM =			1586	272	68	140	285	25	88	58	14	27	12	Ø	Ø	12	15	34	1	5	1	2	7	18	247	86	142	7

Matrix shows frequencies of transitions (second order) of most probable pairs (preceding column) in the following syllable (row). Probability bias is 1%. All pairings occurring less than bias value are ignored by computer.

293

TABLE 16.4

```
N = 1
ONE SYLLABLE - ONE SYLLABLE
FREQUENCIES
                    NEXT ELEMENT
1ST  2ND  3RD  SUM   1    2    3    4    5    6
 1    Ø    Ø   425  337   25   56    7
 2    Ø    Ø   122    3   73   40    6
 3    Ø    Ø   251    9    9  113  120
 4    Ø    Ø   197    2        68  127
 5    Ø    Ø     Ø
 6    Ø    Ø     Ø

SUM =          995  351  107  277  260    Ø    Ø
```

Matrix gives transition frequencies for female only as extracted from duets. Matrix is divided into a dominant upper triangle and a less noisy lower triangle. One syllable will switch most often into itself (indicated by strong diagonal) or less often into syllables with the next higher code values. This linear dependency indicated nonrandom sequence-position dependent endogenous drift through female repertoire.

Analysis of Syllable Sequences

first and second columns describe the pair. The column "sum" shows how many times this pair has been observed. Some pairs are dominant: pair 22,1 has been observed 309 times; pair 24,3 for 123 times; pair 23,2 for 98 times; pair 4,4 for 72 times. It is obvious that these four pairs must have significant positions in the model.

Table 16.1 lists the most probable syllable-to-syllable transitions. This table is used to draw the first approximation model, Fig. 16.1. As described previously, syllables 22 and 6 are merged into the same class.

The most probable transition is syllable 22 into syllable 1, which occurs 12% of the time. By adding syllable 6 we form the starting point in the model 22/6. Table 16.1 shows that 15% of the time (12% + 3%) syllables 6/22 are followed by syllable 1. Hence syllable 1 is the second point in the model. The two points are connected by the line showing the probability of transition (15%).

Now we concentrate on the syllable 1. Table 16.1 shows that syllable 1 is followed most frequently by syllables 22/6, which occurs 12% of the time. Hence we connect the point 1 with point 22/6 by the line showing the probability of transition (12%).

Also 1% of the time syllable 1 is followed by syllable 24. In the model we form a point 24 and a connecting line between points 1 and 24, with the

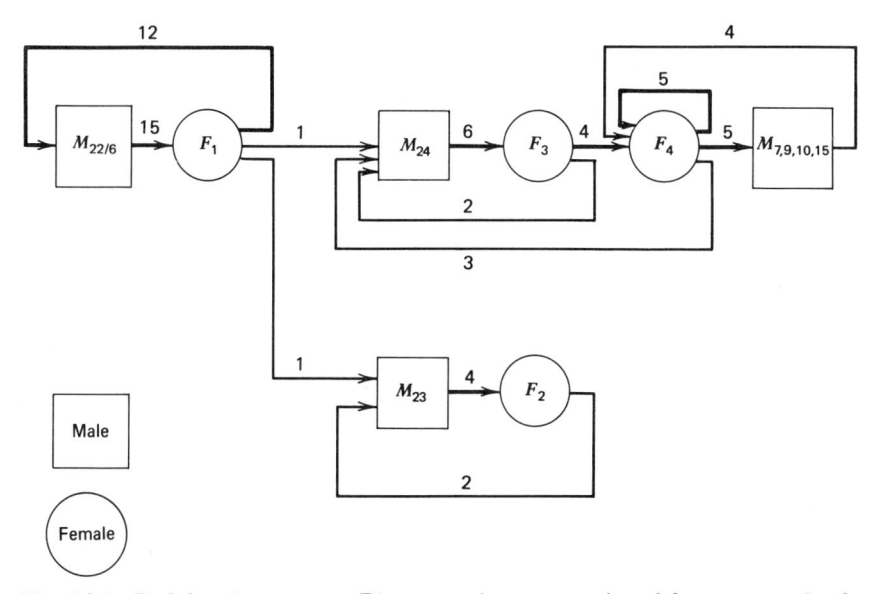

Fig. 16-1 Bird duetting program. Diagrammatic representation of the structure of male (boxes) and female (circles) sequences, based on transitional data from tables. Values shown above the lines are probabilities representing the utilization of sequence pathways.

probability 1%. Somewhat closer examination of Table 16.2 shows that syllable 1 is also followed by syllable 23 in 26 out of 2866 transitions. Because the probability of this transition is close to 1%, we form a new point in the model, 23, and connecting line 1-23, with the probability of approximately 1%. Transitions with less than 1% probability are neglected in this model.

The previously described technique has been used to analyze all the transitions shown in Table 16.1. The final result is the model shown in Fig. 16.1.

16.4 Dependencies and Independencies Controlling Song Sequencies

A brief inspection of the data shows that duets are not simply a random aggregation of syllables. The distribution of entries in the matrices is clustered. The question posed at this point is as follows: how organized are the duets? That is, what dependencies exist between syllables and how do the birds influence each other's songs? The diagrams based on male and female sequences suggest answers about duet structure. They are models of control that (1) summarize the kinds of sequences sung, (2) isolate points of endogenous and exogenous control, (3) predict when syllables will appear in sequences.

Figure 16.1 presents male-female song program. The program could be divided into six major loops:

Loop 22/6-1. This loop presents the starting point of the program. The male appears to control the exit from this loop. Most of the time the male sings 22/6. However, in 1% of the time the male's answer is 24, and the second loop (24-3) starts. Also in 1% of the time the male's answer is 23, and the third loop (23-2) starts.

Loop (24-3). The female appears to control the exit from this loop. After the syllable 3, the female either waits for male's 24 to close the loop again, or she produces the syllable 4.

Loop (24-3-4). This loop is, in fact, an extension of the loop (24-3).

Loops (4-4) or (4-7,9,10,15). Toward the end of the song the female syllable 4 is frequently uttered. Sometimes the male syllables 7, 9, 10, or 15 are found here.

Loop (23-2). This loop, in fact, produces the side branch in the model and presents an alternative to the loop (24-3).

A duet appears to be a chain of discrete signals exchanged between two birds. What information is coded in each syllable? Does the entire sequence convey a message? We have only elementary notions about the communication function of duetting in laughing thrushes. It is our impression that the seven most common syllables in both sexes are not exclusive to the pair of

Discussion of the Model

birds in this study. A preliminary analysis of three other pairs indicates that females and males share these syllables (F_1 to F_4; M_{22} to M_{24}). The remaining male syllables vary from male to male. It may be that rare male syllables identify individuals, whereas shared calls can code for the species. It is clear from the data that the positions of syllables is important. Duets must have a syntax. Each syllable is uttered at a characteristic point in the sequence. Some syllables are given as "questions," some as "replies." At this time it is speculation to suggest what the birds are saying to one another. We have noted that certain syllables are given more often under some conditions. For example, F_2–M_{23} seem to be called up after a disturbance such as a loud noise. This replay loop might mean something like, "I am OK, how about you?" $M_{6/22}$–F_1 may represent a synchronization that "entrains" both birds for more singing. It may signify, "I'm ready to sing, are you?"

The survival value of individual behavioral patterns lies in the fact that they do not appear in a random sequence but rather in a definite progression in time. This is especially true in sequences of patterns in communication. The sequential occurrence of a syllable is related to all the other syllables in the vocabulary. As the sequence progresses, different syllables are called up. When many sequences are observed and displayed in an overlapping fashion, beginning with the first syllable of each sequence, it becomes clear that each individual syllable has a typical position within the overall pattern. Its position can be described as a bell-shaped curve that reaches its peak at the point where, on the average, most of syllables of a given type overlap. The curve declines rapidly to the end with only minor fluctuations. Each syllable's appearance can be summarized by such a sequence-position dependency. An example of this position effect is found in the female's song. From the female matrix (Table 16.4) the following can be observed. The matrix is divided into a dominant upper triangle and less noisy, lower triangle. One syllable will always make a transition most frequently into itself (indicated by the strong descending diagonal) or somewhat less often into the syllables encoded with the next higher values. This is also true in the male for $M_{6/22}$, M_{23}, and M_{24}. A sequence of syllables from each individual progresses linearly; that is, there is a tendency to repeat the previous syllable or to shift to an unuttered one but not to return to former syllables. Hence the appearance of syllables is not equiprobable. The utterance of a syllable is strongly influenced in a position-dependent fashion.

16.5 Discussion of the Model

The model presented in Fig. 16.1 shows that the duetting communication follows a set of rules: syllable 22/6 is always followed by 1; syllable 23 is followed by 2; syllable 24 is followed by 3; syllables 7, 9, 10, and 15 are followed by 4. One could clearly identify different phases in the communication and their characteristic sequential patterns.

The probability that a particular syllable will be uttered seems to depend on (1) the previous syllable given by the partner, (2) the previous syllable given in the birds own song sequence, (3) syllable sequence-position dependent periodicity (endogenous drift). It has been shown that duet sequences are extremely sequence-position dependent and are far from being stationary processes. This means that the probability of a syllable being uttered at sequence-position T depends strongly on T. Hence the probability of transitions from syllable to syllable is strongly event dependent and as a result, transitions early in the sequence cannot be treated with the same statistics as those occurring toward the end of the song. However, one realization of the song is very similar to any other realization and the term nonstationarity is reserved here only to emphasize the sequence-position dependencies inside one song.

The animal communication study could be extended into new areas such as correlation between the communication models, as presented in Fig. 16.1, and particular animal behavior and correlation between the model and neural network activity.

References

1. Salim, Ali and Ripley, S. Dillon: *Handbook of Birds of India and Pakistan*, Vol. 7, Oxford University Press, Bombay and New York, pp. 14–16.
2. Bertram, B.: The Vocal Behaviour fo the Indian Hill Mynah, *Gracula religiosa*, *Anim. Behav. Monog.*, **3** (1970), V2, 79–192.
3. Brockway, B. F.: Stimulation of Ovarian Development and Egg Laying by Male Courtship Vocalization in Budgerigars (*Melopsittacus undulatus*), *Anim. Behav.*, **13** (1965) 575–578.
4. Dilger, W. C.: Duetting in the Crimson-breasted barbet, *Condor*, **55** (1953) 220–221.
5. Harrison, C. J. O.: Allopreening as Agonistic Behaviour, *Behaviour*, **24** (1965) 161–209.
6. Helversen, D. v. and Wickler, W.: Uber den Duettgesang des afrikanishen Drongo *Dicrurus adsimilis* Bechstein, *A. Tierpsychol.*, **29** (1971) 301–321.
7. Isaac, D. and Marler, P.: Ordering of Sequences of Singing Behaviour of Mistle Thrushes in Relationship to Timing, *Anim. Behav.*, **11** (1963) 179–188.
8. Lemon, R. and Chatfield, C.: Organization of Song in Cardinals, *Anim. Behav.*, **19** (1971) 1–17.
9. Lemon, R. and Chatfield, C.: Organization of Song of Rose-Breasted Grosbeaks, *Anim. Behav.*, **21** (1973) 28–44.
10. North, M. and Thorpe, W. H.: Vocal Imitation in the Tropical Bou Bou Shrike *Laniarius aetheopicus major* as a Means of Establishing and Maintaining Social Bonds, *Ibis*, **108** (1966) 432–435.
11. Payne, R. B. and Skinner, N. J.: Temporal Patterns of Duetting in African Barbets, *Ibis*, **112** (1970) 173–183.

References

12. Payne, R. B.: Temporal Pattern of Duetting in the Barbary Shrike *Laniarius Barbarus*, *Ibis*, **112** (1970) 106–108.
13. Soucek, B. and Vencl, F.: Bird communication study using digital computer, *J. Theor. Biol.*, **49** (1975) 147–172.
14. Soucek, B. and Carlson, A. D.: Firefly communication based on time discrimination, sensitivity adjustment and accumulative memory, (to be published).
15. Thimm, F.: Sequentielle und Zeitliche Beziehungen im Reviergesang des Gartenrotschwanzes, *J. Comp. Physiol.*, **84** (1973) 311–334.
16. Thorpe, W. H. and North, M.: Origin and Significance of Vocal Imitation with Special Reference to the Antiphonal Singing of Birds, *Nature*, **208** (1965) 219–222.
17. Thorpe, W. H., Hall-Craggs, J., Hooker, B., Hooker, T., and Hutchison, R.: Duetting and Antiphonal Song in Birds: its extent and significance, *Behaviour Supplement*, XVIII (1972) 189–318.
18. Todt, D.: Zur Steuerung unregelmassiger Verhaltensablaufe, *Kybernetik*, *Kapitel*, **6** (1968) 465–485.
19. Todt, D.: Die Antiphonen Paargesange des Ostafrikanischen Grassangers, *Cisticola hunteri printoides* Neumann, *J. Fur Ornithol.*, **111** (1970a) 332–356.
20. Todt, D.: Zur Ordnung im Gesang der Nachtigall (*Luscinia megarhynchos*), *Ver. Deutch. Zool. Ges.*, **64** (1970b) 249–252.
21. Todt, D.: Gesangliche Reaktionen der Amsel (*Turdus merula* (*L.*) auf ihren Experimentell Reproduzierten Eigengesang, *Z. fur Vergleichende Physiologie*, **66**, 3 (1970c) 294–317.
22. Todt, D.: Aquivalente und Kovalente gesangliche Reaktion einer exgrem regelmassig singenden Nachtigall (*L. megarhynchos*), *Z. vergl. Physiol.*, **71** (1971) 262–285.
23. Tschanz, B.: Trottellumen, *Z. Tierpsychol.*, **4** (1968).
24. Wickler, W.: Aufbau und Paarspezifitat des Gesangsduettes von *Lariarius funebris*, *Z. Tierspsychol.*, **30** (1972) 464–476.
25. Wickler, W. and Uhrig, D.: Bettelrufe, Antworstzeit und Rassenuntershiede im Bergrussungsduett des Schmuckbartvogels *Trachyphonus d'arnaudii*, *Z. Tierpsychol.*, **26** (1969) 651–661.
26. Martin, T. B., Applications of Limited Vocabulary Recognition Systems, in *Speech Recognition. Invited Papers Presented at the 1974 IEEE Symposium*, Academic Press, New York, 1975, pp. 55–57.
27. *Word Recognition System and Voice Data Entry Terminal System*, Scope Electronics, Reston, Virginia, 1975.

Chapter 17

MODELS OF QUANTIZED INFORMATION TRANSMISSION ON NEURAL TERMINALS

Introduction and Survey

Action potential carries the information over the axon until it reaches the neural terminal. The chemical transmitter is contained in vesicles in the presynaptic knot and, on the arrival of the action potential at the neural terminal, some of the vesicles are discharged into the synaptic cleft. An integral number of vesicles are discharged each time. Hence the potential on the other side of junction is built up from an integral number of contributions, or quanta. It is believed that many central synapses mediated by chemical transmitters operate in that way. Quantal transmission was first discovered at the neuromuscular junction (end-plate). This chapter explains in detail the model of the quantized information transmission, following Souček.[1]

A statistical model of the composition of end-plate potentials has been designed and used for detailed study of end-plate potential's amplitude distribution function. The model is based on the present knowledge of synaptic function and takes into account the miniature end-plate potential's amplitude distribution, pulse shape and latency fluctuation as well as the residual potential difference across the membrane, non-linear summation of unit quanta and the mean quantum content. The buildings up of end-plate potentials is a stochastic process in which all of the above factors have significant influence. The computer model for statistical building of end-plate potentials is designed on the basis of the Monte Carlo technique. The end-plate potentials' distribution is not calculated, but is built up in the same way as in a living neuromuscular junction through the addition of miniature end-plate potentials which have random amplitudes and random times of arrival.

Large numbers of experiments have been performed with the model with the mean quantum content in the range from 2 to 200, mean amplitudes of miniature end-plate potentials in the range of 0.2 to 0.8 mv, coefficient of variation of miniature end-plate potentials in the range of 0.1 to 0.2, number of depolarizations per experiment up to 25,000. Latency fluctuation and pulse shape have been taken according to the measurements by Katz & Miledi.[2]

Very good agreement between results produced by the theoretical model and real experiments was obtained. Coefficients of variation of end-plate potentials produced by the model are in good agreement with experimental results for all values of mean quantum content. This is due partly to the fact that not only non-linear summation but also the latency fluctuation diminishes the coefficient of variation.

17.1 Example of the End-Plate Potential

Problem Del Castillo & Katz[3,4] (1954a, b) have shown that the end-plate potential (e.p.p.) at myoneural junction of frog muscle is built up of small all-or-none units ("quanta") which are identical in size and shape with the spontaneously occurring miniature e.p.p.'s (m.e.p.p.). Similar results have been obtained by Boyd & Martin[5] (1956) on mammalian muscle, and have been proved after that on a number of other preparations. It was found that when the average "quantum content", m, of the e.p.p. was small ($m < 5$), the value of m could be obtained from the ratio [mean amplitude of e.p.p.]/[mean amplitude of m.e.p.p.]. The amplitude of the e.p.p. fluctuated in a manner described by Poisson law, suggesting that there is a latent population of excitable units at the junction, each with a small probability of responding to a nerve stimulus. However, at higher and more nearly normal levels ($m > 10$) the fluctuations were much less than expected on this basis.

Martin[6] (1955) has shown that for e.p.p.'s exceeding a few millivolts in size, m cannot be calculated as stated above, but a correction must be applied because, unlike quantal conductance changes, miniature potentials do not add linearly beyond a limited range. When the correction factor was applied to the e.p.p. measurements, the discrepancy between the observed and theoretical values was diminished. The coefficient of variation was in more satisfactory agreement with $m^{-0.5}$, as would be expected according to the Poisson law. (Actually, m.e.p.p.'s add more linearly than indicated by the above correction because of the effect of membrane capacity.)

In a resting terminal, transmitter packets are released at random intervals with a low probability. When the terminal is depolarized by an action potential, the release rate rapidly increases to a high value and then returns to the resting level. Synaptic delays and the time course of acetycholine release at the neuromuscular junction have been measured by Katz & Miledi[2] (1965). This measurement shows that the quantal release at the neuromuscular junction is not instantaneous with the arrival of the action potential, but rather if fluctuates in a random way.

E.p.p.'s are built up of m.e.p.p.'s which are added together in a random fashion. Souček[7] (1971) has developed a general equation for e.p.p. amplitude probability distribution function, and showed that statistical composition of e.p.p.'s can be treated as a transient in stochastic process. The e.p.p.'s probability distribution is a function of the latency distribution, m.e.p.p.'s pulse shape, m.e.p.p.'s amplitude distribution and mean quantal content.

From all the knowledge accumulated up to now, many aspects of neural information transfer are well understood and many of the remaining problems are clearly defined. Two things are, however, obvious:

(a) Experimental investigations of the statistical nature of e.p.p.'s give results for a limited number of cases determined by parameters of measurements. Also, experimental results are smeared by measurement errors, the most serious of which are noise, rough quantization and large, statistical inaccuracies due to the small number of data analyzed.

(b) Analytical investigations of e.p.p.'s ask for the use of sophisticated stochastic process theory. Obtained equations describe the statistical nature of e.p.p.'s in mathematical language, which is not easy for everyday laboratory usage.

To overcome these difficulties, a computer model for statistical building of e.p.p.'s has been designed. The model simulates m.e.p.p.'s with random amplitudes and random times of arrival. The m.e.p.p.'s are added in non-linear fashion, producing e.p.p. Large numbers of simulated "experiments" have been performed for various sets of parameters. The model enables one to make "measurements" of e.p.p.'s distribution functions, analyzing thousands of e.p.p.'s, and controlling quantization steps and noise level. The parameters of simulated experiments have been chosen according to the most important real experiments published up to now. Very good agreements between results produced by theoretical models and real experiments have been obtained. Different models have been investigated. The best results are obtained from the model, which takes into account mean quantum content, m, m.e.p.p.'s pulse shape, m.e.p.p.'s amplitude distribution, latency distribution, and non-linear summation. Coefficients of variation of e.p.p.'s distributions produced by such a model are in very good agreement with experi-

$P(A)$ probability distribution of m.e.p.p.'s amplitudes
v mean amplitude of m.e.p.p.'s
$h(t)$ pulse shape of m.e.p.p.'s
$\alpha(t)$ the release rate of m.e.p.p.'s after depolarization
x fraction of m.e.p.p.'s which is used in summation to form e.p.p.
$g(x)$ distribution function of x
V_0 residual potential difference across the membrane
N number of depolarizations during one experiment.

Figure 17.1 explains the model for statistical building of e.p.p.'s, taking into account facts known up to now.

The e.p.p.'s, on the basis of quantal theory, are composed of a sum of m.e.p.p.'s. It has been shown, experimentally, that the number of m.e.p.p. units composing an e.p.p. is Poisson distributed with the mean quantum content, m (for $m < 10$). Hence, the probability that e.p.p. amplitude, s is composed of 1, 2, 3, k m.e.p.p.'s will be

$$p(k) = \frac{e^{-m}}{k!} m^k. \tag{17.1}$$

The mean quantum content, m, or the average rate of unit release depends upon membrane potential, ion concentrations (particularly calcium) in the bathing medium, the quantity of transmitter available for release, and the history of synaptic use. Here we shall consider only depolarizations applied on a resting terminal.

To simulate an experiment, the experimenter must first decide for the value of m. The program reads m, calculate the Poisson distribution, equation (17.1), and also integral Poisson distribution (Fig. 17.1a),

$$pi(k) = \sum_{i=0}^{k} \frac{e^{-m}}{i!} m^i. \tag{17.2}$$

The distribution $pi(k)$ gives the number of cases when e.p.p. is composed of less (or equal) than k m.e.p.p.'s. The distribution, $pi(k)$, is shown in Figure 17.1f.

In the next step, Figure 17.1b, the program reads in the m.e.p.p.'s amplitude distribution function $P(A)$. This distribution has been measured in many experiments. Usually, $P(A)$ can be presented as Gaussian distribution. The experimenter can choose mean values between 0.2 and 1.0 mv, and coefficient of variation (i.e. standard deviation divided by the mean) between 0 and 0.3. This value depends largely on the experimental conditions and the noise level in the system. The program converts the distribution $P(A)$ into integral form $PI(A)$, which is shown in Figure 17.1h.

Computer Model

mental results for all values of *m*. This is explained through the fact tha only non-linear summation but also the latency fluctuation has the influ on the e.p.p.'s distribution function, and on the coefficient of variation

Method The model for statistical building of e.p.p.'s has been des on the basis of the Monte Carlo technique. The e.p.p.'s distribution calculated, but is built in the same way as in living neuromuscular jun The simulation programs have been written in FORTRAN language for C Data 6600 computer and can be applied on other FORTRAN oriented puters. Before simulating an experiment, the program reads experir data describing m.e.p.p.'s pulse shape, m.e.p.p.'s amplitude distrib latency distribution, and mean quantum content. The experiment choose the number of amplitude-channels for e.p.p.'s distribution (qu tion step). The experimenter can also choose the noise level, or perfc ideal, noiseless experiment. The experimenter also decides on the nun depolarizations during one experiment. The experiments, with few tho of depolarizations, can be simulated, eliminating statistical fluctu which are very large in real experiments because it is difficult to thousands of data and get a preparation to remain absolutely stable fo hours.

The experimental, non-simulated, data used for computer st analysis were obtained from the work of Martin[6] (1955) and his paper be consulted for a complete description of the experimental metho In summary, the usual techniques of intracellular recording of e.p.] m.e.p.p.'s were employed (Fatt & Katz,[8] 1951). The median extensor digitorium IV muscle of the frog (*Rana temporaria*) was used. It was in in an isotonic solution containing $CaCl_2$ and $MgCl_2$ adjusted to red end-plate response to just below threshold for the initiation of a pro muscle action potential. Prostigmine, an anticholinesterase, was incl the solution to increase the amplitude of the spontaneous m.e.p.p.'s. attention has been given to investigate the coefficient of variation as tion of mean quantum content.

17.2 Computer Model

In describing the model and results obtained, the following notation used:

- *m*   mean quantum content
- *s*   amplitude of e.p.p.
- *f(s)* probability distribution of e.p.p.'s amplitudes
- *A*   amplitude of m.e.p.p.

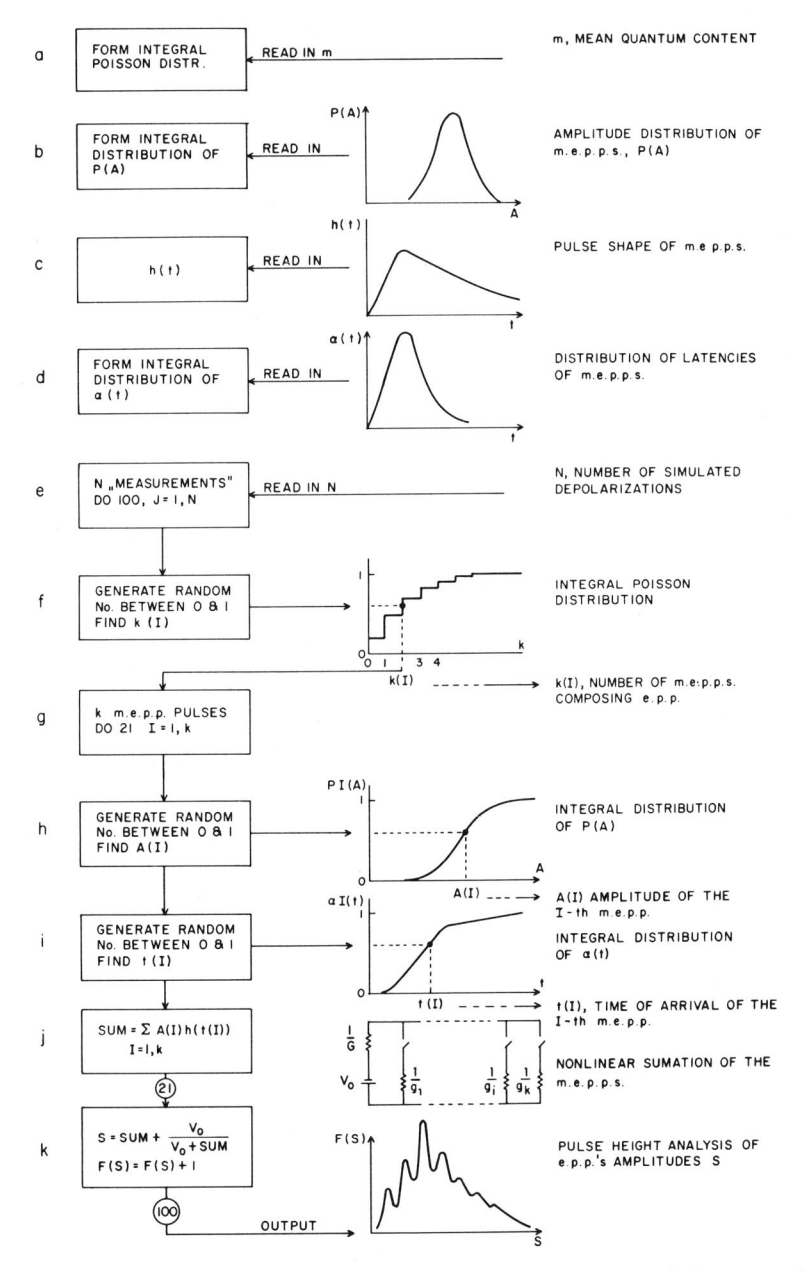

Fig. 17-1 Computer model for the statistical composition of the end-plate potential. Steps a), (b), (c), (d), and (e) present set-up of the experiment. Steps (f), (g), (h), (i), (j), ahd (k) present buildings of one e.p.p.

In the next step, Figure 17.1c, the program reads in the m.e.p.p.'s pulse shape $h(t)$. Its amplitude, rise time, and half-fall time depends on the distance from the end-plate focus on the temperature and the preparation.

In the next step, Figure 17.1d, the program reads the time course of the release rate, or latency distribution $\alpha(t)$. In a resting terminal, transmitter packets are released at random intervals with a low probability. When the terminal is depolarized by an action potential, the release rate, $\alpha(t)$, rapidly increases to a high value and then returns to the resting level. The program converts the distribution, $\alpha(t)$, into integral form, $\alpha I(t)$, which is shown in Figure 17.1i.

In the next step, Figure 17.1e, the program asks for a number of depolarization N during one experiment. For highly accurate statistical analysis, an experiment composed of a few thousand of depolarizations should be simulated.

Steps (a), (b), (c), (d), and (e) present the set-up of the basic experimental conditions.

Steps (f), (g), (h), (i), (j), and (k) simulate one depolarization and are repeated N times (loop no. 100).

In the first of those steps, Figure 17.1f, integral Poisson distribution is used to determine the number of m.e.p.p.'s composing an e.p.p. Random number generator generates the number between 0 and 1. This number is compared with the integral Poisson distribution. K, for which the match is found, presents the number of m.e.p.p.'s. If the random number generator generates all the values between 0 and 1 with equal probability, k will be governed by the Poisson distribution. (For detailed explanation see, e.g. Souček,[9] 1972.)

For each m.e.p.p., steps (g), (h), (i), and (j) are repeated (k repetitions of the loop no. 21).

In the next step, Figure 17.1h, integral distribution, $PI(A)$, is used to determine the random amplitude of one m.e.p.p. The technique is the same as in the previous case, with the use of the random number generator.

In the next step, Figure 17.1i, the integral latency distribution, $\alpha I(t)$, is used to find random time of arrival of m.e.p.p., relative to the action potential.

In the next step, Figure 17.1j, the sum of k m.e.p.p.'s is performed, each having random amplitude and random time of arrival which are not correlated. Figure 17.1j represents also the end-plate membrane during an e.p.p. consisting of k units. To allow for the residual potential difference across a completely short-circuited membrane (the liquid junction potential between myoplasm and the external solution) V_0, which represents the maximum e.m.f. of the e.p.p., is taken to be equal to the recorded resting potential minus 15 mV (see Castillo & Katz,[4] 1954b). At the height of the e.p.p., the potential across the membrane will be $V_0 - s$, where s is the e.p.p. amplitude.

Computer Model

The potential across the resistances, $1/G$, $1/g_1 \ldots 1/g_k$ will be s, $A_1 \ldots A_k$, respectively, where A_1, A_k are m.e.p.p.'s amplitudes.

For $k = 1$

$$\frac{A_1}{V_0 - A_1} = \frac{g_1}{G} \tag{17.3}$$

$$\frac{A_1}{V_0} \approx \frac{g_1}{G} \tag{17.4}$$

and in general

$$\frac{A_i}{V_0} \approx \frac{g_i}{G}. \tag{17.5}$$

For k m.e.p.p.'s we then have

$$\frac{s}{V_0 - s} = \frac{g_1}{G} + \ldots \frac{g_k}{G} = \frac{A_1}{V_0} + \ldots \frac{A_k}{V_0} = \frac{1}{V_0} \cdot \sum_{i=1}^{k} A_i, \tag{17.6}$$

$$s = \frac{\sum_{i=1}^{k} A_i}{1 + (1/V_0) \sum_{i=1}^{k} A_i} = \sum_{i=1}^{k} A_i \cdot \frac{V_0}{V_0 + \sum_{i=1}^{k} A_i}. \tag{17.7}$$

Non-linear summation given by equation (17.7) is shown in Figure 17.1k. If one takes the average of s, $\bar{s}$, and the average of the sum

$$\sum_{i=1}^{k} A_i = m \cdot v,$$

where v is the mean value of m.e.p.p.'s, one gets

$$\bar{s} = \frac{m \cdot v}{1 + (mv/V_0)},$$

$$m = \frac{\bar{s}}{v} \cdot \frac{1}{1 - (\bar{s}/V_0)}. \tag{17.8}$$

Equation (17.8) presents a correction factor, as introduced first by Martin[6] (1955).

In the next step, Figure 17.1k, one count is added to the amplitude channel, s, and the analysis of one e.p.p. is finished.

The similar procedure is repeated N times and the distribution function, $f(s)$, is formed. After N loops, the program plots the curve, $f(s)$ as a result of the experiment.

Next, we shall show the results of a number of simulated experiments. In all of the experiments, the latency distribution, $\alpha(t)$, and m.e.p.p. pulse shape, $h(t)$, are taken on the basis of measurements by Katz & Miledi[2] (1965, Fig. 8).

Figure 17.2 presents the distribution $f(s)$ of e.p.p.'s for the following conditions: $P(A)$ Gaussian distribution, $v = 0.5$ mv; coefficient of variation 0.1; number of depolarizations, $N = 25{,}000$; number of amplitude channels in each measurement is 200; mean quantum content, m, in Figure 17.2 is 10.

In Figure 17.2, one can notice that the distribution $f(s)$ is composed of asymmetrical peaks. For example, the fourth peak has the maximum for $s/v = 3.6$ rather than 4.0. This is explained as a result of imperfect summation of m.e.p.p.'s, due to their different times of arrival. Due to the latency fluctuation, four m.e.p.p.'s are not added with full amplitudes, but rather, they are piling up on the tails of each other.

In Figure 17.2, one can see, going from left to right, that subsequent peaks are broader and after tenth peak one cannot recognize, anymore, separate peaks.

The experiments shown in Figure 17.2 are purposely performed with a fine quantization, low noise level and large number of depolarizations. In this way, a fine structure of the distribution, $f(s)$, is shown.

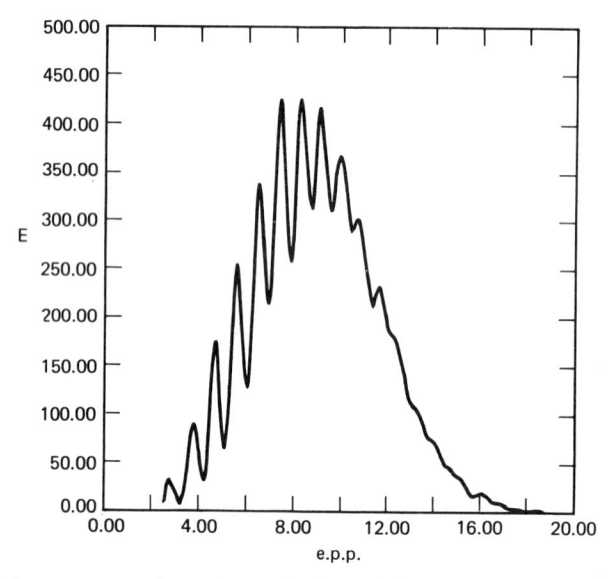

Fig. 17-2 Fine structure of e.p.p.'s amplitude probability distribution. Mean quantum content, m, is 10. One can notice that the distribution is composed of asymmetrical peaks. Abscissa is normalized e.p.p. amplitude s/v, where v is mean amplitude of m.e.p.p.'s.

Computer Model 309

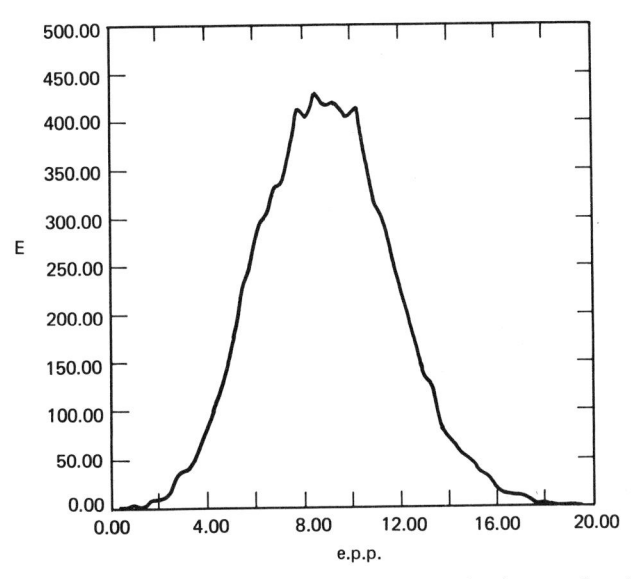

Fig. 17-3 Distortions introduced through rough quantizations and noise. Experimental conditions are the same as in Fig. 17-2, except that the total number of amplitude channels is reduced from 200 to 80 and coefficient of variation of m.e.p.p.'s increased from 0.1 to 0.2. Abscissa is s/v. All fine peaks are lost.

In Figure 17.3, one can see distortions introduced through imperfect measurement. Experimental conditions are the same as in Figure 17.2 except that the total number of amplitude channels is reduced from 200 to 80. Due to the rough quantization, all the fine peaks on the distribution function, $f(s)$, are smeared out and lost. One can notice the difference between distributions in Figures 17.2 and 17.3. In real experiments, the number of amplitude channels is often less than 80, producing significant errors in results.

In Figure 17.4 three different theories for the statistical composition of e.p.p.'s are compared (Poisson; non-linear summation; latency fluctuation). Experimental conditions are as follows: $P(A)$, Gaussian distribution; $v = 0.5$ mv; coefficient of variation is 0.2; mean quantum content, m, is 40; number of depolarizations is $N = 5000$; number of amplitude channels is 80.

Three curves for $f(s)$ are presented: curve A, e.p.p. is composed through linear summation of m.e.p.p.'s. All m.e.p.p.'s are arriving at the same moment [steps (i) and (k) from Figure 17.1 are avoided]; curve B, e.p.p. is composed through non-linear summation of m.e.p.p.'s. All m.e.p.p.'s are arriving at the same moment [step (i) from Figure 17.1 is avoided]; curve C, e.p.p. is composed through non-linear summation of m.e.p.p.'s. M.e.p.p.'s times of arrival are random (the most complete model).

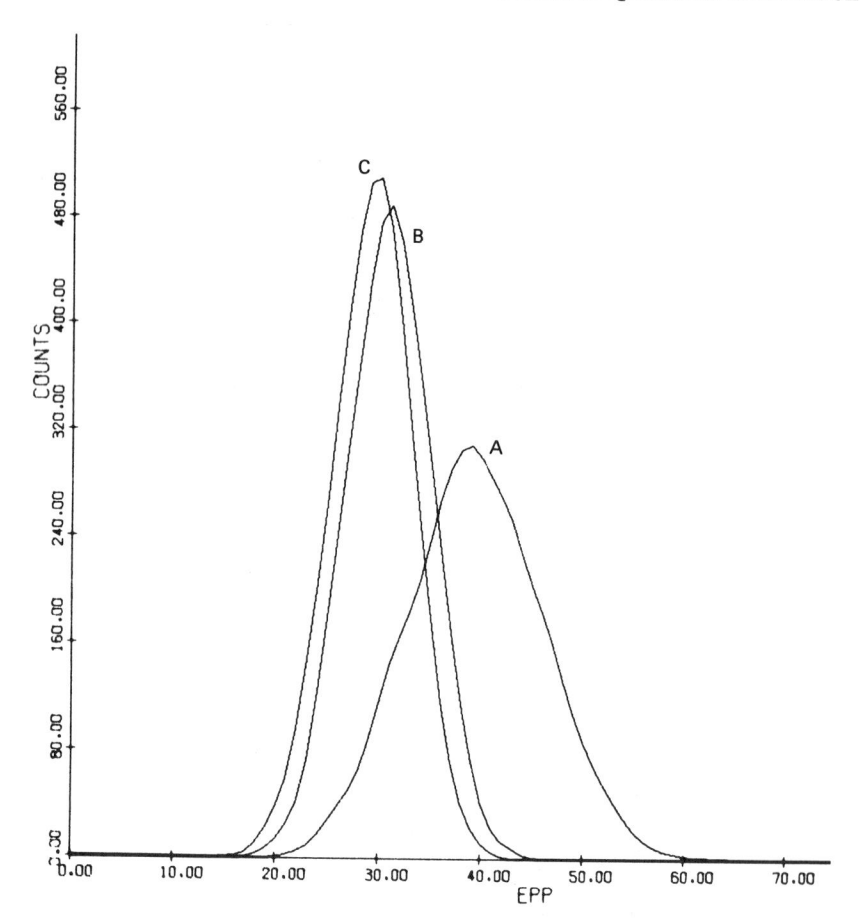

Fig. 17-4 Three models for the statistical composition of e.p.p.'s. Mean quantum content is 40. Curve A, linear summation; curve C, nonlinear summation and influence of latency fluctuation; curve B, non-linear summation.

It is important to compare the three different models with the experimental results. It is well known, up to now, that the first model (curve A) is in approximate agreement with experimental results, only for $m < 10$. Hence, we shall concentrate on the other two models, taking into account non-linear summation and latency fluctuation.

Figure 17.5 presents results of five simulated experiments. Number of depolarizations for each experiment is 2000, number of amplitude channels is 80. Coefficient of variation of m.e.p.p.'s distribution is 0.2.

Computer Model

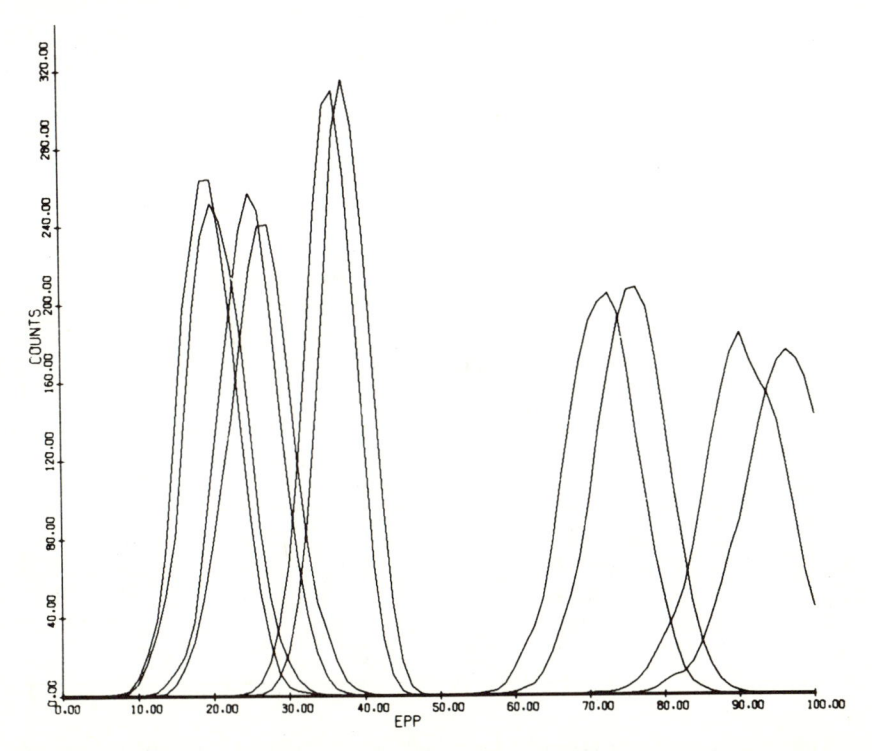

[figure: plot with COUNTS on y-axis (0 to 320.00) and EPP on x-axis (0.00 to 100.00), showing five pairs of peaked curves]

Fig. 17-5 Results of five simulated experiments. Left curve of each pair presents the result of model VL (variable latencies and nonlinear summation). Right curve presents the result of model FL (fixed latencies and nonlinear summation).

The parameters of five simulated experiments are chosen to be the same as in five measurements by Martin[6] (1955), see Table 17.1.

Each experiment is simulated two times: FL, taking into account non-linear summations, but with "fixed latency"; and VL, taking into account both non-linear summation and "variable latency". In Figure 17.5 one can see five pairs of curves for five experiments. The left curve of each pair is the VL curve.

In Table 17.2, coefficients of variations and mean values of e.p.p.'s distributions are presented. One can make a comparison between observed values in five experiments and simulated results for models FL and VL.

In Figure 17.6, coefficients of variations are plotted as a function of mean e.p.p. amplitude $\bar{s}/v$. One can examine five groups of results for five experiments. In all experiments, model VL (variable latency) is in better agreement with observed data than model FL (fixed latency).

TABLE 17.1
Parameters of Five Experiments

Experiment	v(mv)	m
I	0.48	24.8
II	0.44	32.3
III	0.76	62.3
IV	0.34	117.0
V	0.25	143.0

TABLE 17.2
Results of Five Experiments

Experiment	Observed C.V.	$\bar{s}/v$	Model FL C.V.	$\bar{s}/v$	Model VL C.V.	$\bar{s}/v$
I	0.148 ± 0.009	20.8	0.189	20.6	0.189	19.5
II	0.127 ± 0.007	26.4	0.159	26.4	0.158	24.9
III	0.093 ± 0.004	37.8	0.085	37.5	0.091	35.6
VI	0.071 ± 0.003	74.0	0.064	75.5	0.067	71.6
V	0.058 ± 0.003	96.4	0.057	96.4	0.061	91.0

The five experiments just described have different mean values of m.e.p.p.'s, v, and because of that it is difficult to compare results in between experiments. Figure 17.7 presents results of the same group of experiments, but with $v = 0.5$ mv in all five experiments. Figure 17.8 presents coefficients of variations and mean quantum contents plotted as a function of mean e.p.p. amplitude, $\bar{s}/v$, for the above cases.

For small values of $\bar{s}/v$, the two models produce similar m but rather different C.V.

For large values of $\bar{s}/v$, the two models produce similar C.V. but rather different m.

17.3 Fixed and Variable Latency

A model for the statistical composition of end-plate potentials has been designed and used for detailed study of e.p.p.'s amplitude distribution function. The model is based on the present knowledge of synaptic function and takes into account m.e.p.p.'s amplitude distribution, m.e.p.p.'s pulse shape, m.e.p.p.'s latency fluctuation, residual potential differences across the membrane, non-linear summation and mean quantum content. It is shown

Fixed and Variable Latency

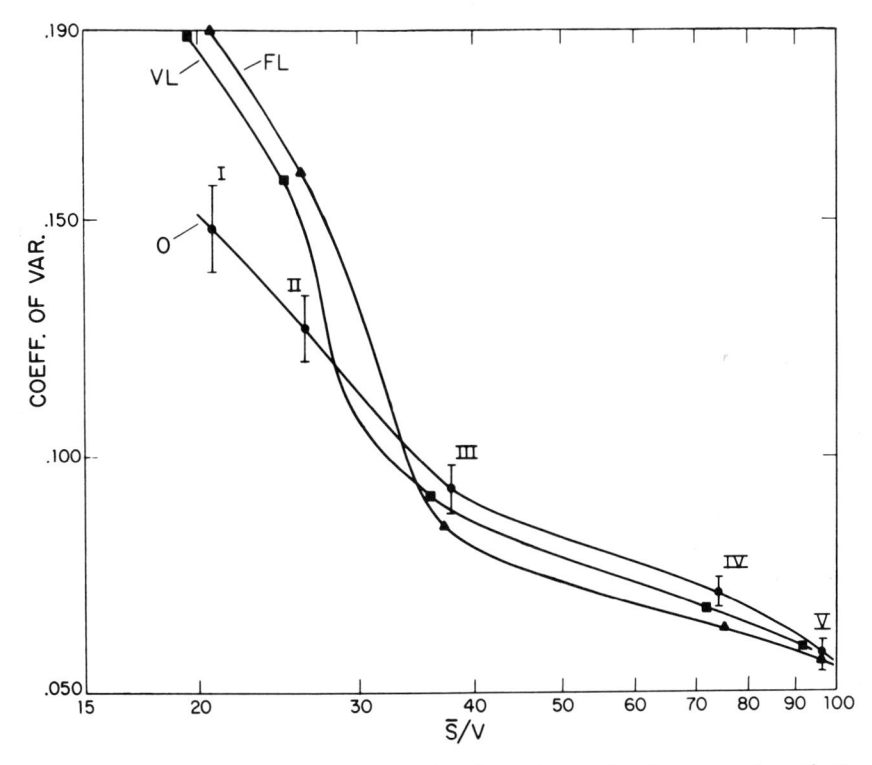

Fig. 17-6 Coefficient of variation as a function of normalized mean value $\bar{s}/v$ for five experiments O, observed data (Martin, 1955); FL, simulated experiment (nonlinear summation); VL, simulated experiment (nonlinear summation and variable latencies).

that the statistical building up of e.p.p.'s is a stochastic process in which all of the above factors have significant influence.

The results provide further support for the view that transmission at a nerve-muscle junction takes place in all-or-none quanta whose sizes are identical with those of spontaneously occurring miniature potentials. The amplitude of the e.p.p. fluctuates in a manner predictable on this basis by Poisson's law.

Unit potential of say 0.25 msec rise time and 0.5 mv size can appear as early as 0.5 msec and as late as 2.6 msec after the arrival of the nerve impulse. This indicates that the nerve impulse is followed by a period of a few milliseconds during which the probability of quantal release is increased.

Due to this latency fluctuation, m.e.p.p.'s are not added with full amplitudes, but rather are piling up on the tails of each other. As a result, the distribution, $f(s)$, of e.p.p.'s is composed of asymmetrical peaks.

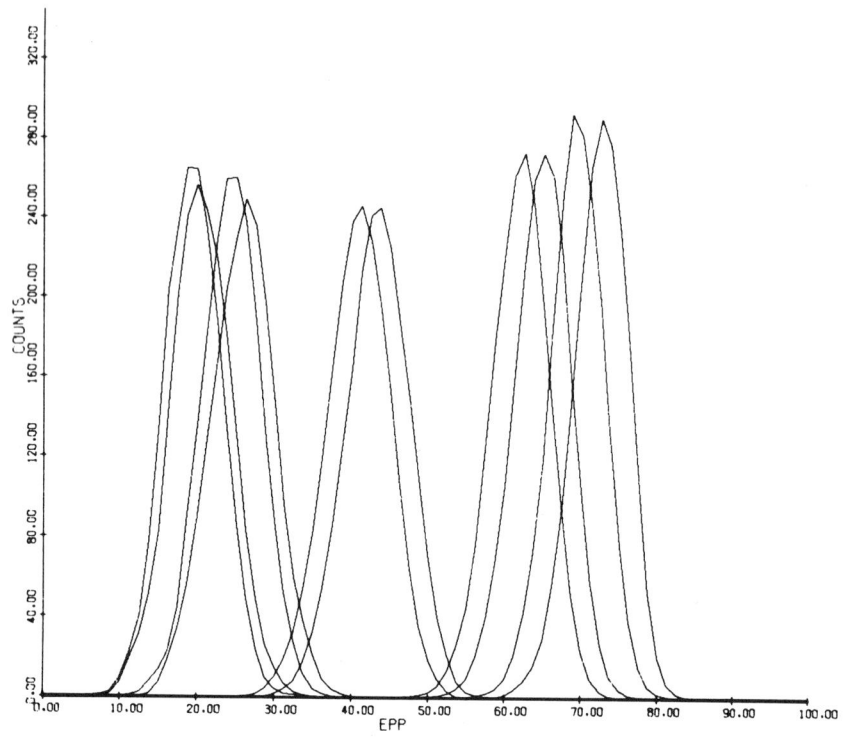

Fig. 17-7 Results of the simulated experiments. Left curve of each pair presents the result of model VL, right curve of model FL. Mean quantal contents in five experiments are 24·8, 32,3, 62·3, 117·143, respectively.

If the noiseless experiment, with fine quantization and large numbers of depolarizations is performed, one can distinguish separate peaks in the distribution, $f(s)$, up to $m = 10$.

For $m > 10$, $f(s)$ distribution has a bell-like shape and its coefficient of variation and mean value can be explained through the influence of the latency fluctuation and non-linear summation of unit components.

In Figure 17.6, comparison is shown between experimentally observed data and data simulated by the "fixed latency" and "variable latency" models. The coefficient of variation is plotted as a function of a mean e.p.p. amplitude, $\bar{s}/v$. Five experiments are presented, as in Table 17.2. In experiment I, observed data gives too small values for C.V. Experiment I has a mean e.p.p. amplitude $\bar{s}/v = 20.8$. According to equation 17.8, we obtain the correction factor, $m/(\bar{s}/v) = 0.864$, giving $m = 24.8$. Poisson distribution with $m = 24.8$ should have coefficient of variation 0.201. Due to the non-

Fixed and Variable Latency

linear summation, this value should be multiplied by the above correction factor, giving the value, 0.175. Allowance must be made for the fact that the m.e.p.p.'s themselves are not of uniform size. This should increase the observed C.V. by a factor $\sqrt{1 + \delta^2}$, where δ is the coefficient of variation of the m.e.p.p.'s amplitudes (in this case $\delta = 0.2$, $\sqrt{1 + \delta^2} = 1.04$). Thus, observed C.V. should be $1.04 \times 0.175 = 0.183$. (Simulated experiment gives the value 0.189.) Observed value, 0.148, is too small and does not agree with either model FL or VL.

Experiment II shows better agreement. In experiments III, IV, and V, curves O and VL are within experimental errors. In all experiments, model VL is in better agreement with experimental data than model FL.

Figure 17.8 shows that for a given mean e.p.p. amplitude, $\bar{s}/v$, model VL always gives a smaller value of C.V. and a larger value of m than model FL.

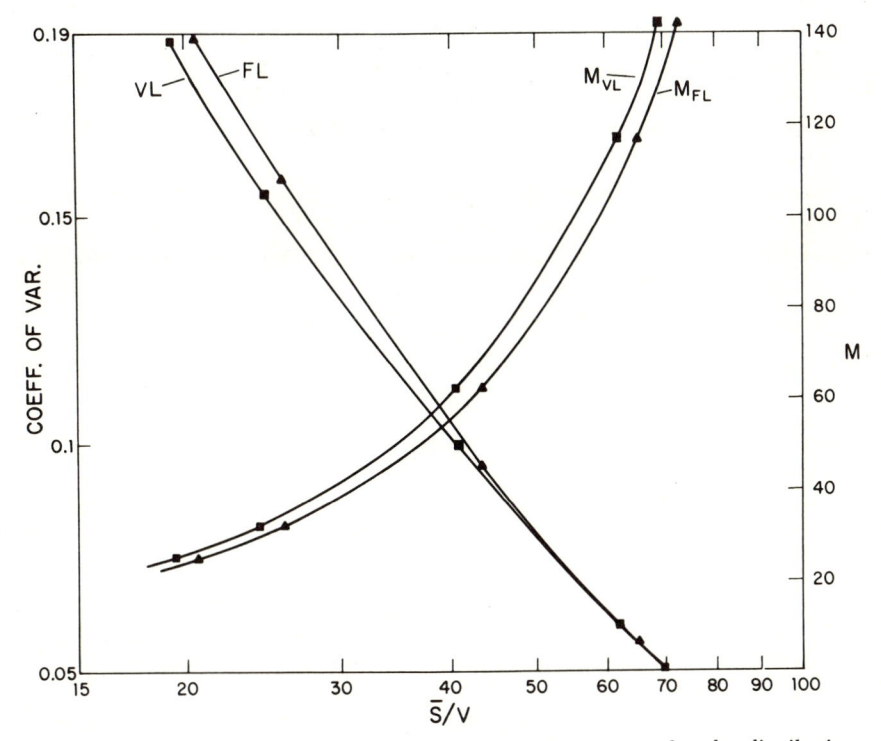

Fig. 17-8 Coefficient of variation and mean quantum content for the distributions of e.p.p.'s, presented in Fig. 17-7. Abscissa is normalized mean amplitude of e.p.p., $\bar{s}/v$. VL and FL curves present coefficient of variation; M_{VL} and M_{FL} present the mean quantum contents for VL and FL models, respectively.

The claim that latency fluctuation reduces the C.V. of the e.p.p.'s, ought to be carefully qualified. If latency fluctuation is introduced into the model while mean quantum content, m, is kept constant, the C.V. is probably slightly increased (Table 17.2: sometimes C.V. is decreased, probably by an amount less than random error). The mean e.p.p. amplitude is decreased by latency fluctuation however, and so, on a model with latency fluctuation, a larger mean quantum content must be taken to arrive at the same e.p.p. amplitude. Consequently, there is a decrease in the C.V. for a given mean e.p.p. amplitude $\bar{s}/v$.

The computer modeling can be especially useful in cases where m.e.p.p.'s amplitudes have skewed distribution.

References

1. Souček, B. *J. Theor. Biol.* **30,** 631, (1971).
2. Katz, B. and Miledi, R.: *Proc. R. Soc. B.* **161,** 483, (1965).
3. Del Castillo, J. and Katz, B.: *J. Physiol., Lond.* **124,** 560, (1950).
4. Del Castillo, J. and Katz, B.: *J. Physiol., Lond.* **125,** 546, (1952).
5. Boyd, I. A. and Martin, A. R.: *J. Physiol., Lond.* **132,** 74, (1950).
6. Martin, A. R.: *J. Physiol. Lond.* **130,** 114, (1955).
7. Souček, B. *Biophys. J.* **11,** 2, 127, (1971).
8. Fatt, P. and Katz, B.: *J. Physiol., Lond.* **115,** 320, (1951).
9. Souček, B.: *Minicomputers in Data Processing and Simulation*, Wiley, New York, 1972.

Appendix

SPECIAL TECHNIQUES

The application of computers in neurobiology and behavior is rapidly spreading. Some special turn key systems and highly specialized languages have also been developed.

On-line Sequential Control of Experiments and the ACT Language

Automated contingency translator (ACT) is a problem-oriented programming language that expresses various contingencies and conditions of behavioral experiments. In many behavioral experiments stimuli are presented to a subject according to a set of prearranged rules. The ACT language could be used to govern the behavioral contingencies of a group of diverse experiments. The ACT compilers exist for a few leading minicomputers. The vocabulary of ACT consists of the pseudowords WHEN, WHILE, GIVE, IF, AFTER, FOLLOWING, GO, PROBABILITY, COUNT, TOTALTIME, and DURATION and the time units UNITS, SEC, MIN, and HR. States are designated by the symbols S, U, or V, and the responses are identified by the letter R. For example, the state declaration might look as follows:

WHEN S1
GIVE U12.5
WHILE V(J)

A complete presentation of the act language may be found elsewhere.[1,2]

317

SNAP: An Interpretive Real-time Language for Biology

Simon[3,4] has published a language called SNAP that has been specifically designed for experimental biology. One may get an idea of the usefulness and power of SNAP from part of the program used to generate a pulse interval histogram. The whole program takes only 11 instructions:

    10   RUBTB   Clear table
    11   U = T + Z   Set U equal to time of last pulse
    12   Y-AN1   Look for positive pulse
    13   BM12Y   If no pulse go back to 12
           and so on

Interpretative languages of this kind are particularly suited for multiple animal-training experiments.

Language Based on Behavioral Notation System

In every behavioral experiment stimuli are presented to a subject according to a set of prearranged rules. Snapper, Knapp, Kushner, and Kadden[5-7] have pointed out that such a sequential system may be formally represented as a finite-state automation. As such, any behavioral experiment may be represented as a sequential system with a finite number of interrelated potential states. A single computer can control in time-sharing mode several independent simultaneous experiments. A new computer language based on the notational system has been developed for this purpose. The simplest format for describing the state graphs to the computer is to type transitions, one per line, in sequential order. Each line includes the number of the state from which the transition will occur, the input causing the transition and any associated outputs, and the number of the state to which the transition leads. Similar information is repeated for all transitions from each state of every state set. Here is an example:

```
Box #(1)   SA (1000)   #STATE SETS 4   1ST RECORDING
                                         ADDRESS (200)
STATE SET 1
  S1----R2/STIM 1 ON---→ S2  <BEGIN SESSION
  S2----60'/STIM 1 OFF---→ S0 <SESSION TIMED OUT
STATE SET 2
  S1---R2--→S2              <DRL STATE SET
  S2----R1/Z1---→S2         <SHORT IRT'S RESET TIMER
                and so on.
```

The present system is built around the minicomputer PDP-8. Other computers and other system requirements could lead to different methods for translating state graphs into actual operating programs. In fact, the notational system has been also used as a base to construct state modules,[5] so that it is possible to program experiments by wiring the modules to mimic the state diagram of the procedure.

Modeling in Neural Systems

Neutral network modeling has been around for many years. With the advent of newer, larger, and faster computers, many new and sophisticated models have been designed. Many models exist to simulate the operation of single neuron, small and large neural networks, behavioral functions, and brain functions. The last group of models overlaps in part with research in artificial intelligence. Research in the above areas is done cooperatively among life scientists, electronic engineers, and computer scientists. For the latest information on neural modeling, one could consult the work of Arbib,[8] Gerstein and Subramanian,[9] Lewis,[10] MacGregor,[11] and Perkel.[12]

References

1. *The ACT Primer: Computer Language and Hardware Interface for Controlling Pyschological Experiments*, Lehigh Valley Electronics, Fogelsville, Pennsylvania, 1969.
2. Millenson, J. R.: On-line Sequential Control of Experiments by an Automated Contingency Translator, in B. Weiss (Ed.), *Digital Computers in the Behavioral Laboratory*, Appleton-Century-Crofts, New York, 1973.
3. Simon, W.: SNAP. An interpretive real-time language for biology, *Med. Biol. Eng.* **5,** (1967), 83–85.
4. Simon, W.: A. method of experimental separation applicable to small computers, *Phys. Med. Biol.*, **15,** (1970), 355–360.
5. Snapper, A. G. and Knapp, J. A.: Multi-purpose logic module for behavior experiments, *Behav. Res. Meth. Instrum.*, **1,** (1969), 264–266.
6. Snapper, A. G., Knapp, J., and Kushner, H.: Mathematical Description of Schedules of Reinforcement, in W. N. Schoenfeld (Ed.), *The Theory of Reinforcement Schedules*. Appleton-Century-Crofts, New York, 1970.
7. Snapper, A. G., and Kadden, R. M.: Time-Sharing in a Small Computer Based on a Behavioral Notation System, in B. Weiss (Ed.), *Digital Computers in the Behavioral Laboratory*, Appleton-Century-Crofts, New York, 1973.
8. Arbib, M. A.: A Top-down approach to brain function, *Neuroscience Abstracts*, 5th Annual Meeting, New York, 1975, 45.
9. Gerstein, G. L. and Subramanian, K. N.: Identification of neural assemblies, ibid, 1013.
10. Lewis, E. R.: A view from the bottom, ibid, 45.
11. MacGregor, R. J.: Models of large neural networks, ibid, 45.
12. Perkel, D. H.: Models of smaller neural networks, ibid, 45.

INDEX

ACCEPT and REJECT statements, 101
Accumulator, 49
Action potential, 187, 194-196f
Analog-to-digital converter, 66, 247
Analog multiplexers, 64
ADC statement, 100
Amplifiers, 62
Amplitude correlation, 158
  histograms, 140
Analyzers, 145
AND gate, 30
Animal communication, 187-189f, 190, 198
  bird duetting, 258-283
  firefly flash communication, 224-257
  katydid communication, 203-223
  receiver, 188, 189f, 198
  signaller, 188, 189f, 198
Animals, birds, 188, 193, 198, 258-283
    chicken, 197
    colonial breeding birds, 192
    duck, 190
    hawk, 188
    herring gull, 190
    ring necked pheasant, 191f
    red-winged blackbird, 190
    turkey, 192
    white crested jay thrush, 258-283
  fish, 197
    goldfish, 195
    stickleback, 190
  frog, 190, 197
  insects, 193, 197
    ant, 190
    bee, 190
    cricket, 190, 197, 198
    firefly, 193, 224-257
    katydid, 203-223
    moth, 190
    social insects, 190, 192
  mammals, 197
    mice, 190
    monkey, 190
    whale, 193
Associative memory, 274
Autocorrelation function, 10

BASIC, 73, 247, 260
Behavioral notation system, 318
Binary addition, 28
  number system, 26
Biocommunication signals, 3
Biomedical signals, 3
Bird duetting, 258-283
  associative memory, 258, 275, 282
  computer program, 260, 262, 265, 268, 269, 275
  frequency matrix, 262, 266
  pair, 259, 277
  probability matrix, 262, 266, 269
  sequential analysis of song, 258, 273, 279
  sonagraph, 260
  syllable, 258-262, 266, 269, 273, 276-282
  transition frequency, 262, 266, 277
  transition matrix, 260, 269, 279, 282

321

tree pattern, 258, 269, 273, 276-282
Bit, 27
Breeding territories, 188, 190
Buffered digital input-output, 64
Byte, 27

Character set, 75
Chemical, odors, 187, 188, 192
Classification of syllables, 287
CLEAR statement, 96
Cocktail party effect, 192
Command code, 54
Communication channel, 37
Communication continuous process, 17
Communication point process, 11
Computer, 190, 199
Computerized experiment, 61
Computer model, 190, 199, 202, 212, 221, 238, 247-257
   firefly communication, 247-257
   katydid communication, 207-223
   logical neurons, 199, 200f
   McCullock-Pitts neuron, 199
   response function, 203, 208-212, 217, 220-222, 224, 227-230, 232
   threshold, 208, 209f, 210, 212, 217
   time function, 208, 209f, 210, 212, 221
   transfer function, 203, 208, 209f, 217, 220, 221
Computer program, 260, 262, 265
   bird duetting, 260, 262, 265, 268, 269, 275
   firefly flash communication, 251-253
   katydid communication, 222
Computer system, 61
Conditional transfer, 51
Control operations, 76
Correlation, 155
   amplitude, 158
   function, 155
   interval, 159
   measurement of, 155
   properties of, 156

Decimal number system, 25
DELAY statement, 96
De Morgan's theorem, 33
Device selection address, 54
Digital-to-analog decoder, 64
DIM statement, 78

Direct memory access, 58
Dwell histogram, 9

End-plate potential, 301
EOR gate, 34
Event scheduling, 104
Evolution, 187

Filters, 62
Firefly flash communication, 224-257
   acceptance, 224, 236, 240
   blocking, 224, 226, 227, 229, 236, 237, 240, 242, 243, 256
   computer model, 247-257
   computer program, 251-253
   courtship flashes, 225
   discrimination level control, 232-234
   double flash interval, 225, 227, 229, 232, 237, 239, 242, 244, 247, 249, 255, 256
   initial inhibition, 224, 228, 236, 240
   inserted or extra male flash, 230, 242, 243, 256
   interflash interval, 225, 239, 240
   interpair interval, 225, 237, 242
   intrapair interval, 225, 240, 242
   locking-in, 234-236, 247, 248, 256
   long-term inhibition, 224, 228, 236, 240
   memory, 224, 225, 227, 232, 233, 237, 247, 254, 256, 257
   motor control, 233, 234, 236, 254
   noise, 224, 229-231, 256
   patrolling flashes, 225
   pulse pattern recognition, 224, 225
   response function, 224, 227-230, 232, 236, 238, 249, 254, 256
   self-adjustment, 227, 232, 233, 237
   sensitivity adjustment, 224, 225, 233-235, 240, 254, 256
   sensory control, 234, 254, 257
   switching, 227, 244-246, 256
   time window, 224, 231, 236, 249, 254-256
Flip-flop, 35
FOR-NEXT statements, 79
FORTRAN, 195
Frequency spectrum, 175
   cosine wave, 177
   exponential wave, 178
   pulse waveform, 178

Index

Fourier analysis, 173

Generator potential, 195
Genes, 198
GOSUB-RETURN statements, 82
GO-TO statement, 77

Harmonic oscillator, 119

IF-THEN statement, 77
Information, theory of, 39
Information content, 37
Input-Output instruction, 54
  programming, 83
  transfer, 51
INPUT statement, 83
Instinction register, 49
Interfacing, 45
Interrupt, 52
Interrupt line, 57
Interval correlation, 159
Invertor, 30

Katydid communication, 203-223
  aggression, 203, 223
  aggressive sounds, 204, 204f, 207, 208
  alternation, 203, 204, 206-217, 221, 223
  calling song, 204
  chirp interval, 205, 205f, 207
  chirp period, 205, 205f, 207-212, 214, 216-221
  disturbance signals, 204, 204f
  follower, 204-206f, 208-212, 214-220, 222, 223
  leader, 204-206f, 208-212, 214-220, 222, 223
  partial delay, 203, 206f-208, 210, 211, 213-221, 223
  response period and interval, 203, 205, 205f, 207, 209, 211, 212, 214-223
  sliding sequence, 217
  solo, 204f, 206, 206f-213, 216-218, 220, 223
  solo overlap, 203, 206f-208, 210-221
  stable operating point, 216, 217, 219, 220, 223
  stimulus period and interval, 203, 205f, 206, 209-212, 214-223

Labelling, 76

Language, human, 188, 288
Latency, fixed, 312
  variable, 315
Logical operations, 29
Logical neuron, 37
Loops, 78

Memory, 48
Memory data register, 49
Message, 187, 188, 190, 192
  distance, 41
  unit, 281
Microcomputer, 45
Minicomputer, 45
Minicomputer-analyzer, 147
Monte Carlo techniques, 127, 223, 247
Muscle, 194
  muscle fibers, 194, 196f
  endplate, 194, 196f

Nervous system, 187, 188, 194, 198, 202
Neural continuous process, 14
Neural point processes, 6
Neural terminals, 300
Neuron, 194, 198, 199, 202
  action potential, 187, 194-196f
  axon, 194, 196f, 197f
  dendrites, 194, 196f, 197f
  endplate, 194, 196f
  giant manthner cell, 195, 197f
  interneurons, 194
  local potentials, 194, 195
  logical, 199, 200f
  McCullock-Pitts, 199
  motor neuron, 194-196f
  oscillator, 205
  pacemaker, 205
  postsynaptic, 194
  synapse, 194, 196f
  threshold, 199
  transmitter, 194
Noise, 188, 192, 211, 212, 218, 220, 222-224, 229-231
NOR gate, 31
Numbers, 75

Observer, 188
Octal number system, 28
Operation code, 54
OR gate, 31

Index

Pairs, 262
Party-line, 53
Pattern recognition, frequencies, 286
Pheromones, 190
PL/M language, 113
PLOT statement, 94
Poststimulus histogram, 71
Power spectra, 173
Predator-prey, 192, 198
PRINT statement, 84
Priority, 104
Probability distributions, 122
   Gaussian, 126, 136
   Poisson, 127, 133
   uniform, 126, 130
Pulse-time code, 187, 195

Quantized information transmission, 300

Random data, 122
Random time intervals, 133
READ and DATA statements, 86
Real-time BASIC, 93
   clock, 69
   executive, 103
   programming, 93
   scheduling, 107
   statement, 100
Recognition of messages, 286
Redundancy, 41
Reproductive process, 188, 190, 198
   courtship process, 190
Response function, 203, 209, 224, 227-230,
   232, 236, 238, 249, 254, 256

Sample-hold circuit, 64
Schmitt trigger, 71
Selection, 198
Sensory receptors, 187, 188, 192, 194, 195,
   197
   filters, 188, 195
   generator potential, 195
   transducer, 188, 195
Sequential control of experiment, 317
Sequential pattern, 258
SET CLOCK statement, 99
SET RATE statement, 98
Signals, 187, 190, 192, 197

Signal modalities, 187, 188, 191, 192
   auditory, sound, vocalization, 187, 188,
   190, 192, 193
   chemical, odors, 187, 188, 192, 194
   tactile, 187, 188, 192
   visual, visual displays, 187, 188, 190,
   192, 193
Simulation, 117
   deterministic data, 117
   experimental data, 131
   random data, 122
Sine-wave generation, 117
SNAP language, 318
Sonagraph machine, 184
Song sequences, 296
Sound, 193
   bird duetting, 258-283
   frequency and amplitude, 188, 193
   katydid sounds, 203-223
   spectral energy distribution, 193
   temporal pattern, 193
Species specificity, 192, 198, 225
Stimuli, 188, 191, 195, 198, 207
Subroutines, 81
Syllable, 259
Syllable sequences, 290

TAB statement, 85
Task definition, 104
Time-to-digital conversion, 68
Time scheduling, 104
TIM statement, 99
Transducer, 188, 195
Transfer function, 203, 208, 209, 214,
   217, 220, 221
Transformation of random variable,
   128
Trees, 269
Triplets, 269

Unconditional transfer, 51
USE statement, 97

Variables, 75
Visual modality, 193
Voice input to the computer, 288

Word, 27